Military and Veteran Issues

snova
New York

Military and Veteran Issues

Title of Book Goes Here in Roboto Condensed Bold 11Pt Font
Name(s) of author(s) or editor(s) of book goes here in Roboto 10pt font
ISBN:
eBook ISBN:

Title of Book Goes Here in Roboto Condensed Bold 11Pt Font
Name(s) of author(s) or editor(s) of book goes here in Roboto 10pt font
ISBN:
eBook ISBN:

Title of Book Goes Here in Roboto Condensed Bold 11Pt Font
Name(s) of author(s) or editor(s) of book goes here in Roboto 10pt font
ISBN:
eBook ISBN:

Title of Book Goes Here in Roboto Condensed Bold 11Pt Font
Name(s) of author(s) or editor(s) of book goes here in Roboto 10pt font
ISBN:
eBook ISBN:

Title of Book Goes Here in Roboto Condensed Bold 11Pt Font
Name(s) of author(s) or editor(s) of book goes here in Roboto 10pt font
ISBN:
eBook ISBN:

Title of Book Goes Here in Roboto Condensed Bold 11Pt Font
Name(s) of author(s) or editor(s) of book goes here in Roboto 10pt font
ISBN:
eBook ISBN:

More information about this series can be found at
https://novapublishers.com/product-category/series/georgian-classics/

Jerry Shields
Editor

Emerging Military Technologies

Directed-Energy Weapons, Hypersonic Weapons and Lethal Autonomous Weapon Systems

snova
New York

Library of Congress Cataloging-in-Publication Data

ISBN: 979-8-89530-613-0 (Hardcover)
ISBN: 979-8-89530-637-6 (e-Book)

Published by Nova Science Publishers, Inc. † New York

Contents

Preface

The U.S. military has long relied upon technological superiority to ensure its dominance in conflict and to underwrite U.S. national security. In recent years, however, technology has both rapidly evolved and rapidly proliferated—largely as a result of advances in the commercial sector. This development has threatened to erode the United States' traditional sources of military advantage. The Department of Defense (DOD) has undertaken a number of initiatives to arrest this trend. This book provides an overview of selected emerging military technologies including: directed energy weapons; hypersonic weapons and lethal autonomous weapons.

Chapter 1

Emerging Military Technologies: Background and Issues for Congress*

Kelley M. Sayler

Summary

Members of Congress and Pentagon officials are increasingly focused on developing emerging military technologies to enhance U.S. national security and keep pace with U.S. competitors. The U.S. military has long relied upon technological superiority to ensure its dominance in conflict and to underwrite U.S. national security. In recent years, however, technology has both rapidly evolved and rapidly proliferated—largely as a result of advances in the commercial sector. As former Secretary of Defense Chuck Hagel observed, this development has threatened to erode the United States' traditional sources of military advantage. The Department of Defense (DOD) has undertaken a number of initiatives to arrest this trend. For example, in 2014, DOD announced the Third Offset Strategy, an effort to exploit emerging technologies for military and security purposes as well as associated strategies, tactics, and concepts of operation. In support of this strategy, DOD established a number of organizations focused on defense innovation, including the Defense Innovation Unit and the Defense Wargaming Alignment Group.

More recently, the 2018 National Defense Strategy echoed the underpinnings of the Third Offset Strategy, noting that U.S. national security will likely be

> affected by rapid technological advancements and the changing character of war.... New technologies include advanced computing, "big data" analytics, artificial intelligence, autonomy, robotics, directed energy, hypersonics, and biotechnology—the very technologies that ensure we will be able to fight and win the wars of the future.

Similarly, the 2022 National Defense Strategy notes that artificial intelligence, quantum science, autonomy, biotechnology, and space technologies have the potential to change warfighting. The United States is the leader in developing many of these technologies. However, China and Russia—key strategic competitors—are making steady progress in developing advanced military technologies. As these technologies are integrated into foreign and domestic military forces and deployed, they could hold significant implications for the future of international security writ large, and will have to be a significant focus for Congress, both in terms of funding and program oversight.

---

* This is an edited, reformatted and augmented version of Congressional Research Service Publication No. R46458, updated February 22, 2024.

In: Emerging Military Technologies
Editor: Jerry Shields
ISBN: 979-8-89530-613-0

This report provides an overview of selected emerging military technologies in the United States, China, and Russia:

- artificial intelligence,
- lethal autonomous weapons,
- hypersonic weapons,
- directed energy weapons,
- biotechnology, and
- quantum technology.

It also discusses relevant initiatives within international institutions to monitor or regulate these technologies, considers the potential implications of emerging military technologies for warfighting, and outlines associated issues for Congress. These issues include the level and stability of funding for emerging technologies, the management structure for emerging technologies, the challenges associated with recruiting and retaining technology workers, the acquisitions process for rapidly evolving and dual-use technologies, the protection of emerging technologies from theft and expropriation, and the governance and regulation of emerging technologies. Such issues could hold implications for congressional authorization, appropriation, oversight, and treaty-making.

Introduction

Members of Congress and Pentagon officials are increasingly focused on developing emerging military technologies to enhance U.S. national security and keep pace with U.S. competitors. The U.S. military has long relied upon technological superiority to ensure its dominance in conflict and to underwrite U.S. national security. In recent years, however, technology has both rapidly evolved and rapidly proliferated—largely as a result of advances in the commercial sector. As former Secretary of Defense Chuck Hagel has observed, this development has threatened to erode the United States' traditional sources of military advantage.[1] The Department of Defense (DOD) has undertaken a number of initiatives in recent years in an effort to arrest this trend. For example, in 2014, DOD announced the Third Offset Strategy, an effort to exploit emerging technologies for military and security purposes as well as associated strategies, tactics, and concepts of operation.[2] In support of this strategy, DOD established a number of organizations focused on defense innovation, including the Defense Innovation Unit and the Defense Wargaming Alignment Group.

More recently, the 2018 National Defense Strategy has echoed the underpinnings of the Third Offset Strategy, noting that U.S. national security will likely be affected by rapid technological advancements and the changing character of war…. New technologies include advanced computing, "big data" analytics, artificial intelligence, autonomy, robotics, directed

---

[1] Remarks as delivered by Secretary of Defense Secretary of Defense Chuck Hagel, "Defense Innovation Days Opening Keynote," September 3, 2014, at https://www.defense.gov/ Newsroom/Speeches/Speech/Article /605602/.

[2] The Third Offset Strategy is a strategy for maintaining U.S. military superiority. It succeeds the First and Second Offsets—nuclear weapons and the precision-guided munitions regime, respectively. Remarks as prepared for delivery by Deputy Secretary of Defense Bob Work, "National Defense University Convocation," August 5, 2014, at https://www.defense.gov/ Newsroom/Speeches/Speech/Article/605598/.

energy, hypersonics, and biotechnology—the very technologies that ensure we will be able to fight and win the wars of the future.[3]

Similarly, the 2022 National Defense Strategy notes that artificial intelligence, quantum science, autonomy, biotechnology, and space technologies have the potential to change warfighting.[4] Although the United States is the leader in developing many of these technologies, China and Russia—key strategic competitors—are making steady progress in developing advanced military technologies. As they are integrated into foreign and domestic military forces and deployed, these technologies could hold significant implications for congressional considerations and the future of international security writ large.

This report provides an overview of selected emerging military technologies in the United States, China, and Russia:

- artificial intelligence,
- lethal autonomous weapons,
- hypersonic weapons,
- directed energy weapons,
- biotechnology,
- and quantum technology.

It also discusses relevant initiatives within international institutions to monitor or regulate these technologies, considers the potential implications of emerging military technologies, and outlines associated issues for Congress. Such issues could hold implications for congressional authorization, appropriation, oversight, and treaty-making.

Artificial Intelligence (AI)[5]

Although the U.S. government has no official definition of artificial intelligence, policymakers generally use the term AI to refer to a computer system capable of human-level cognition. AI is further divided into three categories: narrow AI, general AI, and artificial superintelligence.

Narrow AI systems can perform only the specific task that they were trained to perform, while general AI systems would be capable of performing a broad range of tasks, including those for which they were not specifically trained. Artificial superintelligence refers to a system "that greatly exceeds the cognitive performance of humans in virtually all domains of interest."[6] General AI systems and artificial superintelligence do not yet—and may never—exist.[7]

---

[3] Department of Defense, "Summary of the 2018 National Defense Strategy of The United States of America," 2018, p. 3, at https://dod.defense.gov/Portals/1/Documents/pubs/2018-National-Defense-Strategy-Summary.pdf.

[4] Department of Defense, *2022 National Defense Strategy of The United States of America*, October 2022, p. 19, at https://media.defense.gov/2022/Oct/27/2003103845/-1/-1/1/2022-NATIONAL-DEFENSE-STRATEGY-NPR- MDR.PDF.

[5] For more information about artificial intelligence, see CRS Report R45178, *Artificial Intelligence and National Security*, by Kelley M. Sayler.

[6] Nick Bostrom, Superintelligence: Paths, Dangers, Strategies (Oxford, United Kingdom: Oxford University Press, 2014).

[7] For a discussion of narrow versus general artificial intelligence, as well as a range of expert opinions about the future of general artificial intelligence, see Nick Bostrom, *Superintelligence: Paths, Dangers, Strategies* (Oxford, United Kingdom: Oxford University Press, 2014).

Narrow AI is currently being incorporated into a number of military applications by both the United States and its competitors. Such applications include but are not limited to intelligence, surveillance, and reconnaissance;[8] logistics; cyber operations; command and control; and semi- autonomous and autonomous vehicles. These technologies are intended in part to augment or replace human operators, freeing them to perform more complex and cognitively demanding work. In addition, AI-enabled systems could (1) react significantly faster than systems that rely on operator input; (2) cope with an exponential increase in the amount of data available for analysis; and (3) enable new concepts of operations, such as swarming (i.e., cooperative behavior in which unmanned vehicles autonomously coordinate to achieve a task) that could confer a warfighting advantage by overwhelming adversary defensive systems.

Narrow AI, however, could introduce a number of challenges. For example, such systems may be subject to algorithmic bias as a result of their training data or models. Researchers have repeatedly discovered instances of racial bias in AI facial recognition programs due to the lack of diversity in the images on which the systems were trained, while some natural language processing programs have developed gender bias.[9] Such biases could hold significant implications for AI applications in a military context. For example, incorporating undetected biases into systems with lethal effects could lead to cases of mistaken identity and the unintended killing of civilians or noncombatants.

Similarly, narrow AI algorithms can produce unpredictable and unconventional results that could lead to unexpected failures if incorporated into military systems. In a commonly cited demonstration of this phenomenon (illustrated in *Figure 1*), researchers combined a picture that an AI system correctly identified as a panda with random distortion that the computer labeled "nematode." The difference in the combined image is imperceptible to the human eye, but it resulted in the AI system labeling the image as a gibbon with 99.3% confidence. Such vulnerabilities could be exploited intentionally by adversaries to disrupt AI-reliant or -assisted target identification, selection, and engagement. This could, in turn, raise ethical concerns—or, potentially, lead to violations of the law of armed conflict—if it results in the system selecting and engaging a target or class of targets that was not approved by a human operator.

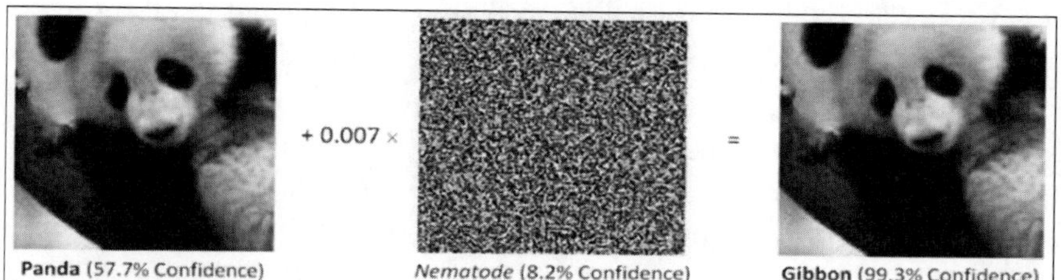

Source: Andrew Ilachinski, AI, Robots, and Swarms, Issues Questions, and Recommended Studies, Center for Naval Analyses, January 2017, p. 61.

Figure 1. AI Failure in Image Recognition.

[8] For a discussion of intelligence, surveillance, and reconnaissance, see CRS Report R46389, *Intelligence, Surveillance, and Reconnaissance Design for Great Power Competition*, by Nishawn S. Smagh.

[9] Brian Barrett, "Lawmakers Can't Ignore Facial Recognition's Bias Anymore," Wired, July 26, 2018, at https://www.wired.com/story/amazon-facial-recognition-congress-bias-law-enforcement/; and Will Knight, "How to Fix Silicon Valley's Sexist Algorithms," MIT Technology Review, November 23, 2016, at https://www.technology review.com/s/ 602950/how-to-fix-silicon-valleys-sexist-algorithms/.

Finally, recent news reports and analyses have highlighted the role of AI in enabling increasingly realistic photo, audio, and video digital forgeries, popularly known as "deep fakes." Adversaries could deploy this AI capability as part of their information operations in a "gray zone" conflict.[10] Deep fake technology could be used against the United States and its allies to generate false news reports, influence public discourse, erode public trust, and attempt blackmail of government officials. For this reason, some analysts argue that social media platforms—in addition to deploying deep fake detection tools—may need to expand the means of labeling and authenticating content.[11] Doing so might require that users identify the time and location at which the content originated or properly label content that has been edited. Other analysts have expressed concern that regulating deep fake technology could impose an undue burden on social media platforms or lead to unconstitutional restrictions on free speech and artistic expression.[12] These analysts have suggested that existing law is sufficient for managing the malicious use of deep fakes and that the focus should be instead on the need to educate the public about deep fakes and minimize incentives for creators of malicious deep fakes.

United States

DOD's unclassified investments in AI have grown from just over $600 million in FY2016 to approximately $1.8 billion in FY2024, with the department maintaining over 685 active AI projects.[13] Pursuant to the FY2019 National Defense Authorization Act (NDAA; P.L. 115-232), DOD established the Joint Artificial Intelligence Center (JAIC, pronounced "jake") to coordinate DOD projects of over $15 million; the JAIC was granted acquisition authority by Section 808 of the FY2021 NDAA (P.L. 116-283).[14] The JAIC has undertaken a number of National Mission Initiatives for AI, including predictive maintenance,[15] humanitarian aid and disaster relief, warfighter health, and business process transformation. In addition, the JAIC maintains the Joint Common Foundation, a "secure cloud-based AI development and experimentation environment" intended to support the testing and fielding of department-wide

---

10 "Gray zone" conflicts are those that occur below the threshold of formally declared war. For more information about information operations, see CRS In Focus IF10771, *Defense Primer: Operations in the Information Environment*, by Catherine A. Theohary.

11 Some social media platforms such as Twitter have established rules for labeling and removing certain types of synthetic or manipulated media. See Yoel Roth and Ashita Achuthan, "Building rules in public: Our approach to synthetic & manipulated media," *Twitter*, February 4, 2020, at https://blog.twitter.com/en_us/topics/company/2020/ new-approach-to-synthetic-and-manipulated-media.html.

12 Jessica Ice, "Defamatory Political Deepfakes and the First Amendment," *Case Western Reserve Law Review*, 2019, at https://scholarlycommons.law.case.edu/caselrev/ vol70/iss2/12.

13 The amount listed as the FY2024 investment reflects DOD's FY2024 unclassified budget request for AI. DOD's actual investments in AI in FY2024 may be higher; based on historical precedent, they are unlikely to be lower. Office of the Under Secretary of Defense (Comptroller)/Chief Financial Officer, *United States Department of Defense Fiscal Year 2024 Budget Request*, March 2023, p. 15, at hhttps://comptroller. defense.gov/Portals/45/Documents/defbudget/FY2024/FY2024_Budget_Request.pdf; and Government Accountability Office, *Artificial Intelligence: Status of Developing and Acquiring Capabilities for Weapon Systems*, February 2022, at https://www.gao.gov/assets/gao-22-104765.pdf.

14 P.L. 115-232, Section 2, Division A, Title II, §1051; and P.L. 116-283, Section 2, Division A, Title VIII, §808.

15 Predictive maintenance uses AI "to predict the failure of critical parts, automate diagnostics, and plan maintenance based on data and equipment condition." Department of Defense, "Summary of the 2018 Department of Defense Artificial Intelligence Strategy," February 12, 2019, p. 11, at https://media.defense.gov/2019/Feb/12 /2002088963/-1/-1/1/SUMMA RY-OF-DOD-AI-STRATEGY.PDF.

AI capabilities.[16] In December 2021, Deputy Secretary of Defense Kathleen Hicks directed the establishment of the Chief Digital and Artificial Intelligence Office, which is to "serve as the successor organization to the JAIC, reporting directly to the Deputy Secretary of Defense."[17]

The FY2019 NDAA also directed DOD to publish a strategic roadmap for AI development and fielding, as well as to develop guidance on "appropriate ethical, legal, and other policies for the Department governing the development and use of artificial intelligence enabled systems and technologies in operational situations."[18] In support of this mandate, the Defense Innovation Board (DIB), an independent federal advisory committee to the Secretary of Defense, drafted recommendations for the ethical use of artificial intelligence.[19] Based on these recommendations, DOD then adopted five ethical principles for AI based on the DIB's recommendations: responsibility, equitability, traceability, reliability, and governability.[20] On May 26, 2021, Deputy Secretary of Defense Kathleen Hicks issued a memorandum providing guidance on the implementation of Responsible Artificial Intelligence (RAI), in keeping with the ethical principles.[21] DOD then released an RAI strategy and implementation pathway in June 2022.[22] Section 1544 of the FY2024 NDAA (P.L. 118-31) directs DOD to assess whether a given AI technology operationalizes the RAI principles, "to report and remediate any artificial intelligence technology that is determined not to be in compliance with the [RAI] framework," and to discontinue the use of noncompliant technology "until effective remediation is achievable." E.O. 14110, Executive Order (E.O.) on the Safe, Secure, and Trustworthy Development and Use of Artificial Intelligence, additionally provides guidance on standards for AI safety and security.[23]

Finally, Section 1051 of the FY2019 NDAA established a National Security Commission on Artificial Intelligence (NSCAI) to conduct a comprehensive assessment of militarily relevant AI technologies and to provide recommendations for strengthening U.S.

---

[16] Joint Artificial Intelligence Center, "Joint Common Foundation," at https://www.ai.mil/ jcf.html.

[17] Department of Defense, "Establishment of the Chief Digital and Artificial Intelligence Officer," December 8, 2021, at https://media.defense.gov/2021/Dec/08/2002906075/-1/-1/1/MEMORANDUM-ON-ESTABLISHMENT-OF-THE-CHIEF-DIGITAL-AND-ARTIFICIAL-INTELLIGENCE-OFFICER.PDF.

[18] P.L. 115-232, Section 2, Division A, Title II, §238.

[19] For a discussion of DOD's rationale for developing principles for ethical AI, as well as DOD's existing ethical commitments related to AI, see Defense Innovation Board, "AI Principles: Recommendations on the Ethical Use of Artificial Intelligence by the Department of Defense," October 31, 2019, at https://media.defense.gov/ 2019/Oct/31/ 2002204458/-1/-1/0/DIB_AI_PRINCIPLES_PRIMARY_DOCUMENT.PDF.

[20] For definitions of these principles, see Department of Defense, "DOD Adopts Ethical Principles for Artificial Intelligence," February 24, 2020, at https://www.defense.gov/Newsroom/Releases/Release/Article/ 2091996/dod- adopts-ethical-principles-for-artificial-intelligence/.

[21] RAI is to focus on RAI governance, warfighter trust, AI product and acquisition lifecycle, requirements validation, responsible AI ecosystem, and AI workforce. For additional information about RAI, see Kathleen H. Hicks, "Implementing Responsible Artificial Intelligence in the Department of Defense," May 26, 2021, at https://media.defense.gov/2021/May/27/2002730593/-1/-1/0/IMPLEMENTING-RESPONSIBLE-ARTIFIC IAL-INTELLIGENCE-IN-THE-DEPARTMENT-OF-DEFENSE.PDF.

[22] Department of Defense, *U.S. Department of Defense Responsible Artificial Intelligence Strategy and Implementation Pathway*, June 2022, at https://media.defense.gov/2022/ Jun/22/2003022604/-1/-1/0/Department-of-Defense- Responsible-Artificial-Intelligence-Strategy-and-Implementation-Pathway.PDF. The Defense Innovation Unit (DIU) has additionally issued *Responsible AI Guidelines in Practice: Lessons Learned from the DIU Portfolio* to operationalize DOD's Ethical Principles for AI within DIU's commercial prototyping and acquisitions programs. Jared Dunnmon et al., *Responsible AI Guidelines in Practice: Lessons Learned from the DIU Portfolio*, November 14, 2021, at https://assets.ctfassets.net/3nanhbfkr0pc/ acoo1Fj5uungnGNPJ3QWy/3a1dafd64f22efcf8f27380aafae9789/ 2021_RAI_Report-v3.pdf.

[23] For additional information, see CRS Report R47843, Highlights of the 2023 Executive Order on Artificial Intelligence for Congress, by Laurie Harris and Chris Jaikaran; and CRS Insight IN12286, The AI Executive Order and Its Potential Implications for DOD, by Alexandra G. Neenan and Kelley M. Sayler.

competitiveness. The commission's final report to Congress was delivered in March 2021 and generally offers recommendations along five key lines of effort: (1) investing in research and development, (2) applying AI to national security missions, (3) training and recruiting AI talent, (4) protecting and building upon U.S. technology advantages, and (5) marshalling global AI cooperation.[24] A number of NSCAI's recommendations have been passed into law.[25] In addition, Section 247 of the FY2022 NDAA (P.L. 117-81) directs the Secretary of Defense to submit to the congressional defense committees annual status reports on the department's intent to implement NSCAI recommendations, as well as any associated timelines for implementation.[26]

China

China is widely viewed as the United States' closest competitor in the international AI market.[27] China's 2017 "Next Generation AI Development Plan" describes AI as a "strategic technology" that has become a "focus of international competition."[28] Recent Chinese achievements in the field demonstrate China's potential to realize its goals for AI development. In particular, China has pursued language and facial recognition technologies, many of which it plans to integrate into the country's domestic surveillance network. Such technologies could be used to counter espionage and aid military targeting. In addition to developing various types of air, land, sea, and undersea autonomous military vehicles, China is actively pursuing swarm technologies, which could be used to overwhelm adversary missile defense interceptors. Moreover, open-source publications indicate that China is developing a suite of AI tools for cyber operations.[29]

China's management of its AI ecosystem stands in stark contrast to that of the United States.[30] In general, few boundaries exist between Chinese commercial companies, university research laboratories, the military, and the central government. China's National Intelligence Law, for example, requires companies and individuals to "support, assist, and cooperate with national intelligence work."[31] As a result, the Chinese government has a direct means of guiding military AI development priorities and accessing technology developed for civilian purposes.

---

[24] National Security Commission on Artificial Intelligence, Final Report, March 2021, at https://www.nscai.gov/wp-content/uploads/2021/03/Full-Report-Digital-1.pdf. Pursuant to Section 238 of the FY2019 NDAA, RAND Corporation, a federally funded research and development center, additionally conducted a review of DOD's posture for AI. See Danielle C. Tarraf et al., The Department of Defense Posture for Artificial Intelligence: Assessment and Recommendations, RAND Corporation, 2019, https://www.rand.org/pubs/research_reports/RR4229.html.

[25] See, for example, Section 216 ("Improvements relating to Steering Committee on Emerging Technology and National Security Threats"), Section 228 ("Executive Education on Emerging Technologies for Senior Civilian and Military Leaders"), Section 903 ("Enhanced role for the Undersecretary of Defense for Research and Engineering on the Joint Requirements Oversight Council"), Section 909 ("Digital Talent Recruiting Officer"), and Section 1118 ("Occupational series for digital career fields") of the FY2022 NDAA.

[26] This requirement is to expire on December 27, 2023.

[27] See, for example, Kai-Fu Lee, *AI Superpowers: China, Silicon Valley, and the New World Order* (Boston, MA: Houghton Mifflin Co., 2018).

[28] China State Council, "A Next Generation Artificial Intelligence Development Plan," p. 2.

[29] Elsa Kania, *Battlefield Singularity: Artificial Intelligence, Military Revolution, and China's Future Military Power*, Center for a New American Security, November 28, 2017, p. 27.

[30] Ibid., p. 6.

[31] Arjun Kharpal, "Huawei says it would never hand data to China's government. Experts say it wouldn't have a choice," *CNBC*, March 5, 2019.

Russia

Russian president Vladimir Putin has stated that "whoever becomes the leader in [AI] will become the ruler of the world."[32] At present, however, Russian AI development lags significantly behind that of the United States and China. As part of Russia's effort to close this gap, Russia has released a national strategy that outlines 5- and 10-year benchmarks for improving the country's AI expertise, educational programs, datasets, infrastructure, and legal regulatory system.[33] Russia has indicated it will continue to pursue its 2008 defense modernization agenda, which called for robotizing 30% of the country's military equipment by 2025.[34]

The Russian military has been researching a number of AI applications, with a heavy emphasis on semiautonomous and autonomous military vehicles. Russia has also reportedly built a combat module for unmanned ground vehicles that may be capable of autonomous target identification— and, potentially, target engagement—and it plans to develop a suite of AI-enabled autonomous systems.[35] In addition, the Russian military plans to incorporate AI into unmanned aerial, naval, and undersea vehicles and is reportedly developing swarming capabilities.[36] These technologies could reduce both cost and manpower requirements, potentially enabling Russia to field more systems with fewer personnel. Russia is also exploring innovative uses of AI for remote sensing and electronic warfare, which could in turn reduce an adversary's ability to effectively communicate and navigate on the battlefield.[37] Finally, Russia has made extensive use of AI technologies for domestic propaganda and surveillance, as well as for information operations directed against the United States and U.S. allies.[38]

Despite Russia's aspirations, analysts argue that it may be difficult for Russia to make significant progress in AI development. For example, some analysts note that Russian academics have produced few research papers on AI—ranking 22nd in AI-related publications globally[39]—and that the Russian technology industry has yet to produce AI applications on par

---

[32] "'Whoever leads in AI will rule the world': Putin to Russian children on Knowledge Day," *RT.com*, September 1, 2017, at https://www.rt.com/news/401731-ai-rule-world-putin/.

[33] Office of the President of the Russian Federation, "Decree of the President of the Russian Federation on the Development of Artificial Intelligence in the Russian Federation" (Center for Security and Emerging Technology, Trans.), October 10, 2019, at https://cset. georgetown.edu/research/decree-of-the-president-of-the-russian-federation-on-the-develop ment-of-artificial-intelligence-in-the-russian-federation/.

[34] Tom Simonite, "For Superpowers, Artificial Intelligence Fuels New Global Arms Race," *Wired*, August 8, 2017.

[35] Tristan Greene, "Russia is Developing AI Missiles to Dominate the New Arms Race," *The Next Web*, July 27, 2017, at https://thenextweb.com/artificial-intelligence/2017/07/27 /russia-is-developing-ai-missiles-to-dominate-the-new- arms-race/; and Kyle Mizokami, "Kalashnikov Will Make an A.I.-Powered Killer Robot," Popular Mechanics, July 19, 2017, at https://www.popularmechanics.com/military/weapons/news/a27393/kalashnikov-to-make-ai-directed-machine-guns/.

[36] Samuel Bendett, "Red Robots Rising: Behind the Rapid Development of Russian Unmanned Military Systems," *The Strategy Bridge*, December 12, 2017.

[37] Jill Dougherty and Molly Jay, "Russia Tries to Get Smart about Artificial Intelligence"; *The Wilson Quarterly*, Spring 2018; and Margarita Konaev and Samuel Bendett, "Russian AI-Enabled Combat: Coming to a City Near You?" *War on the Rocks*, July 31, 2019, at https://warontherocks.com/2019/07/russian-ai-enabled-combat-coming-to-a-city-near-you/.

[38] Alina Polyakova, "Weapons of the Weak: Russia and AI-driven Asymmetric Warfare," Brookings Institution, November 15, 2018, at https://www.brookings.edu/research/ weapons-of-the-weak-russia-and-ai-driven-asymmetric-warfare/; and Chris Meserole and Alina Polyakova, "Disinformation Wars," *Foreign Policy*, May 25, 2018, at https://foreign policy.com/2018/05/25/disinformation-wars/.

[39] Margarita Konaev et al., *Headline or Trend Line? Evaluating Chinese-Russian Collaboration in AI*, Center for Security and Emerging Technology, August 2021, p. 9.

with those produced by the United States and China.[40] Other analysts counter that such factors may be irrelevant, arguing that while Russia has never been a leader in internet technology, it has managed to become a notably disruptive force in cyberspace.[41] Russia may also be able to draw upon its growing technological cooperation with China.[42]

International Institutions

A number of international institutions have examined issues surrounding AI, including the Group of Seven (G7), the Asia-Pacific Economic Cooperation (APEC), and the Organisation for Economic Co-operation and Development (OECD), which developed the first intergovernmental set of principles for AI.[43] These principles are intended to "promote AI that is innovative and trustworthy and that respects human rights and democratic values."[44] The United States is one of 42 countries—including the OECD's 36 member countries, Argentina, Brazil, Colombia, Costa Rica, Peru, and Romania—to have adopted the OECD AI Principles. These principles serve as the foundation for the Group of Twenty's (G20's) June 2019 Ministerial Statement on human-centered AI.[45] In addition, the OECD established the AI Policy Observatory in 2019 to develop policy options that will "help countries encourage, nurture, and monitor the responsible development of trustworthy AI systems for the benefit of society."

Similarly, in October 2021, the North Atlantic Treaty Organization (NATO) released its first AI strategy.[46] According to NATO Secretary General Jens Stoltenberg, the strategy is to "set standards for responsible use of artificial intelligence, in accordance with international law, outline how [NATO] will accelerate the adoption of artificial intelligence in what [it does], set out how [NATO] will protect this technology, and address the threats posed by the use of

---

[40] Leon Bershidsky, "Take Elon Musk Seriously on the Russian AI Threat," *Bloomberg*, September 5, 2017, at https://www.bloomberg.com/view/articles/2017-09-05/take-elon-musk-seriously-on-the-russian-ai-threat; and Alina Polyakova, "Weapons of the Weak: Russia and AI-driven Asymmetric Warfare," Brookings Institution, November 15, 2018, at https://www.brookings.edu/research/weapons-of-the-weak-russia-and-ai-driven-asymmet ric-warfare/.

[41] Gregory C. Allen, "Putin and Musk Are Right: Whoever Masters AI Will Run the World," *CNN*, September 5, 2017.

[42] Samuel Bendett and Elsa Kania, *A New Sino-Russian High-tech Partnership*, Australian Strategic Policy Institute, October 29, 2019, at https://www.aspi.org.au/report/new-sino-russian-high-tech-partnership. Some analysts have cautioned, however, that "the extent and scope of Chinese-Russian collaboration in AI may be overstated by both Chinese and Russian sources as well as U.S. observers." Margarita Konaev et al., *Headline or Trend Line? Evaluating Chinese-Russian Collaboration in AI*, Center for Security and Emerging Technology, August 2021, p. 9.

[43] In May 2020, the United States joined the G7's Global Partnership on AI, which is "to guide the responsible adoption of AI based on shared principles of 'human rights, inclusion, diversity, innovation and economic growth.'" Matt O'Brien, "US joins G7 artificial intelligence group to counter China," Associated Press, May 28, 2020.

[44] Organisation for Economic Co-operation and Development, "OECD Principles on AI," June 2019, at https://www.oecd.org/going-digital/ai/principles/.

[45] "G20 Ministerial Statement on Trade and Digital Economy," June 9, 2019, at https:// www.mofa.go.jp/files/000486596.pdf.

[46] North Atlantic Treaty Organization, "Summary of the NATO Artificial Intelligence Strategy," October 22, 2021, at https://www.nato.int/cps/en/natohq/official_texts_187617.htm?mc_ cid=8f2b5c99db&mc_eid=5a3e8d2b43.

artificial intelligence by adversaries."[47] NATO has additionally established AI test centers and a data and AI review board to "ensure the 'operationalization' of the AI strategy."[48]

Finally, in September 2021, the United States announced that it had formed a trilateral security partnership with Australia and the United Kingdom.[49] This partnership, known as AUKUS, includes defense cooperation on artificial intelligence and autonomy, as well as hypersonic weapons, quantum technologies, and other advanced capabilities.[50]

Potential Questions for Congress

- What measures is DOD taking to implement its ethical principles for artificial intelligence? Are such measures sufficient to ensure DOD's adherence to the principles?
- How is DOD testing and evaluating AI systems to ensure that they have not been exploited by adversaries?
- Do DOD and the intelligence community have adequate information about the state of foreign military AI applications and the ways in which such applications may be used to harm U.S. national security?
- How has the establishment of the Chief Digital and Artificial Intelligence Office affected DOD's ability to adopt AI applications?
- How should national security considerations with regard to deep fakes be balanced with free speech protections, artistic expression, and beneficial uses of the underlying technologies? What efforts, if any, should the U.S. government undertake to ensure that the public is educated about deep fakes?

Lethal Autonomous Weapon Systems (LAWS)[51]

Although there is no internationally agreed definition of lethal autonomous weapon systems, Department of Defense Directive (DODD) 3000.09 defines LAWS as a class of weapon systems capable of both independently identifying a target and employing an onboard weapon to engage and destroy the target without manual human control. This concept of autonomy is

---

[47] Vivienne Machi, "NATO ups the ante on disruptive tech, artificial intelligence," *C4ISRNET*, November 3, 2021, at https://www.c4isrnet.com/digital-show-dailies/feindef/2021/11/03/nato-ups-the-ante-on-disruptive-tech-artificial- intelligence/.

[48] Ibid. For additional information, see NATO, "NATO's Data and Artificial Intelligence Review Board," October 17, 2022, at https://www.nato.int/cps/en/natohq/official_texts_208374. htm; and NATO, "DIANA Network: Test Centers," July 20, 2023, at https://www. diana.nato.int/resources/site1/general/maps/diana-test-centres-en_v5.pdf.

[49] The White House, "Joint Leaders Statement on AUKUS," September 15, 2021, at https://www.whitehouse.gov/briefing-room/statements-releases/2021/09/15/joint-leaders-statement-on-aukus/.

[50] The White House, "Fact Sheet: Implementation of the Australia – United Kingdom – United States Partnership (AUKUS)," April 5, 2022, at https://www.whitehouse.gov/briefing-room/statements-releases/2022/04/05/fact-sheet-implementation-of-the-australia-united-kingdom-united-states-partnership-aukus/. For additional information about AUKUS and advanced capability development, see CRS Report R47599, *AUKUS Pillar 2: Background and Issues for Congress*, by Patrick Parrish and Luke A. Nicastro.

[51] For additional information about LAWS, see CRS Report R44466, *Lethal Autonomous Weapon Systems: Issues for Congress*, by Nathan J. Lucas.

also known as "human out of the loop" or "full autonomy." The directive contrasts LAWS with human- supervised, or "human on the loop," autonomous weapon systems, in which operators have the ability to monitor and halt a weapon's target engagement. Another category is semi-autonomous, or "human in the loop," weapon systems that "only engage individual targets or specific target groups that have been selected by a human operator."[52]

LAWS would require computer algorithms and sensor suites to classify an object as hostile, make an engagement decision, and guide a weapon to the target. Although these systems are not yet in widespread development,[53] it is believed they would enable military operations in communications-degraded or -denied environments where traditional systems may not be able to operate. Some analysts have noted that LAWS could additionally "allow weapons to strike military objectives more accurately and with less risk of collateral damage" or civilian casualties.[54]

Others, including approximately 30 countries and 165 nongovernmental organizations, have called for a preemptive ban on LAWS due to ethical concerns such as a perceived lack of accountability for use and a perceived inability to comply with the proportionality and distinction requirements of the law of armed conflict. Some analysts have also raised concerns about the potential operational risks posed by lethal autonomous weapons.[55] These risks could arise from "hacking, enemy behavioral manipulation, unexpected interactions with the environment, or simple malfunctions or software errors."[56] Although such risks could be present in automated systems, they could be heightened in autonomous systems, in which the human operator would be unable to physically intervene to terminate engagements—potentially resulting in wider-scale or more numerous instances of fratricide, civilian casualties, or other unintended consequences.[57]

---

[52] Department of Defense Directive 3000.09, "Autonomy in Weapon Systems," Updated January 25, 2023, at https://www.esd.whs.mil/portals/54/documents/dd/issuances/dodd/3000 09p.pdf.

[53] Some analysts have argued that certain loitering munitions such as the Israeli Harpy meet the United States' definition of LAWS. See, for example, *Defense Innovation Board, AI Principles: Recommendations on the Ethical Use of Artificial Intelligence by the Department of Defense - Supporting Document*, October 2019, p. 12, at https://media. defense.gov/2019/Oct/31/2002204459/-1/-1/0/ DIB_AI_PRINCIPLES_SUPPORTING_ DOCUMENT.PDF. In addition, while a United Nations report concluded that Turkey's deployment of the STM *Kargu-2* constitutes the first use of a lethal autonomous weapon system in combat, the U.N. described the *Kargu-2* as being "*programmed* to attack targets" [emphasis added]. For this reason, it is unlikely that the *Kargu-2* meets the U.S. definition of LAWS. United Nations Security Council, "Letter dated 8 March 2021 from the Panel of Experts on Libya established pursuant to resolution 1973 (2011) addressed to the President of the Security Council," March 8, 2021, p. 17, at https://undocs.org/S/2021/229.

[54] U.S. Government, "Humanitarian Benefits of Emerging Technologies in the Area of Lethal Autonomous Weapons," March 28, 2018, at https://ogc.osd.mil/Portals/99/Law%20of% 20War/Practice%20Documents/US %20Working%20Paper%20-%20Humanitarian %20benefits%20of%20emerging%20technologies%20in%20the%20area%20of%20LAWS%20-%20CCW_GGE.1_2018_WP.4_E.pdf?ver=O0lg6BIxsFt57nrOuz3xHA%3D%3D.

[55] See, for example, Paul Scharre, "Autonomous Weapons and Operational Risk," Center for a New American Security, February 2016, at https://s3.amazonaws.com/files.cnas.org/ documents/CNAS_Autonomous-weapons-operational- risk.pdf.

[56] Ibid.

[57] Ibid.

United States

The United States is not known to currently have LAWS in its inventory; however, there is no prohibition on the development, fielding, or employment of LAWS. DODD 3000.09 establishes DOD guidelines for the future development and fielding of LAWS to ensure that they comply with "the law of war, applicable treaties, weapon system safety rules, and applicable rules of engagement."[58] This directive includes a requirement that LAWS be designed to "allow commanders and operators to exercise appropriate levels of human judgment over the use of force."[59] "Human judgment over the use of force" does not require manual human "control" of the weapon system, as is often reported, but instead requires broader human involvement in decisions about how, when, where, and why the weapon will be employed.

In addition, DODD 3000.09 requires that the software and hardware of all systems, including lethal autonomous weapons, be tested and evaluated to ensure they

> [f]unction as anticipated in realistic operational environments against adaptive adversaries; complete engagements in a timeframe consistent with commander and operator intentions and, if unable to do so, terminate engagements or seek additional human operator input before continuing the engagement; and are sufficiently robust to minimize failures that could lead to unintended engagements or to loss of control of the system to unauthorized parties.

Any changes to a system's operating state—for example, due to machine learning—would require the system to be retested and reevaluated to ensure that it has retained its safety features and ability to operate as intended. In addition to the standard weapons review process, a secondary senior-level review is required for covered autonomous and semi-autonomous systems. This review requires the Under Secretary of Defense for Policy (USD[P]), the Vice Chairman of the Joint Chiefs of Staff (VCJCS), and the Under Secretary of Defense for Research and Engineering (USD[R&E]) to approve the system before formal development. USD(P), VCJCS, and the Under Secretary of Defense for Acquisition and Sustainment (USD[A&S]) must then approve the system before fielding. In the event of "urgent military need," this senior-level review may be waived by the Deputy Secretary of Defense. DODD 3000.09 additionally establishes the Autonomous Weapon System Working Group—composed of representatives of USD(P); USD(R&E); USD(A&S); DOD General Counsel; the Chief Digital and AI Officer; the Director, Operational Test and Evaluation; and the Chairman of the Joint Chiefs of Staff—to support and advise the senior-level review process.[60]

Per Section 251 of the FY2024 (P.L. 118-31), the Secretary of Defense is to notify the defense committees of any changes to DODD 3000.09 within 30 days. The Secretary is directed to provide a description of the modification and an explanation of the reasons for the modification.

---

58 Department of Defense Directive 3000.09, "Autonomy in Weapon Systems," Updated January 25, 2023, at https://www.esd.whs.mil/portals/54/documents/dd/issuances/dodd/300009p. pdf. For an explanation of this directive, see CRS In Focus IF11150, *Defense Primer: U.S. Policy on Lethal Autonomous Weapon Systems*, by Kelley M. Sayler.

59 Department of Defense Directive 3000.09, "Autonomy in Weapon Systems," Updated January 25, 2023, at https://www.esd.whs.mil/portals/54/documents/dd/issuances/dodd/30000 9p.pdf.

60 Ibid.

China

According to former U.S. Secretary of Defense Mark Esper, some Chinese weapons manufacturers, such as Ziyan, have advertised their weapons as having the ability to select and engage targets autonomously.[61] It is unclear whether these claims are accurate; however, China has no prohibition on the development of LAWS, which it has characterized as weapons that exhibit—at a minimum—five attributes:

> The first is lethality, which means sufficient pay load (charge) and for means [*sic*] to be lethal. The second is autonomy, which means absence of human intervention and control during the entire process of executing a task. Thirdly, impossibility for termination, meaning that once started there is no way to terminate the device. Fourthly, indiscriminate effect, meaning that the device will execute the task of killing and maiming regardless of conditions, scenarios and targets. Fifthly evolution, meaning that through interaction with the environment the device can learn autonomously, expand its functions and capabilities in a way exceeding human expectations.[62]

Russia

Russia has proposed the following definition of LAWS: "unmanned technical means other than ordnance that are intended for carrying out combat and support missions without any involvement of the operator" beyond the decision of whether and how to deploy the system.[63] Russia has noted that LAWS could "ensure the increased accuracy of weapon guidance on military targets, while contributing to lower rate of unintentional strikes against civilians and civilian targets."[64] Although Russia has not publicly stated that it is developing LAWS, Russian weapons manufacturer Kalashnikov has reportedly built a combat module for unmanned ground vehicles capable of autonomous target identification and, potentially, target engagement.[65]

International Institutions

Since 2014, the United States has participated in international discussions of LAWS under the auspices of the United Nations Convention on Certain Conventional Weapons (U.N. CCW). The U.N. CCW has considered proposals by states parties to issue political declarations about LAWS, as well as proposals to regulate or ban them. At the U.N. CCW, the United States and Russia have opposed a preemptive ban on LAWS, while China has supported a ban on the use—but not development—of LAWS, which it defines as weapon systems that are inherently

---

[61] Patrick Tucker, "SecDef: China is Exporting Killer Robots to the Mideast," *Defense One*, November 5, 2019.
[62] U.N. CCW, "China: Position Paper," April 11, 2018, p. 1, at https://unog.ch/80256EDD006B8954/(httpAssets)/E42AE83BDB3525D0C125826C0040B262/$file/CCW_GGE.1_2018_WP.7.pdf.
[63] U.N. CCW, "Russian Federation: Potential opportunities and limitations of military uses of lethal autonomous weapons systems," 2019, at https://unog.ch/80256EDD006B8954/(httpAssets)/B7C992A51A9FC8BFC12583BB00637BB9/$file/CCW.GGE.1.2019.WP.1_R+E.pdf.
[64] Ibid.
[65] Kyle Mizokami, "Kalashnikov Will Make an A.I.-Powered Killer Robot," *Popular Mechanics*, July 19, 2017.

indiscriminate and thus in violation of the law of war.[66] The United States has instead proposed the development of a nonbinding Code of Conduct, which "would help States promote responsible behavior and compliance with international law."[67]

The United States has additionally issued the "Political Declaration on Responsible Military Use of Artificial Intelligence and Autonomy," which notes that "military use of AI must be in compliance with applicable international law" and occur within a "responsible human chain of command and control."[68] To date, 50 states have endorsed the declaration.[69]

Potential Questions for Congress

- To what extent are potential U.S. adversaries developing LAWS? How, if at all, should this affect U.S. LAWS research and development?
- What role should the United States play in U.N. CCW discussions of LAWS? Should the United States support the status quo or advocate regulation of or a ban on LAWS?
- If the United States chooses to develop LAWS, are current weapons review processes and legal standards for their employment in conflict sufficient?

Hypersonic Weapons[70]

A number of countries, including the United States, Russia, and China, are developing hypersonic weapons—those that fly at speeds of at least Mach 5, or five times the speed of sound. In contrast to ballistic missiles, which also travel at hypersonic speeds, hypersonic weapons do not follow a parabolic ballistic trajectory and can maneuver en route to their destination, making defense against them difficult.

There are two categories of hypersonic weapons:

- *Hypersonic glide vehicles* are launched from a rocket before gliding to a target.[71]
- *Hypersonic cruise missiles* are powered by high-speed engines throughout the duration of their flight.

---

[66] For additional information about U.N. CCW discussions on LAWS, see CRS In Focus IF11294, *International Discussions Concerning Lethal Autonomous Weapon Systems*, by Kelley M. Sayler.

[67] U.S. Delegation to the U.N. CCW, "Opening Statement, as delivered by Joshua Dorosin," December 2, 2021, at https://geneva.usmission.gov/2021/12/03/convention-on-ccw-group-of-governmental-experts-on-emerging-technologies-in-the-area-of-laws/.

[68] Department of State, "Political Declaration on Responsible Military Use of Artificial Intelligence and Autonomy," November 9, 2023, at https://www.state.gov/political-declaration-on-responsible-military-use-of-artificial-intelligence- and-autonomy-2/.

[69] For a list of endorsing states, see https://www.state.gov/political-declaration-on-responsible-military-use-of-artificial- intelligence-and-autonomy/.

[70] For additional information about hypersonic weapons, see CRS Report R45811, *Hypersonic Weapons: Background and Issues for Congress*, by Kelley M. Sayler; and CRS In Focus IF11459, *Defense Primer: Hypersonic Boost-Glide Weapons*, by Kelley M. Sayler and Amy F. Woolf.

[71] When hypersonic glide vehicles are mated with their rocket booster, the resulting weapon system is often referred to as a hypersonic boost-glide weapon.

Analysts disagree about the strategic implications of hypersonic weapons. Some have identified two factors that could hold significant implications for strategic stability: (1) the weapon's short time-of-flight, which, in turn, compresses the timeline for response, and (2) its unpredictable flight path, which could generate uncertainty about the weapon's intended target and therefore heighten the risk of miscalculation or unintended escalation in the event of a conflict.[72]

Other analysts have argued that the strategic implications of hypersonic weapons are minimal because U.S. competitors such as China and Russia already possess the ability to strike the United States with intercontinental ballistic missiles, which, when launched in salvos, could overwhelm U.S. missile defenses.[73] Furthermore, these analysts argue that in the case of hypersonic weapons, traditional principles of deterrence hold: "it is really a stretch to try to imagine any regime in the world that would be so suicidal that it would even think threating to use—not to mention to actually use—hypersonic weapons against the United States ... would end well."[74]

United States

The Pentagon requested $4.7 billion in FY2023 for hypersonic weapons and $225.5 million for hypersonic defense programs.[75] The Pentagon declined to provide a breakout of funding for hypersonic-related research in FY2024, but requested $11 billion for long-range fires—a category that includes hypersonic weapons.[76] DOD is currently developing hypersonic weapons under the Navy's Conventional Prompt Strike program, which is intended to provide the U.S. military with the ability to strike hardened or time-sensitive targets with conventional warheads, as well as through several Air Force, Army, and DARPA programs.[77] Analysts who support these development efforts argue that hypersonic weapons could enhance deterrence, as well as provide the U.S. military with an ability to defeat capabilities such as mobile missile launchers and advanced air and missile defense systems that form the foundation of U.S.

---

[72] See, for example, Richard H. Speier et al., *Hypersonic Missile Proliferation: Hindering the Spread of a New Class of Weapons*, RAND Corporation, 2017, at https://www.rand.org /pubs/research_reports/RR2137.html.

[73] David Axe, "How the U.S. Is Quietly Winning the Hypersonic Arms Race," The Daily Beast, January 16, 2019, at https://www.thedailybeast.com/how-the-us-is-quietly-winning-the-hypersonic-arms-race. See also Mark B. Schneider, "Moscow's Development of Hypersonic Missiles," p. 14.

[74] Jyri Raitasalo, "Hypersonic Weapons are No Game-Changer," The National Interest, January 5, 2019, at https://nationalinterest.org/blog/buzz/hypersonic-weapons-are-no-game-changer-40632.

[75] Office of the Under Secretary of Defense (Comptroller)/Chief Financial Officer, *Defense Budget Overview: United States Department of Defense Fiscal Year 2023 Budget Request*, April 2022, p. 4-6, at https://comptroller.defense.gov/ Portals/45/Documents/defbudget/ FY2023/FY2023_Budget_Request_Overview_Book.pdf. For additional information about hypersonic missile defense, see CRS In Focus IF11623, *Hypersonic Missile Defense: Issues for Congress*, by Kelley M. Sayler.

[76] CRS correspondence with DOD Comptroller on October 12, 2023; November 27, 2023; and January 4, 2024.

[77] In a June 2018 memorandum, DOD announced that the Navy would lead the development of a common glide vehicle for use across the services. The services coordinate efforts on a Common Hypersonic Glide Body Board of Directors with rotating chairmanship. Sydney J. Freedberg, Jr., "Army Ramps Up Funding for Laser Shield, Hypersonic Sword," Breaking Defense, February 28, 2020, at https://breakingdefense.com/2020/02/army-ramps-up-funding-for-laser- shield-hypersonic-sword/. For a full history of U.S. hypersonic weapons programs, see CRS Report R41464, *Conventional Prompt Global Strike and Long-Range Ballistic Missiles: Background and Issues*, by Amy F. Woolf.

competitors' anti- access/area denial strategies.[78] Others have argued that hypersonic weapons confer little to no additional warfighting advantage and note that the U.S. military has yet to identify any mission requirements or concepts of operation for hypersonic weapons.[79]

The United States is unlikely to field an operational hypersonic weapon before FY2025; however, in contrast to Russia and China, the United States is not known to be developing hypersonic weapons for potential use with a nuclear warhead. As a result, the United States is seeking to develop hypersonic weapons that can attack targets with greater accuracy, which could be more technically challenging to develop than nuclear-armed—and less accurate—Russian and Chinese systems.

Section 237 of the FY2023 NDAA (P.L. 117-263) directs the Secretary of Defense to both assess DOD's capacity to test and evaluate hypersonic capabilities and "[identify] test facilities outside the Department of Defense that have potential to be used to expand [DOD] capacity ... including test facilities of other departments and agencies of the Federal Government, academia, and commercial test facilities." Section 218 of the FY2024 NDAA (P.L. 118-31) directs the Secretary to update this assessment at least once every two years. It additionally directs the Secretary to conduct a study to evaluate at least two possible locations in the United States that "have potential to be used as additional corridors for long-distance hypersonic system testing" and to submit to the congressional defense committees an annual report on DOD funding and investments in hypersonic capabilities.

China

According to Tong Zhao, a fellow at the Carnegie-Tsinghua Center for Global Policy, "most experts argue that the most important reason to prioritize hypersonic technology development [in China] is the necessity to counter specific security threats from increasingly sophisticated U.S. military technology" such as U.S. regional missile defenses.[80] China's pursuit of hypersonic weapons, like Russia's, reflects a concern that U.S. hypersonic weapons could enable the United States to conduct a preemptive, decapitating strike on China's nuclear arsenal and supporting infrastructure. U.S. missile defense deployments could then limit China's ability to conduct a retaliatory strike against the United States.[81]

China has developed the DF-41 intercontinental ballistic missile (ICBM), which, according to a 2014 report by the U.S.-China Economic and Security Review Commission, could carry a

---

[78] Roger Zakheim and Tom Karako, "China's Hypersonic Missile Advances and U.S. Defense Responses," remarks at the Hudson Institute, March 19, 2019. See also Department of Defense Fiscal Year (FY) 2020 Budget Estimates, Army Justification Book of Research, Development, Test and Evaluation, Volume II, Budget Activity 4, p. 580.

[79] See, for example, Valerie Insinna, "Air Force's top civilian hints at changes to hypersonic weapons programs," *Defense News*, September 22, 2021, at https://www. defensenews. com/air/2021/09/22/air-forces-top-civilian-hints-at-changes-to-hypersonic-weapons-programs/.

[80] Tong Zhao, "Conventional Challenges to Strategic Stability: Chinese Perceptions of Hypersonic Technology and the Security Dilemma," Carnegie-Tsinghua Center for Global Policy, July 23, 2018, at https://carnegietsinghua.org/2018/ 07/23/conventional-challenges-to-strategic-stability-chinese-perceptions-of-hypersonic-technology-and-security- dilemma-pub-76894.

[81] Ibid.; and Lora Saalman, "China's Calculus on Hypersonic Glide," August 15, 2017, Stockholm International Peace Research Institute, at https://www.sipri.org/commentary/topical-backgrounder/2017/chinas-calculus-hypersonic-glide.

nuclear hypersonic glide vehicle.[82] General Terrence O'Shaughnessy, then-commander of U.S. Northern Command, seemed to confirm this assessment in February 2020, when he testified that "China is testing a [nuclear-capable] intercontinental-range hypersonic glide vehicle … which is designed to fly at high speeds and low altitudes, complicating our ability to provide precise warning."[83]

Reports indicate that China may have tested a nuclear-capable HGV[84]—launched by a Long March rocket—in August 2021.[85] In contrast to the ballistic missiles that China has previously used to launch HGVs, the Long March, a fractional orbital bombardment system (FOBS), launches the HGV into orbit before the HGV de-orbits to its target. This could provide China with a space-based global strike capability and further reduce the amount of target warning time prior to a strike.[86]

China has additionally tested the DF-ZF hypersonic glide vehicle at least nine times since 2014.

U.S. defense officials have reportedly identified the range of the DF-ZF as approximately 1,200 miles and have stated that the missile may be capable of performing evasive maneuvers during flight.[87] Although unconfirmed by intelligence agencies, some analysts believe the DF-ZF could have become operational as early as 2020.[88] In addition, in August 2018 China successfully tested Starry Sky-2, a nuclear-capable hypersonic vehicle prototype.[89] Some reports indicate that the Starry Sky-2 could be operational by 2025.[90] U.S. officials have declined to comment on the program.[91]

---

[82] *U.S.-China Economic and Security Review Commission 2014 Annual Report,* p. 292, at https://www.uscc.gov/sites/default/files/annual_reports/Complete%20Report.PDF.

[83] General Terrence J. O'Shaughnessy, "Statement before the Senate Armed Services Committee," February, 13, 2020, at https://www.armed-services.senate.gov/imo/media/ doc/OShaughnessy_02-13-20.pdf.

[84] It is not clear if this nuclear-capable HGV is the same model as that referenced by General O'Shaughnessy.

[85] Demetri Sevastopulo and Kathrin Hille, "China tests new space capability with hypersonic missile," October 16, 2021, at https://www.ft.com/content/ba0a3cde-719b-4040-93cb-a486e1f843fb. China's Foreign Ministry Spokesperson Zhao Lijian has stated that "this was a routine test of [a] space vehicle," rather than a test of a nuclear-capable HGV. Zhao Lijian, "Remarks at Regular Press Conference," Ministry of Foreign Affairs of the People's Republic of China, October 18, 2021, at https://www.fmprc.gov.cn/mfa_eng/xwfw_665399/s2510_665401/t1915130.shtml.

[86] Greg Hadley, "Kendall: China Has Potential to Strike Earth From Space," *Air Force Magazine*, September 20, 2021, at https://www.airforcemag.com/global-strikes-space-china-frank-kendall/.

[87] "Gliding missiles that fly faster than Mach 5 are coming," *The Economist*, April 6, 2019, at https://www.economist.com/science-and-technology/2019/04/06/gliding-missiles-that-fly-faster-than-mach-5-are-coming; and Franz-Stefan Gady, "China Tests New Weapon Capable of Breaching US Missile Defense Systems," *The Diplomat*, April 28, 2016, at https://thediplomat.com/2016/04/china-tests-new-weapon-capable-of-breaching-u-s- missile-defense-systems/.

[88] *U.S.-China Economic and Security Review Commission 2015 Annual Report*, p. 20, at https://www.uscc.gov/sites/default/files/annual_reports/2015%20Annual%20Report%20to%20Congress.PDF.

[89] Jessie Yeung, "China claims to have successfully tested its first hypersonic aircraft," *CNN*, August 7, 2018, at https://www.cnn.com/2018/08/07/china/china-hypersonic-aircraft-intl/index.html. See also *U.S.-China Economic and Security Review Commission 2018 Annual Report*, p. 220, at https://www.uscc.gov/sites/default/files/annual_reports/2018% 20Annual%20Report%20to%20Congress.pdf.

[90] *U.S.-China Economic and Security Review Commission Report 2015,* p. 20.

[91] Bill Gertz, "China Reveals Test of New Hypersonic Missile," *The Washington Free Beacon*, August 10, 2018, at https://freebeacon.com/national-security/chinas-reveals-test-new-hyper sonic-missile/.

Russia

Although Russia has conducted research on hypersonic weapons technology since the 1980s, it accelerated its efforts in response to U.S. missile defense deployments in both the United States and Europe, and in response to the U.S. withdrawal from the Anti-Ballistic Missile Treaty in 2002.[92] Detailing Russia's concerns, President Putin stated in 2018 that "the US is permitting constant, uncontrolled growth of the number of anti-ballistic missiles, improving their quality, and creating new missile launching areas. If we do not do something, eventually this will result in the complete devaluation of Russia's nuclear potential. Meaning that all of our missiles could simply be intercepted."[93] Russia thus seeks hypersonic weapons, which can maneuver as they approach their targets, as an assured means of penetrating U.S. missile defenses and restoring its sense of strategic stability.[94]

Russia is pursuing two nuclear-capable hypersonic weapons: the Avangard and the 3M22 Tsirkon (or Zircon).[95] Avangard is a hypersonic glide vehicle launched from an ICBM, giving it "effectively 'unlimited' range."[96] Reports indicate that Avangard is currently deployed on the SS- 19 Stiletto ICBM, though Russia plans to eventually launch the vehicle from the Sarmat ICBM. Sarmat reportedly entered combat duty in September 2023.[97] Tsirkon has a range of between approximately 250 and 600 miles and can be fired from the vertical launch systems mounted on cruisers Admiral Nakhimov and Pyotr Veliky, Project 20380 corvettes, Project 22350 frigates, and Project 885 Yasen-class submarines, among other platforms.[98] Russian news sources assert that Tsirkon was successfully launched from a Project 22350 frigate in January, October, and December 2020 and May 2022, and from a Project 885 Yasen-class

---

[92] United Nations Office of Disarmament Affairs, *Hypersonic Weapons: A Challenge and Opportunity for Strategic Arms Control*, February 2019, at https://www.un.org/disar mament/publications/more/hypersonic-weapons-a-challenge-and-opportunity-for-strategic-arms-control/.

[93] Vladimir Putin, "Presidential Address to the Federal Assembly," March 1, 2018, at http://en.kremlin.ru/events/president/news/56957.

[94] In this instance, "strategic stability" refers to a "bilateral nuclear relationship of mutual vulnerability." See Tong Zhao, "Conventional Challenges to Strategic Stability: Chinese Perceptions of Hypersonic Technology and the Security Dilemma," Carnegie-Tsinghua Center for Global Policy, July 23, 2018, at https://carnegietsinghua.org/2018/07/23/ conventional-challenges-to-strategic-stability-chinese-perceptions-of-hypersonic-technology-and-security-dilemma-pub-76894.

[95] Although it is frequently characterized as a "hypersonic weapon"—including by U.S. defense officials, a third weapon, the Kinzhal, is a maneuvering air-launched ballistic missile. See, for example, CSIS Missile Defense Project, "Kh-47M2 Kinzhal," Updated March 19, 2022, at https://missilethreat.csis.org/missile/kinzhal/.

[96] Steve Trimble, "A Hypersonic Sputnik?" *Aviation Week*, January 14-27, 2019, p. 20.

[97] Al Jazeera, "Russia puts advanced Sarmat nuclear missile system on 'combat duty,'" September 2, 2023, at https://www.aljazeera.com/news/2023/9/2/russia-puts-advanced-sarmat-nuclear-missile-system-on-combat-duty#:~:text=Russia%20test%2Dfired% 20the% 20Sarmat,in%20Russia's%20far%20east%20region. Sarmat could reportedly accommodate at least three Avangard vehicles. See Malcolm Claus, "Russia unveils new strategic delivery systems," *Jane's* (subscription required), at https://janes.ihs.com/ Janes/Display/FG_899127-JIR.

[98] "Russia makes over 10 test launches of Tsirkon seaborne hypersonic missile," *TASS*, December 21, 2018, at http://tass.com/defense/1037426. See also *Russia Military Power: Building a Military to Support Great Power Aspirations*, Defense Intelligence Agency, 2017, p. 79, at https://www.dia.mil/portals/27/documents/news/ military%20power% 20publications/russia%20military%20power%20report%202017.pdf.

submarine in October 2021.[99] Russia reportedly deployed Tsirkon on the Project 22350 frigate Admiral of the Fleet of the Soviet Union Gorshkov in January 2023.[100]

International Institutions

No international treaty or agreement is dedicated to overseeing the development of hypersonic weapons. Although the New START Treaty—a strategic offensive arms treaty between the United States and Russia—does not specifically limit hypersonic weapons, it does limit ICBMs, which could be used to launch hypersonic glide vehicles.[101] Because Russia has deployed its Avangard hypersonic glide vehicle on an SS-19 ICBM, it has agreed that missiles equipped with Avangard count under New START. Furthermore, Article V of the treaty states that "when a Party believes that a new kind of strategic offensive arm is emerging, that Party shall have the right to raise the question of such a strategic offensive arm for consideration in the Bilateral Consultative Commission (BCC)." Accordingly, some legal experts hold that it would be possible to negotiate provisions that would count additional types of hypersonic weapons under the New START limits.[102] However, because New START is due to expire in 2026, this may be a short-term solution.[103] In addition, the treaty would not cover hypersonic weapons developed in countries other than the United States and Russia.[104]

Finally, some analysts have noted that if any parties to the Outer Space Treaty were to launch a nuclear-armed HGV on a fractional orbital bombardment system, they would likely be in violation of Article IV of the treaty, which prohibits the placement of "any objects carrying nuclear weapons or any other kinds of weapons of mass destruction" into orbit.[105]

Potential Questions for Congress

- What mission(s) will hypersonic weapons be used for? Are hypersonic weapons the most cost-effective means of executing these potential missions?

99 "TASS: Russia Conducts First Ship-Based Hypersonic Missile Test," *Reuters*, February 27, 2020, at https://www.voanews.com/europe/tass-russia-conducts-first-ship-based-hyper sonic-missile-test; and Samuel Cranny- Evans, "Russia conducts first submarine test launches of Tsirkon hypersonic missile," *Jane's* (subscription required), October 4, 2021.

100 Guy Faulconbridge, "Putin deploys new Zircon hypersonic cruise missiles to Atlantic," Reuters, January 4, 2023, at https://www.reuters.com/world/europe/putin-sends-off-frigate-armed-with-new-hypersonic-cruise-missile-2023-01-04/.

101 For example, Russia's Avangard hypersonic glide vehicle is reportedly launched by an intercontinental ballistic missile. See Rachel S. Cohen, "Hypersonic Weapons: Strategic Asset or Tactical Tool?," *Air Force Magazine*, May 7, 2019, at https://www.airforcemag. com/hypersonic-weapons-strategic-asset-or-tactical-tool/.

102 James Acton notes: "during [New START] negotiations, Russia argued that boost-glide weapons might constitute 'a new kind of strategic offensive arm,' in which case they would trigger bilateral discussions about whether and how they would be regulated by the treaty—a position [then] rejected by the United States." James M. Acton, *Silver Bullet?: Asking the Right Questions about Conventional Prompt Global Strike*, Carnegie Endowment for International Peace, 2013, p. 139, at https://carnegieendowment.org/files/cpgs.pdf.

103 CRS Report R41219, The New START Treaty: Central Limits and Key Provisions, by Amy F. Woolf.

104 In addition to China, countries with hypersonic weapons development programs include Australia, India, France, Germany, South Korea, North Korea, and Japan. For additional information, see CRS Report R45811, *Hypersonic Weapons: Background and Issues for Congress*, by Kelley M. Sayler.

105 Jeffrey Lewis, "China's Orbital Bombardment System Is Big, Bad News—but Not a Breakthrough," *Foreign Policy*, October 18, 2021, at https://foreignpolicy.com/2021/ 10/18/hypersonic-china-missile-nuclear-fobs/.

- Given the lack of defined mission requirements for hypersonic weapons, how should Congress evaluate funding requests for hypersonic weapons programs or the balance of funding requests for hypersonic weapons programs, enabling technologies, and supporting test infrastructure?
- How, if at all, will the fielding of hypersonic weapons affect strategic stability? Is there a need for risk-mitigation measures, such as expanding New START, negotiating new multilateral arms control agreements, or undertaking transparency and confidence-building activities?

Directed Energy (DE) Weapons[106]

DOD defines directed energy (DE) weapons as those using concentrated electromagnetic energy, rather than kinetic energy, to "incapacitate, damage, disable, or destroy enemy equipment, facilities, and/or personnel."[107] DE weapons could be used by ground forces in short-range air defense (SHORAD), counter-unmanned aircraft systems (C-UAS), or counter-rocket, artillery, and mortar (C-RAM) missions.[108] DE weapons could offer low costs per shot and—assuming access to a sufficient power supply[109]—nearly limitless magazines that, in contrast to existing conventional systems, could enable an efficient and effective means of defending against missile salvos or swarms of unmanned systems. Theoretically, DE weapons could also provide options for boost-phase missile intercept, given their speed-of-light travel time; however, as in the case of hypersonic missile defense, experts disagree on the affordability, technological feasibility, and utility of this application.[110]

High-powered microwave weapons, a subset of DE weapons, could be used as a nonkinetic means of disabling electronics, communications systems, and improvised explosive devices, or as a nonlethal "heat ray" system for crowd control.

United States

Although the United States has been researching directed energy since the 1960s, some experts have observed that "actual directed energy programs … have frequently fallen short of expectations," with DOD investing billions of dollars in programs that were ultimately

---

[106] For additional information about directed energy weapons, see CRS Report R46925, *Department of Defense Directed Energy Weapons: Background and Issues for Congress*, coordinated by Kelley M. Sayler.

[107] Joint Chiefs of Staff, *Joint Electromagnetic Spectrum Operations, Joint Publication 3-85*, May 22, 2020, GL-6.

[108] For more information about the role of DE weapons in C-UAS missions, see CRS In Focus IF11426, *Department of Defense Counter-Unmanned Aircraft Systems*, by John R. Hoehn and Kelley M. Sayler.

[109] Although research has been conducted on chemically fueled lasers, most countries are now pursuing solid state lasers, which are fueled by electrical power. As a result, the cost per shot is equivalent to the cost of the electrical power required to fire the shot. See Ariel Robinson, "Directed Energy Weapons: Will They Ever Be Ready?," *National Defense*, July 1, 2015, at https://www.nationaldefensemagazine.org/articles/2015/7/1/2015july-directed-energy-weapons-will-they-ever-be-ready.

[110] See, for example, James N. Miller and Frank A. Rose, "Bad Idea: Space-Based Interceptors and Space-Based Directed Energy Systems," Center for Strategic and International Studies, December 13, 2018, at https://defense360.csis.org/bad-idea-space-based-interceptors-and-space-based-directed-energy-systems/; and Justin Doubleday, "Pentagon punts MDA's laser ambitions, shifts funding toward OSD-led 'laser scaling,'" *Inside Defense*, February 19, 2020, at https://insidedefense.com/daily-news/pentagon-punts-mdas-laser-ambitions-shifts-funding- toward-osd-led-laser-scaling.

cancelled.[111] Others contend that developments in commercial lasers could be leveraged for military applications.[112] Directed energy weapons programs continue, however, to face questions about their technological maturity, including questions about the ability to improve beam quality and control to militarily useful levels and the ability to meet power, cooling, and size requirements for integration into current platforms.[113]

The U.S. Navy fielded the first operational U.S. DE weapon, the Laser Weapon System (LaWS), in 2014 aboard the USS Ponce. LaWS was a 30-kilowatt (-kW) laser prototype that "was capable of blinding enemy forces as a warning, shooting down drones, disabling boats, or damaging helicopters."[114] In addition, the Navy installed its 60-kW laser, HELIOS, on the USS Preble in FY2022 and plans to continue at-sea testing through at least the end of FY2028.[115] Similarly, the Army delivered a platoon of 50-kW Directed Energy Mobile Short-Range Air Defense System prototypes to the 4th Battalion, 60th Air Defense Artillery Regiment (Fort Sill, Oklahoma) in 2023.[116] The Air Force is currently conducting field assessments of several counter-UAS DE systems, including both laser and high-powered microwave systems.[117]

Overall, DOD requested approximately $1 billion for directed energy weapons programs in FY2024.[118] Many of these programs are intended to support DOD's Directed Energy Roadmap. According to OUSD(R&E), DOD's DE roadmap outlines DOD's plans to increase power levels of HEL weapons from around 150 kilowatt (kW), as is currently feasible, to 500 kW class—with reduced size and weight—by FY2025. DOD seeks "to further reduce size and weight and increase power to MW [megawatt] levels by FY2026."[119]

---

111 Paul Scharre, Preface to "Directed-Energy Weapons: Promise and Prospects," Center for a New American Security, April 2015, p. 4.

112 See Ariel Robinson, "Directed Energy Weapons: Will They Ever Be Ready?," *National Defense*, July 1, 2015, at https://www.nationaldefensemagazine.org/articles/2015/7/1/2015july-directed-energy-weapons-will-they-ever-be- ready.

113 Ibid.

114 Kyle Mizokami, "The U.S. Army Plans To Field the Most Powerful Laser Weapon Yet," *Popular Mechanics*, August 7, 2019. Kilowatts are units of power. For example, 1 kilowatt is equal to 1,000 watts.

115 DOD, *Department of Defense, Fiscal Year (FY) 2024 Budget Estimates, Navy Justification Book Volume 2 of 5*, Research, Development, Test & Evaluation, Navy, March 2023, p. 979, at https://www.secnav.navy.mil/fmc/fmb/Documents/24pres/RDTEN_BA4_Book. pdf.

116 Venetia Gonzales, "Groundbreaking laser prototype systems delivered to 4-60th Air Defense Artillery Regiment," U.S. Army, September 21, 2023, at https://www.army.mil/article/270134/groundbreaking_laser_prototype_systems_delivered_to_4_60th_air_defense_artil lery_regiment.

117 For a more in-depth discussion of U.S. DE programs, see CRS Report R46925, *Department of Defense Directed Energy Weapons: Background and Issues for Congress*, coordinated by Kelley M. Sayler.

118 CRS conversation with Principal Director for Directed Energy Dr. Frank Peterkin, May 17, 2023. See also Government Accountability Office, *Directed Energy Weapons: DOD Should Focus Transition on Planning*, April 2023, p. 1.

119 CRS correspondence with the Office of the Under Secretary of Defense for Research and Engineering, September 8, 2022. For reference, although no consensus exists regarding the precise power level that would be needed to neutralize different target sets, DOD briefing documents suggest that a laser of approximately 100 kW could engage UASs, rockets, artillery, and mortars, whereas a laser of around 300 kW could additionally engage small boats and cruise missiles flying in certain profiles (i.e., flying across—rather than at—the laser). Lasers of 1 MW could potentially neutralize ballistic missiles and hypersonic weapons. Dr. Jim Trebes, "Advancing High Energy Laser Weapon Capabilities: What is OUSD (R&E) Doing?," presentation at IDGA, October 21, 2020; and CRS conversation with then-Principal Director for Directed Energy Modernization Dr. Jim Trebes, November 17, 2020. Required power levels could be affected by additional factors such as adversary countermeasures and atmospheric conditions and effects.

China

According to the US-China Economic and Security Review Commission, China has been developing DE weapons since at least the 1980s and has made steady progress in developing HPM and increasingly powerful HELs.[120] China has reportedly developed a 30-kilowatt road-mobile DE system, LW-30, designed to engage unmanned aerial vehicles and precision-guided weapons.[121] Reports indicate that China is also developing an airborne DE weapon pod and has used or proposed using DE weapons to interfere with U.S. and allied military aircraft and to disrupt U.S. freedom of navigation operations in the Indo-Pacific.[122]

According to the Defense Intelligence Agency, China is additionally pursuing DE weapons

> to disrupt, degrade, or damage satellites and their sensors and possibly already has a limited capability to employ laser systems against satellite sensors. China likely will field a ground-based laser weapon that can counter low-orbit space-based sensors by 2020, and by the mid-to-late 2020s, it may field higher power systems that extend the threat to the structures of non-optical satellites.[123]

Russia

Russia has been conducting DE weapons research since the 1960s, with a particular emphasis on HELs. Russia has reportedly deployed the Peresvet, a mobile, ground-based HEL, with several mobile intercontinental ballistic missile units. Although little is publicly known about Peresvet, including its power level, some analysts assert it is to dazzle satellites and provide point defense against unmanned aircraft systems.[124] Russia's deputy defense minister Alexei Krivoruchko has stated that efforts are underway to increase Peresvet's power level and to deploy it on military aircraft.[125] Reports suggest that Russia may also be developing HPMs as well as additional HELs capable of performing antisatellite missions.

---

[120] US-China Economic and Security Review Commission (USCC), *USCC 2017 Annual Report*, November 2017, p. 563, at https://www.uscc.gov/sites/default/files/2019-09/2017_ Annual_Report_to_Congress.pdf.

[121] Nikolai Novichkov, "Airshow China 2018: CASIC's LW-30 laser weapon system breaks cover," *Jane's Defence Weekly*, November 9, 2018.

[122] Andrew Tate, "China aiming to procure airborne laser-based weapon pod," *Jane's Defence Weekly*, January 8, 2020; and Patrick M. Cronin and Ryan D. Neuhard, "Countering China's Laser Offensive," The Diplomat, April 2, 2020, at https://thediplomat. com/2020/04/countering-chinas-laser-offensive/.

[123] Defense Intelligence Agency, *Challenges to Security in Space*, February 2019, p. 20, at https://www.dia.mil/Portals/27/Documents/News/Military%20Power%20Publications/Space_Threat_V14_020119_sm.pdf.

[124] Defense Intelligence Agency, *Challenges to Security in Space*, February 2019, p. 23, at https://www.dia.mil/Portals/27/Documents/News/Military%20Power%20Publications/Space_Threat_V14_020119_sm.pdf; and "Putin hails new Russian laser weapons," *Associated Press*, May 17, 2019, at https://apnews.com/ff03960c48a6440bacc1c2512 a7c197a.

[125] Bart Hendrickx, "Peresvet: a Russian mobile laser system to dazzle enemy satellites," *The Space Review*, June 5, 2020, at https://www.thespacereview.com/article/3967/1.

International Institutions

DE weapons "are not authoritatively defined under international law, nor are they currently on the agenda of any existing multilateral mechanism."[126] However, certain applications of DE weapons are prohibited. For example, Protocol I of the CCW "Protocol on Blinding Lasers" prohibits the employment of "laser weapons specifically designed, as their sole combat function or as one of their combat functions, to cause permanent blindness to unenhanced vision."[127] Some analysts have suggested that multilateral agreements should be considered. For example, Congress may consider prohibitions on nonlethal anti-personnel uses of DE weapons—such as "heat rays" or lasers intended to cause temporary visual impairment—or on certain military applications of DE weapons—such as aircraft interference—in peacetime.[128] Other analysts have argued that DE weapons could be considered more humane than conventional weapons because their accuracy could potentially reduce collateral damage and because they could provide a nonlethal anti- personnel capability in circumstances in which lethal force might otherwise be used.[129]

Potential Questions for Congress

- Does the technological maturity of DE weapons warrant current funding levels? To what extent, if at all, can advances in commercial lasers be leveraged for military applications?
- How successful have U.S. field tests of DE weapons been? Are any changes to operational concepts, rules of engagement, or tactics required to optimize the use of DE weapons or deconflict the use of DE weapons with other U.S. military operations?
- In what circumstances and for what purposes should the U.S. military's use of DE weapons be permissible? What, if any, regulations, treaties, or other measures should the United States consider with regard to the use of DE weapons in both war and peacetime?

---

126 "Directed Energy Weapons: Discussion paper for the Convention on Certain Conventional Weapons (CCW)," Article 36, November 2017.

127 The protocol does not cover the development, procurement, or possession of such weapons, nor does it prohibit the employment of laser weapons that may cause blindness "as an incidental or collateral effect." *Additional Protocol to the Convention on Prohibitions or Restrictions on the Use of Certain Conventional Weapons Which May Be Deemed to Be Excessively Injurious or to Have Indiscriminate Effects*, Vienna, October 13, 1995, United Nations, Treaty Series, vol. 1380, p. 370, at https://treaties.un.org/doc/Treaties/ 1995/10/19951013%2001-30%20AM/Ch_XXVI_02_ap.pdf. For additional information about the protocol and its relationship to DE weapons programs, see Appendix I of CRS Report R41526, *Navy Shipboard Lasers for Surface, Air, and Missile Defense: Background and Issues for Congress*, by Ronald O'Rourke.

128 Patrick M. Cronin and Ryan D. Neuhard, "Countering China's Laser Offensive," The Diplomat, April 2, 2020, at https://thediplomat.com/2020/04/countering-chinas-laser-offen sive/.

129 See, for example, Mark Gunzinger and Chris Dougherty, *Changing the Game: The Promise of Directed-Energy Weapons*, Center for Strategic and Budgetary Assessments, April 19, 2021, at https://csbaonline.org/uploads/documents/CSBA_ChangingTheGame_ereader. pdf.

Biotechnology

Biotechnology leverages life sciences for technological applications. A number of developments in biotechnology hold potential implications for the U.S. military and for international security writ large. As a 2018 Government Accountability Office report notes, the Departments of Defense, State, and Homeland Security, and the Office of the Director of National Intelligence assess that biotechnologies, such as the low-cost gene-editing tool CRISPR,[130] have the potential to

> alter genes or create DNA to modify plants, animals, and humans. Such biotechnologies could be used to enhance [or degrade] the performance of military personnel. The proliferation of synthetic biology—used to create genetic code that does not exist in nature—may increase the number of actors that can create chemical and biological weapons.[131]

Similarly, the U.S. intelligence community's 2016 Worldwide Threat Assessment cited genome editing as a potential weapon of mass destruction.[132]

In addition, biotechnology could be used to create adaptive camouflage, cloaking devices, or lighter, stronger, and—potentially—self-healing body and vehicle armor.[133] Concerns have been raised that U.S. competitors may not hold the same ethical standards in the research and application of biotechnologies, particularly regarding biological weapons, genome editing, or more invasive forms of human performance modification.[134]

United States

Overall, DOD requested $1.3 billion for biotechnology efforts in FY2023; the department did not provide a topline breakout of its request for biotechnology in FY2024.[135] Unclassified U.S. biotechnology programs with military applications center primarily on improving "readiness,

---

[130] For a general overview of CRISPR, see CRS Report R44824, *Advanced Gene Editing: CRISPR-Cas9*, by Marcy E. Gallo et al.

[131] Government Accountability Office, *National Security: Long-Range Emerging Threats Facing the United States as Identified by Federal Agencies*, December 2018, at https://www. gao.gov/assets/700/695981.pdf.

[132] James R. Clapper, "Statement for the Record: Worldwide Threat Assessment of the US Intelligence Community," delivered before the U.S. Senate Committee on Armed Services, February 9, 2016.

[133] Patrick Tucker, "The US Army Is Making Synthetic Biology a Priority," *Defense One*, July 1, 2019; and "Army scientists explore synthetic biology potential," U.S. Army, June 24, 2019, at https://www.army.mil/article/223495/ rmy_scientists_explore_synthetic_biology_ potential.

[134] James R. Clapper, "Statement for the Record: Worldwide Threat Assessment of the US Intelligence Community," delivered before the U.S. Senate Committee on Armed Services, February 9, 2016; and Daniel R. Coats, "Statement for the Record: Worldwide Threat Assessment of the US Intelligence Community," delivered before the U.S. Senate Committee on Armed Services, March 6, 2018. Although the U.S. military has long used certain drugs such as caffeine, modafinil, dextroamphetamine, and various sleep aids to enhance soldier performance, it bans other performance-enhancing drugs and techniques such as anabolic steroids and blood doping. See Paul Scharre and Lauren Fish, *Human Performance Enhancement*, Center for a New American Security, November 7, 2018, at https://www.cnas.org/ publications/reports/human-performance-enhancement-1.

[135] Office of the Under Secretary of Defense (Comptroller)/Chief Financial Officer, Defense Budget Overview: United States Department of Defense Fiscal Year 2023 Budget Request, April 2022, p. 4-7, at https://comptroller.defense.gov/Portals/45/Documents/defbudget/ FY2023/FY2023_Budget_Request_Overview_Book. pdf.

resilience, and recovery." DARPA, for example, has a number of biotechnology programs devoted to battlefield medicine, diagnostics, and prognostics. It is also exploring options for mitigating the effects of traumatic brain injury, treating neuropsychiatric illnesses such as depression and post-traumatic stress, and protecting against infectious diseases and bio-engineered threats to the U.S. food supply. In addition, DARPA's Safe Genes program seeks "to [protect] service members from accidental or intentional misuse of genome editing technologies."[136]

DOD is also exploring "advanced development and operational testing of biomanufacturing capabilities to provide alternative sources for critical supply chain materials" and other "products needed for non-medical supply chain resiliency," and is to invest $1 billion in bioindustrial domestic manufacturing infrastructure over five years.[137] Similarly, DOD has announced the establishment of a Tri-Service Biotechnology for a Resilient Supply Chain program, which is to include "more than $270 million investment over five years … to support the advanced development of bio-based materials for defense supply chains, such as fuels, fire-resistant composites, polymers and resins, and protective materials."[138] In addition, Section 215 of the FY2023 NDAA directs the Secretary of Defense, subject to the availability of appropriations, to provide support for the development of a network of bioindustrial manufacturing facilities.[139] In support of these efforts, DOD released its Biomanufacturing Strategy in March 2023.[140]

Some reports suggest that the United States is also researching or has previously researched biotechnology and neuroscience applications to increase soldier lethality, including applications to make soldiers "stronger, smarter, [and] more capable, and … give them more endurance than other humans."[141] Some groups have expressed ethical concerns about this

---

136 See Defense Advanced Research Projects Agency, "Our Research: Biological Technologies Office," at https://www.darpa.mil/our-research?tFilter=&oFilter=1.

137 Office of the Under Secretary of Defense (Comptroller)/Chief Financial Officer, *Defense Budget Overview: United States Department of Defense Fiscal Year 2023 Budget Request*, April 2022, p. 4-7, at https://comptroller.defense.gov/Portals/45/Documents/defbudget/FY2023/FY2023_Budget_Request_Overview_Book.pdf; and The White House, "FACT SHEET: The United States Announces New Investments and Resources to Advance President Biden's National Biotechnology and Biomanufacturing Initiative," September 14, 2022, at https://www.whitehouse.gov/briefing- room/statements-releases/2022/09/14/fact-sheet-the-united-states-announces-new-investments-and-resources-to-advance-president-bidens-national-biotechnology-and-biomanufacturing-initiative/.

138 The White House, "FACT SHEET: The United States Announces New Investments and Resources to Advance President Biden's National Biotechnology and Biomanufacturing Initiative," September 14, 2022, at https://www.whitehouse.gov/briefing-room/statements-releases/2022/09/14/fact-sheet-the-united-states-announces-new-investments-and-resources-to-advance-president-bidens-national-biotechnology-and-biomanufacturing-initiative/.

139 P.L. 117-263, Section 2, Division A, Title II, §215. DOD's FY2024 budget request notes that, following a $300 million congressional add in support of Section 215, the BioMADE Manufacturing Innovation Institute—a DOD- sponsored public-private partnership headquartered at the University of Minnesota in St. Paul—is to initiate the development of the network. DOD, *Department of Defense Fiscal Year (FY) 2024 Budget Estimates, Office of the Secretary of Defense, Defense-Wide Justification Book Volume 3 of 5 Research, Development, Test & Evaluation,* March 2023, p. 296, at https://comptroller.defense. gov/Portals/45/Documents/defbudget/fy2024/budget_justification/pdfs/03_RDT_and_E/OS D_PB2024.pdf.

140 Department of Defense, *Biomanufacturing Strategy*, March 21, 2023, at https://media.defense.gov/2023/Mar/22/2003184301/-1/-1/1/BIOMANUFACTU RING-STRATEGY.PDF.

141 Annie Jacobsen, *The Pentagon's Brain: An Uncensored History of DARPA, America's Top-Secret Military Research Agency* (New York: Little, Brown and Company, 2015). See also Michael Joseph Gross, "The Pentagon's Push to Program Soldiers' Brains," *The Atlantic*, November 2018, at https://www.theatlantic.com/magazine/archive/2018/11/the-pentagon-wants-to-weaponize-the-brain-what-could-go-wrong/570841/.

research; although the United States had a series of presidential bioethics commissions between 1974 and 2017, there is no current national framework for examining ethical concerns.[142]

Although there does not appear to be a DOD-specific biotechnology research strategy, the Biden Administration released the *National Biodefense Strategy and Implementation Plan for Countering Biological Threats, Enhancing Pandemic Preparedness, and Achieving Global Health Security*—as well as an associated *National Security Memorandum on Countering Biological Threats, Enhancing Pandemic Preparedness, and Achieving Global Health Security (NSM-15)*— in October 2022.[143] These documents outline "how the U.S. Government will manage its activities to more effectively assess, prevent, prepare for, respond to, and recover from biological threats, coordinating its biodefense efforts with those of [state, local, tribal, and territorial] entities, international partners, industry, academia, nongovernmental entities, and the private sector."[144] The Administration notes that, while it is currently "implementing key actions in the Strategy with existing funding," full implementation "will require the support of Congress to provide additional resources, including the President's $88 billion request over five years for pandemic preparedness and biodefense."[145]

Congress has expressed an interest in conducting oversight of the military applications of emerging biotechnologies. For example, per Section 263 of the FY2020 NDAA (P.L. 116-92), DOD is to conduct "a review of the military understanding and relevancy of applications of emerging biotechnologies to national security requirements of the Department of Defense" and provide recommendations for future legislative and administrative activities."[146] Section 278 of the FY2021 NDAA (P.L. 116-283) directs DOD to "conduct an assessment and direct comparison of capabilities in emerging biotechnologies for national security purposes ... between the capabilities of the United States and the capabilities of adversaries of the United States."[147] Similarly, Section 1312 of the FY2024 NDAA (P.L. 118-31) directs the Secretary of Defense to conduct "an analysis to determine if any biotechnology entity, or any subsidiary,

---

[142] For a history of these commissions, see Presidential Commission for the Study of Bioethical Issues, "History of Bioethics Commissions," archived January 15, 2017, at https://bioethics archive.georgetown.edu/pcsbi/history.html.

[143] The White House, National Biodefense Strategy and Implementation Plan for Countering Biological Threats, Enhancing Pandemic Preparedness, and Achieving Global Health Security, October 2022, at https://www.whitehouse.gov/wp-content/uploads/2022/10/ National-Biodefense-Strategy-and-Implementation-Plan- Final.pdf; and The White House, National Security Memorandum on Countering Biological Threats, Enhancing Pandemic Preparedness, and Achieving Global Health Security, October 18, 2022, at https://www.whitehouse.gov/briefing-room/presidential-actions/2022/10/18/national-security-memorandum-on-countering-biological-threats-enhancing-pandemic-prepared ness-and-achieving-global-health-security/.See also Diane Dieuliis, "Biotechnology for the Battlefield: In Need of a Strategy," War on the Rocks, November 27, 2018.

[144] The White House, National Biodefense Strategy and Implementation Plan for Countering Biological Threats, Enhancing Pandemic Preparedness, and Achieving Global Health Security, October 2022, at https://www.whitehouse.gov/wp-content/uploads/2022/10/ National-Biodefense-Strategy-and-Implementation-Plan- Final.pdf.

[145] The White House, "Fact Sheet: Biden-Harris Administration Releases Strategy to Strengthen Health Security and Prepare for Biothreats," October 18, 2022, at https://www.whitehouse. gov/briefing-room/statements-releases/2022/10/18/fact-sheet-biden-harris-administration-releases-strategy-to-strengthen-health-security-and-prepare-for-biothreats/. This $88 billion request includes funds for the Department of Health and Human Services, the Office of the Assistant Secretary for Preparedness and Response, the Centers for Disease Control and Prevention, the National Institutes of Health, the Food and Drug Administration, the Department of State, and the U.S. Agency for International Development; it does not include funds for DOD.

[146] P.L. 116-92, Section 2, Division A, Title II, §263.

[147] P.L. 116-283, Section 2, Division A, Title II, §278.

parent, affiliate, or successor of such an entity, should be identified as a Chinese military company or a military- civil fusion contributor and included on [DOD's 1260H] list."[148]

Finally, Section 1091 of the FY2022 NDAA (P.L. 117-81) establishes the National Security Commission on Emerging Biotechnology, which is to "consider the methods, means, and investments necessary to advance and secure the development of biotechnology, biomanufacturing, and associated technologies by the United States to comprehensively address the national security and defense needs of the United States." The commission delivered an interim report to Congress in December 2023 and is to deliver its final report in December 2024.[149]

China

Motivated by an aging population and growing health care needs, China has been particularly interested in conducting biotechnology research. Biotechnology is cited as a key strategic priority within China's Made in China 2025 initiative and is additionally highlighted within China's current five-year development plan.[150] In particular, China is aggressively pursuing biotechnologies for genetic testing and precision medicine. In 2016, Chinese scientists became the first to use the CRISPR gene-editing tool on humans, and in 2018, a Chinese scientist produced—perhaps with the approval of the Chinese government—the first "gene-edited babies."[151] In addition, China maintains one of the world's largest repositories of genetic information, the National Genebank, which includes U.S. genetic data. Such information could be used to develop personalized disease treatment plans or, potentially, precision bioweapons.[152]

Open-source information about China's research into specific military applications of biotechnology is limited; however, China's policy of military-civil fusion would enable the Chinese military to readily leverage developments in civilian biotechnology.[153] Furthermore, reports indicate that China's Central Military Commission "has funded projects on military brain science, advanced biomimetic systems, biological and biomimetic materials, human performance enhancement, and 'new concept' biotechnology," while the Chinese military's medical institutions have conducted extensive research on CRISPR gene editing.[154]

---

[148] For additional information about DOD's 1260H list, see Section 1260H of the FY2021 NDAA (P.L. 116-283).

[149] See National Security Commission on Emerging Biotechnology, *Interim Report*, December 2023, at https://www.biotech.senate.gov/press-releases/interim-report/.

[150] "Outline of the People's Republic of China 14th Five-Year Plan for National Economic and Social Development and Long-Range Objectives for 2035," *Xinhua News Agency*, March 12, 2021, Translated by Etcetera Language Group, Inc., at https://cset.georgetown.edu/wp-content/uploads/t0284_14th_Five_Year_Plan_EN.pdf.

[151] Amidst international outcry, China later sentenced the scientist to three years in jail and termed his work "extremely abominable in nature." Michael Standaert, "'Extremely abominable': Chinese gene-editing scientist faces law," *Al Jazeera*, November 26, 2018. See also, Elsa Kania, "Weaponizing Biotech: How China's Military Is Preparing for a 'New Domain of Warfare,'" *Defense One*, August 14, 2019.

[152] David J. Lynch, "Biotechnology: the US-China dispute over genetic data," *Financial Times*, July 31, 2017. See also Elsa Kania and Wilson VornDick, "China's Military Biotech Frontier: CRISPR, Military-Civil Fusion, and the New Revolution in Military Affairs," *The Jamestown Foundation*, October 8, 2019, at https://jamestown.org/program/chinas-military-biotech-frontier-crispr-military-civil-fusion-and-the-new-revolution-in-military-affairs/.

[153] Elsa Kania and Wilson VornDick, "Weaponizing Biotech: How China's Military Is Preparing for a 'New Domain of Warfare,'" *Defense One*, August 14, 2019, at https://www. defenseone.com/ideas/2019/08/chinas-military-pursuing- biotech/159167/.

[154] Ibid.

Russia

Although Russia released BIO2020—a whole-of-government strategy for improving the standing of Russia's biotechnology sector—in 2012, biotechnology research in Russia continues to lag behind that of the United States and China.[155] BIO2020 identifies Russia's priority areas for biotechnology research as biopharmaceutics and biomedicine, industrial biotechnology and bioenergetics, agricultural and food biotechnology, forest biotechnology, environmental protection biotechnology, and marine biotechnology.[156]

Little information is publicly available on how Russia might employ such dual-use technologies within a military or national security context. However, the accusation that the country attempted to assassinate a former double agent for the United Kingdom using a Novichok nerve agent—in violation of the 1992 Chemical Weapons Convention—suggests that it may be similarly unrestrained in weaponizing biological agents, including those derived from synthetic biology.[157] Indeed, the Soviet Union is known to have maintained an extensive, long-standing biological weapons program, Biopreparat, in violation of the 1972 Biological Weapons Convention.[158] Furthermore, in August 2020, the End-User Review Committee (ERC)—composed of representatives of the U.S. Departments of Commerce, State, Defense, Energy, and, where appropriate, Treasury—stated that it has "reasonable cause" to believe that three Russian research institutes are associated with the Russian biological weapons program.[159]

International Institutions

Only the weaponization of biotechnology is prohibited under international law.[160] Some international institutions have demonstrated interest in considering broader implications of biotechnologies. For example, since 1983, ASEAN has maintained a subcommittee on biotechnology that facilitates coordination of regional biotechnology projects. Similarly, since 1993, the OECD has maintained an Internal Co-ordination Group for Biotechnology that monitors developments in biotechnology and facilitates coordination among various sectors involved in biotechnology research (e.g., agriculture, science and technology, environment,

---

[155] Russian Federation, "BIO2020: Summary of the State Coordination Program for the Development of Biotechnology in the Russian Federation," 2012.

[156] Ibid.

[157] Mark Urban, "Salisbury attack 'evidence' of Russian weapon stockpile," *BBC*, March 4, 2019. For a full assessment of the potential national security threats posed by synthetic biology, see the Committee on Strategies for Identifying and Addressing Potential Biodefense Vulnerabilities Posed by Synthetic Biology Consensus Report: *Biodefense in the Age of Synthetic Biology*, National Academy of Sciences, 2018, at http://nap.edu/24890.

[158] Lukas Trakimavičius "Is Russia Violating the Biological Weapons Convention?," Atlantic Council, May 23, 2018, at https://www.atlanticcouncil.org/blogs/new-atlanticist/is-russia-violating-the-biological-weapons-convention/.

[159] The ERC added these research institutes to the Entity List, which identifies entities acting "contrary to the national security or foreign policy interests of the United States." Department of Commerce, "Addition of Entities to the Entity List, and Revision of Entries on the Entity List," August 27, 2020, https://www.federalregister.gov/documents/2020/08/ 27/2020-18909/addition-of-entities-to-the-entity-list-and-revision-of-entries-on-the-entity-list.

[160] The United States, China, and Russia have ratified the 1972 Biological Weapons Convention, which is a legally binding treaty that bans the development and production of biological weapons.

industry). In addition, the United Nations Convention on Biological Diversity is charged with governing the development and use of genetically modified organisms.[161] These entities are not, however, focused specifically on military applications of biotechnology.

In terms of potential militarization, the 1972 Biological Weapons Convention requires review conferences, which every five years assess both the implementation of the treaty and ongoing developments in biotechnology. Annual meetings are held between review conferences to informally consider relevant topics, as well as to address national bilateral and multilateral efforts to enhance biosecurity. Some analysts have argued that an international framework should be established to consider the militarization of biotechnologies and discuss potential regulation of or limits on certain applications.[162]

Potential Questions for Congress

- Is a DOD biotechnology strategy or organization needed to identify research priorities and coordinate department-wide research? What, if any, resources or organizational changes would be required to fully implement a national biodefense strategy?
- What military applications of biotechnologies are U.S. competitors developing? Is the U.S. military appropriately balancing the potential warfighting utility of biotechnologies with ethical considerations?
- What, if any, national and international frameworks are needed to consider the ethical, moral, and legal implications of military applications of biotechnologies such as synthetic biology, genome editing, and human performance modification?

Quantum Technology[163]

Quantum technology translates the principles of quantum physics into technological applications.[164] In general, quantum technology has not yet reached maturity; however, it could hold significant implications for the future of military sensing, encryption, and communications. GAO reports that DOD, State, DHS, and ODNI have assessed that "quantum communications could enable adversaries to develop secure communications that U.S. personnel would not be able to intercept or decrypt. Quantum computing may allow adversaries

---

[161] The United States is not a party to this convention or its associated protocols.

[162] See, for example, Brett Edwards, "We've got to talk: The militarization of biotechnology," *Bulletin of the Atomic Scientists*, August 4, 2017, at https://thebulletin.org/2017/08/weve-got-to-talk-the-militarization-of-biotechnology/.

[163] See also CRS In Focus IF11836, *Defense Primer: Quantum Technology*, by Kelley M. Sayler.

[164] These principles include superposition—in which "a quantum system can exist in two or more states at once"—and entanglement—in which "two or more quantum objects in a system can be intrinsically linked such that measurement of one dictates the possible measurement outcomes for another, regardless of how far apart the two objects are." Emily Grumbling and Mark Horowitz, eds., *Quantum Computing: Progress and Prospects,* National Academy of Sciences, 2019, at https://www.nap.edu/read/25196/chapter/1. For additional information about quantum technology, see CRS Report R45409, *Quantum Information Science: Applications, Global Research and Development, and Policy Considerations*, by Patricia Moloney Figliola.

to decrypt [unclassified, classified, or sensitive] information, which could enable them to target U.S. personnel and military operations."[165]

Quantum technology could have other military applications, such as quantum sensing, which could theoretically enable significant improvements in submarine detection, rendering the oceans "transparent."[166] This could, in turn, compromise the survivability of the U.S. sea-based nuclear deterrent. Quantum sensing could also provide alternative positioning, navigation, and timing options that could in theory allow militaries to continue to operate at full performance in GPS- degraded or GPS-denied environments.

Military application of such technologies could be constrained, however, by the fragility of quantum states, which can be disrupted by minute movements, changes in temperature, or other environmental factors. As physicist Mikkel Hueck has explained, "if future devices that use quantum technologies [continue to] require cooling to very cold temperatures, then this will make them expensive, bulky, and power hungry."[167] As a result, widespread adoption will likely require significant advances in materials science and fabrication techniques.

United States

According to a Defense Science Board Task Force on Applications of Quantum Technologies assessment, three applications of quantum technologies demonstrate the most promise for the U.S. military: quantum sensing, quantum computing, and quantum communications.[168] The task force notes that quantum sensing could "dramatically improve" DOD's ability to conduct certain missions, providing precision navigation and timing options in environments in which GPS is degraded or denied; that quantum computers could "potentially give DOD substantial computation power" for decryption, signals processing, and AI; and that quantum communications could improve networking technologies.[169] The task force concludes that "quantum sensing applications are currently poised for mission use whereas quantum computing and communications are in earlier stages of development.... Quantum radar will not provide upgraded capability to DOD."[170] Both DARPA and the services fund an array of quantum technology programs across these and other research areas.

---

[165] Government Accountability Office, *National Security: Long-Range Emerging Threats Facing the United States as Identified by Federal Agencies*, December 2018, at https://www.gao.gov/assets/700/695981.pdf. Significant advances in quantum computing will likely be required to break current encryption methods. Indeed, some analysts believe that a quantum computer with around 20 million qubits—shorthand for "quantum bits," or computing units that leverage the principle of superposition—would be required to break these methods; the most advanced quantum computers today have around 256 qubits. See Siobhan Roberts, "This new startup has built a record-breaking 256-qubit quantum computer," *MIT Technology Review*, November 17, 2021, at https://www.technology review.com/2021/11/17/1040243/quantum-computer-256-bit-startup/.

[166] Michael J. Biercuk and Richard Fontaine, "The Leap into Quantum Technology: A Primer for National Security Professionals," *War on the Rocks*, November 17, 2017, at https://waront herocks.com/2017/11/leap-quantum-technology-primer-national-security-professionals/.

[167] U.S. Army Research Lab Public Affairs, "Army researchers see path to quantum computing at room temperature," April 30, 2020, at https://www.army.mil/article/235127/ army_ researchers_see_path_to_quantum_computing_at_room_temperature#:~:text=%E2 %80%9CIf%20future%20devices%20that%20use,power%20hungry%2C%E2%80% 9D%20Heuck%20 said.

[168] Defense Science Board, *Applications of Quantum Technologies: Executive Summary*, October 2019, at https://dsb.cto.mil/reports.htm.

[169] Ibid.

[170] Ibid.

In addition, some analysts believe that an initial quantum computer prototype capable of breaking current encryption methods could be developed in the 2030 to 2040 timeframe.[171] For this reason, the United States is investing in post-quantum cryptography (also known as quantum-resistant cryptography). In May 2022, the Biden administration released National Security *Memorandum on Promoting United States Leadership in Quantum Computing While Mitigating Risks to Vulnerable Cryptographic Systems (NSM-10),* which "directs specific actions for agencies to take as the United States begins the multi-year process of migrating vulnerable computer systems to quantum-resistant cryptography."[172] NSM-10 notes that the Director of the National Institute of Standards and Technology and the Director of the National Security Agency (NSA) are developing and expected to publicly release by 2024 technical standards for quantum-resistant cryptography. In September 2022, NSA issued a cybersecurity advisory stating that it "expects the transition to [quantum-resistant] algorithms for [national security systems] to be complete by 2035 in line with NSM-10."[173]

Per Section 234 of the FY2019 NDAA (P.L. 115-232), the Secretary of Defense—acting through the Under Secretary of Defense for Research and Engineering—is tasked with coordinating quantum technology programs and providing "for interagency cooperation and collaboration on quantum information science and technology research and development between the Department of Defense and other departments and agencies of the United States and appropriate private sector entities."[174] In addition, Section 220 of the FY2020 NDAA (P.L. 116-92) authorizes the Secretary of each military department to establish Quantum Information Science (QIS) Research Centers that may "engage with appropriate public and private sector organizations" to advance quantum research. To date, the Navy has designated the Naval Research Laboratory as its QIS Research Center, while the Air Force has designated the Air Force Research Laboratory as a QIS Research Center for both the Air Force and Space Force. The Army does not plan to establish a QIS Research Center at this time.[175]

Finally, Section 214 of the FY2021 NDAA (P.L. 116-283) directs the services to compile and annually update a list of technical challenges that quantum computers could potentially address within the next one to three years.[176] It also directs the services to establish programs with small and medium businesses to provide quantum computing capabilities to government,

---

171 See, for example, "Quantum computers will break the encryption that protects the internet," *The Economist*, October 10, 2018, at https://www.economist.com/science-and-technology/ 2018/10/20/quantum-computers-will-break-the-encryption-that-protects-the-internet.

172 The White House, *National Security Memorandum on Promoting United States Leadership in Quantum Computing While Mitigating Risks to Vulnerable Cryptographic Systems*, May 4, 2022, at https://www.whitehouse.gov/briefing-room/statements-releases/2022/05/04/ national-security-memorandum-on-promoting-united-states-leadership-in-quantum-computing-while-mitigating-risks-to-vulnerable-cryptographic-systems/.

173 National Security Agency, "Cybersecurity Advisory: Announcing the Commercial National Security Algorithm Suite 2.0," September 7, 2022, at https://media.defense.gov/2022/Sep/ 07/2003071834/-1/-1/0/ CSA_CNSA_2.0_ALGORITHMS_.PDF.

174 Section 219 of the FY2024 NDAA amended this provision to authorize the Secretary of Defense to "carry out a program of fellowships in quantum information science and technology research and development for individuals who have a graduate or postgraduate degree." Similarly, Section 220 of the FY2024 NDAA (P.L. 118-31) directs the Secretary of Defense to "to establish public-private talent exchange programs with private-sector entities working on quantum information sciences and technology research applications."

175 CRS correspondence with the Office of the Under Secretary of Defense for Research and Engineering, April 5, 2021.

176 On behalf of the services, the Office of the Under Secretary of Defense for Research and Engineering identified these challenges as quantum chemistry, optimization, and machine learning. CRS correspondence with the Office of the Under Secretary of Defense for Research and Engineering, March 25, 2022.

industry, and academic researchers working on these challenges. Section 1722 directs DOD to conduct an assessment of the risks posed by quantum computers, as well as current standards for post- quantum cryptography. In addition, Section 229 of the FY2022 NDAA (P.L. 117-81) directs the Secretary of Defense to establish activities to "to accelerate the development and deployment of dual-use quantum capabilities," while Section 511 expands the grant program for science, technology, engineering, and math education in the Junior Research Officers' Training Corps to include quantum information sciences.

Finally, Section 231 of the FY2024 NDAA (P.L. 118-31) authorizes the Secretary of Defense to carry out a pilot program on near-term quantum computing applications.

China

China has increasingly prioritized quantum technology research within its development plans.[177] Indeed, President Xi has cited quantum communications and quantum computing as key research initiatives "prioritized for major breakthroughs by 2030," an objective that is also cited in the country's National Science and Technology Innovation Program.[178] China is already a world leader in quantum technology. In 2016, China launched the world's first quantum satellite (Micius) to provide a "global quantum encrypted communications capability."[179] In 2017, China hosted the first quantum-secured intercontinental videoconference.[180] Furthermore, China is investing heavily in terrestrial quantum communications networks. It completed construction of a 2,000 kilometer (approximately 1,250 miles) Beijing-Shanghai quantum network in 2016 and plans to expand that network nationwide in the years to come.[181] While such advances in quantum technology have been driven primarily by academia, China has expressed its intent to leverage them for military applications in the country's Thirteenth Five-Year S&T Military-Civil Fusion Special Projects Plan.

---

[177] For a history of China's quantum technology research and development initiatives, see Elsa B. Kania and John Costello, *Quantum Hegemony?: China's Ambitions and the Challenge to U.S. Innovation Leadership*, Center for a New American Security, September 2018, p. 8, at https://s3.amazonaws.com/files.cnas.org/documents/CNASReport-Quantum-Tech_ FINAL.pdf?mtime=20180912133406.

[178] Ibid., p. 6.

[179] Office of the Secretary of Defense, *Annual Report to Congress: Military and Security Developments Involving the People's Republic of China 2019*, May 2, 2019, p. 146, at https://media.defense.gov/2019/May/02/2002127082/-1/-1/1/ 2019_CHINA_MILITARY _POWER_REPORT.pdf. This satellite—as well as the other communications networks discussed in this section—employ quantum key distribution (QKD), a subset of quantum communications. QKD enables secure communications that cannot be covertly intercepted during transmission. QKD communications can, however, be intercepted at the relay stations currently required for long-distance transmissions.

[180] Office of the Secretary of Defense, *Annual Report to Congress: Military and Security Developments Involving the People's Republic of China 2019*, May 2, 2019, p. 101, at https://media.defense.gov/2019/May/02/2002127082/-1/-1/1/2019_CHINA_MILITARY_ POWER_REPORT.pdf.

[181] Elsa B. Kania and John Costello, *Quantum Hegemony?: China's Ambitions and the Challenge to U.S. Innovation Leadership*, p. 14. This terrestrial network is reportedly connected to Micius, which provides a satellite link "spanning 2,600 km [1,600 miles] between two observatories—one east of Beijing and the other just a few hundred kilometers from China's border with Kazakhstan." See Hamish Johnston, "Quantum cryptography network spans 4600 km in China," *Physics World*, January 7, 2021, at https://physicsworld.com/a/ quantum-cryptography-network-spans-4600- km-in-china/.

Russia

Russian development of quantum technology, as with artificial intelligence, lags significantly behind that of the United States and China, with some analysts noting that Russia is likely "5 to 10 years behind" in quantum computing.[182] In an effort to spur development, Russia announced plans in December 2019 to invest $790 million in quantum research over the next five years and adopted a five-year Russian Quantum Technologies Roadmap.[183] These initiatives are not military-specific, however, and limited information is available in open sources about how Russia might apply them to its military.

International Institutions

No major international institutions have formal initiatives devoted to monitoring or regulating military or other applications of quantum technology.

Potential Questions for Congress

- Does the maturity of military applications of quantum technology warrant current funding levels? To what extent, if at all, can advances in commercial quantum technology be leveraged for military applications?
- Are adequate measures being taken to develop quantum-resistant encryption and to protect data that has been encrypted using current methods?
- How mature are U.S. competitor efforts to develop military applications of quantum technologies? To what extent, if at all, could such efforts threaten advanced U.S. military capabilities such as submarines and fifth-generation stealth aircraft?

Potential Implications of Emerging Technologies for Warfighting

The implications of emerging technologies for warfighting and strategic stability are difficult—if not impossible—to predict, as they will be a function of many factors, including the rate of technological advancement in both the United States and competitor nations, the manner in which emerging technologies are integrated into existing military forces and concepts of operation, the interactions between emerging technologies, and the extent to which national policies and international law enable or inhibit their development, integration, and use.

Nonetheless, many emerging technologies exhibit characteristics that could potentially affect the future character of war. For example, developments in technologies such as AI, big data analytics, and lethal autonomous weapons could diminish or remove the need for a human

---

182 Quirin Schiermeier, "Russia joins race to make quantum dreams a reality," *Nature*, December 17, 2019, at https://www.nature.com/articles/d41586-019-03855-z.

183 For comparison, the U.S. National Quantum Initiative Act (P.L. 115-368), signed into law in December 2018, commits the United States to investing $1.25 billion in quantum research over five years.

operator. This could, in turn, increase combat efficiency and accelerate the pace of combat—potentially with destabilizing consequences.

Emerging technologies such as low-cost drones could shift the balance between quality—upon which U.S. military forces have traditionally relied—and quantity, as well as between offense and defense. For example, swarms of coordinated, unmanned vehicles could overwhelm defensive systems, providing a greater advantage to the attacker, while directed energy weapons that provide a low-cost means of neutralizing such attacks, could favor the defender. Thus, emerging technologies could shift the offense-defense balance multiple times over the coming decades.

Interactions among emerging technologies could also improve existing military capabilities or enable new capabilities—with unforeseen consequences for warfighting and strategic stability. For example, an enabling technology like AI could be paired with quantum computing to produce more powerful methods of machine learning, potentially leading to improvements in image recognition and target identification and enabling more sophisticated autonomous weapons.

Similarly, AI could be paired with 5G communications technologies to enable virtual training environments or with biotechnology in a "brain-computer interface" to enhance human cognition or control prosthetics or robotic systems.[184] Such developments could, in turn, require new strategies, tactics, and concepts of operation.[185]

Emerging military technologies—particularly complex systems such as AI and LAWS—could additionally produce unintended consequences if they fail to perform as anticipated. These consequences could range from system failure to violations of the law of armed conflict. As analyst Paul Scharre has noted, "in the most extreme case, an autonomous weapon could continue engaging inappropriate targets until it exhausts its magazine, potentially over a wide area."[186] This could, in turn, result in mass fratricide or civilian casualties—a possibility that has led some analysts to call for a pre-emptive ban on LAWS.

Finally, emerging military technologies could raise an array of ethical considerations. For example, some analysts have argued that the use of LAWS would be inherently immoral—regardless of whether the weapon could be used legally—because a human operator would not make specific target selection and engagement decisions.[187] Others have countered that human operators would continue to exercise "appropriate levels of human judgement over the use of force" and would remain accountable for ensuring that the deployment of LAWS conforms to the requirements of the laws of armed conflict.[188] Those supporting a pre-emptive ban on LAWS have additionally appealed to the Martens Clause, which appears in the1899 Hague Convention preamble and states that weapons usage should conform to the "principles of humanity and the

---

[184] For additional information about military applications of 5G, see CRS In Focus IF11251, *National Security Implications of Fifth Generation (5G) Mobile Technologies*, by John R. Hoehn and Kelley M. Sayler.

[185] For a discussion of these and other military and security implications—including implications for deterrence, crisis stability, force posture, and military roles and missions—see Robert O. Work and Shawn Brimley, *20YY: Preparing for War in the Robotic Age*, Center for a New American Century, January 22, 2014, pp. 31-35, at https://www.cnas.org/ publications/reports/20yy-preparing-for-war-in-the-robotic-age.

[186] Paul Scharre, "Autonomous Weapons and Operational Risk," Center for a New American Security, February 2016, at https://s3.amazonaws.com/files.cnas.org/documents/CNAS_ Autonomous-weapons-operational-risk.pdf.

[187] See, for example, Bonnie Docherty, *Heed the Call: A Moral and Legal Imperative to Ban Killer Robots*, Human Rights Watch, August 21, 2018, at https://www.hrw.org/report/2018/ 08/21/heed-call/moral-and-legal-imperative-ban- killer-robots.

[188] Department of Defense Directive 3000.09, "Autonomy in Weapon Systems," Updated May 8, 2017, at https://www.esd.whs.

dictates of the public conscience."[189] These analysts believe that LAWS contravene that requirement; however, others have noted that the Martens Clause has not been used previously to ban a weapons system and, furthermore, that the legal status of the Martens Clause is questionable and instead constitutes "merely a recognition of 'customary international law'."[190] Similarly, some analysts have raised ethical concerns about applications of biotechnology that involve human testing or modification as well as the weaponization of biotechnology, which could potentially be used for targeted genetic attacks.[191]

Issues for Congress

Congress has previously demonstrated interest in conducting oversight of emerging military technologies beyond technology-specific activities. In Section 247 of the FY2019 NDAA (P.L. 115-232), Congress specified "a set of classified reports that set forth a direct comparison between the capabilities of the United States in emerging technology areas and the capabilities of adversaries of the United States."[192] These areas include hypersonic weapons, AI, quantum technology, directed energy weapons, and other relevant technologies as determined by the Secretary of Defense. Similarly, Section 1251 of the FY2022 NDAA (P.L. 117-81) directs the Under Secretary of Defense for Research and Engineering, in coordination with the Director of the Office of Net Assessment, to "conduct a comparative analysis assessment of the efforts of the United States Government and the Government of the People's Republic of China to develop and deploy" emerging technologies such as directed energy weapons, hypersonic weapons, biotechnology, and quantum technology. Section 225 of the FY2019 NDAA additionally tasked the Under Secretary of Defense for Research and Engineering with generating procedures for developing "technologies that are urgently needed to react to a technological development of an adversary of the United States or to respond to a significant and urgent emerging technology [that are] not receiving appropriate research funding or attention from the Department of Defense."

Furthermore, Section 232 of the FY2020 NDAA (P.L. 116-92) tasked the Secretary of Defense with developing "a process to ensure that the policies of the Department of Defense relating to emerging technology are formulated and updated continuously as such technology

---

189 See, for example, Bonnie Docherty, *Heed the Call: A Moral and Legal Imperative to Ban Killer Robots*, Human Rights Watch, August 21, 2018, at https://www.hrw.org/report/ 2018/08/21/heed-call/moral-and-legal-imperative-ban- killer-robots.

190 Paul Scharre, *Army of None: Autonomous Weapons and the Future of War* (New York: W.W. Norton & Company, 2018), pp. 263-266.

191 For a more in-depth discussion of ethical considerations related to biotechnology, see CRS Report R44824, *Advanced Gene Editing: CRISPR-Cas9*, by Marcy E. Gallo et al. See also Elsa Kania and Wilson VornDick, "China's Military Biotech Frontier: CRISPR, Military-Civil Fusion, and the New Revolution in Military Affairs," *The Jamestown Foundation*, October 8, 2019, at https://jamestown.org/program/chinas-military-biotech-frontier-crispr- military-civil-fusion-and-the-new-revolution-in-military-affairs/.

192 Each report is to include the following elements: "(1) an evaluation of spending by the United States and adversaries on such technology, (2) an evaluation of the quantity and quality of research on such technology, (3) an evaluation of the test infrastructure and workforce supporting such technology, (4) an assessment of the technological progress of the United States and adversaries on such technology, (5) descriptions of timelines for operational deployment of such technology, [and] (6) an assessment of the intent or willingness of adversaries to use such technology."

is developed by the Department,"[193] while Section 236 of the FY2021 NDAA (P.L. 116-283) granted the Secretary the authority to establish a Steering Committee tasked with developing assessments of and a strategy for emerging technology and national security threats.

As Congress continues to review the Pentagon's plans for emerging military technologies during the annual authorization and appropriations process, it might consider issues surrounding funding considerations, management, personnel, acquisition, technology protection, governance and regulation, and oversight.

Funding Considerations

A number of emerging military technologies, including hypersonic weapons and directed energy weapons, have experienced fluctuations in funding over the years. According to a U.S. government interagency task force on the defense industrial base, such "fluctuations challenge the viability of suppliers within the industrial base by diminishing their ability to hire and retain a skilled workforce, [achieve] production efficiencies, and in some cases, [stay] in business."[194]

Other analysts have noted that such fluctuations are often due to unavoidable tradeoffs between technology investment priorities or to questions about a given technology's feasibility or maturity.[195]

Some analysts have suggested that, given the potential for technological surprise, funding for overall research and development is inadequate. Summarizing such views, technology expert Martijn Rasser notes that reducing overall research and development in order to enable "big bets" or heavy investments in a particular technology or technologies, can be a risky approach because "we just don't know where the next breakthroughs will come from."[196]

Management

In general, DOD manages each of the aforementioned emerging military technologies separately due to the distinct expertise required. For example, within the Office of the Under Secretary of Defense for Research and Engineering (USD[R&E])/Chief Technology Officer (CTO), there are separate principal directors for artificial intelligence and autonomy, hypersonic weapons, directed energy, biotechnology, and quantum technology—among other

---

[193] Section 232 defines emerging technology as "technology determined to be in an emerging phase of development by the Secretary of Defense, including quantum computing, technology for the analysis of large and diverse sets of data (commonly known as 'big data analytics'), artificial intelligence, autonomous technology, robotics, directed energy, hypersonics, biotechnology, and such other technology as may be identified by the Secretary."

[194] Interagency Task Force in Fulfillment of Executive Order 13806, *Assessing and Strengthening the Manufacturing and Defense Industrial Base and Supply Chain Resiliency of the United States*, September 2018, p. 21, at https://media.defense.gov/2018/Oct/ 05/2002048904/-1/-1/1/ASSESSING-AND-STRENGTHENING-THE-MANUFACTURING-AND%20DEFENSE-INDUSTRIAL-BASE-AND-SUPPLY-CHAIN-RESILIENCY.PDF.

[195] See, for example, Ariel Robinson, "Directed Energy Weapons: Will They Ever Be Ready?," *National Defense*, July 1, 2015, at https://www.nationaldefensemagazine.org/articles/2015/7/1/2015july-directed-energy-weapons-will-they- ever-be-ready.

[196] See, for example, Will Knight, "Trump Proposes a Cut in Research Spending, but a Boost for AI," *Wired*, February 11, 2020, at https://www.wired.com/story/trump-proposes-cut-research-spending-boost-ai/. For more information about federal R&D funding, including a discussion of DOD R&D funding, see CRS Report R46341, *Federal Research and Development (R&D) Funding: FY2021*, coordinated by John F. Sargent Jr.

technology areas. Development of each of these technologies is guided by a standalone technology roadmap and, in the case of AI, a classified strategy. Although the USD(R&E)/CTO has oversight over emerging military technologies, some analysts have suggested that there is a need for a more holistic approach to portfolio management that better considers how such technologies might be combined and integrated.[197]

Furthermore, prior to May 2022, these principal directors reported to the USD(R&E)/CTO through a singular Director for Modernization, who was "responsible for managing the capability analysis and investments for the modernization priorities outlined in the National Defense Strategy," including the technologies discussed in this report.[198] In May 2022, DOD reorganized the office of the USD(R&E)/CTO and created separate reporting structures for "enabling technology"—including the Principal Director for AI and autonomy—and "applied technology"—including the Principal Directors for directed energy and hypersonic weapons (see Figure 2). The Directors for Enabling Technology and Applied Technology report to the Deputy CTO for Critical Technologies. In contrast, the Principal Directors for biotechnology and quantum science report through the Director of Science and Technology Futures to a different Deputy CTO—the Deputy CTO for Science and Technology.[199] Congress may consider the ways in which this organizational change affects the oversight and integration of emerging technologies.

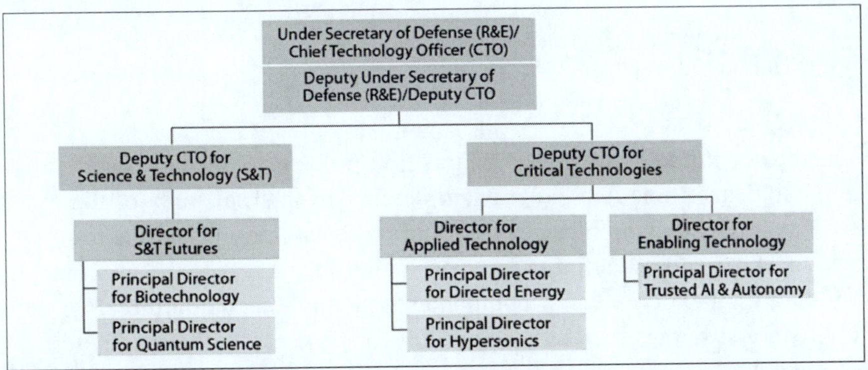

Source: CRS image, adapted from https://www.cto.mil/wp-content/uploads/2022/05/usdre_org_chart_09may2022_distro_a.pdf.

Notes: Chart depicts only the reporting structure for the OUSD(R&E) principal directors associated with the technologies discussed in this report. For a full OUSD(R&E) organizational chart, see "Office of the Under Secretary of Defense for Research and Engineering Organization," May 9, 2022, at https://www.cto. mil/wp-content/uploads/2022/05/usdre_org_chart_09may2022_ distro_a.pdf.

Figure 2. Selected Reporting Structure of the Office of the Under Secretary of Defense for Research and Engineering (R&E).

---

197 See, for example, Government Accountability Office, *Weapon System Acquisitions: Opportunities Exist to Improve the Department of Defense's Portfolio Management*, August 2015, at https://www.gao.gov/assets/680/672205.pdf; and Pete Modigliani, *After the divorce: How the Pentagon can position itself for speed, agility, and innovation in the new era of acquisitions*, MITRE, March 2019, at https://www.mitre.org/sites/default/files/publications/pr-18-03404-3-after- the-divorce-white-paper.pdf.

198 CRS In Focus IF10834, Defense Primer: Under Secretary of Defense for Research and Engineering, by Marcy E. Gallo.

199 See Department of Defense, "Organizational Improvements to the Office of the Under Secretary of Defense for Research and Engineering," May 10, 2022, at https://www.defense.gov/News/Releases/Release/Article/3026367/organizational-improvements-to-the-office-of-the-under-secretary-of-defense-for/.

Finally, senior leaders do not always agree on the priorities among emerging military technologies—both in terms of effort and funding—and such priorities can shift frequently. This fluctuation has led some analysts to suggest that DOD should adopt a technology strategy "to set spending priorities that can be sustained over time, outlasting individual leaders."[200]

Personnel

Some reports indicate that DOD and the defense industry have difficulty recruiting and retaining personnel with expertise in emerging technologies because research funding and salaries significantly lag behind those of commercial companies.[201] Other reports suggest that such challenges stem from quality-of-life factors, as well as from a belief among many technology workers that "they can achieve large-scale change faster and better outside the government than within it."[202] DOD faces additional challenges in training and educating its standing workforce. Examples of recommendations for addressing this set of challenges include increasing technology education opportunities at military academies, enhancing partnerships between DOD and research universities, creating government fellowships and accelerated promotion tracks for technology workers, and improving the technology literacy of human resource teams.[203]

Acquisition

DOD may need to continue adjusting its acquisition process to account for rapidly evolving dual- use technologies such as AI.[204] For example, a 2017 internal study of the process found that it takes an average of 81 months for information technology programs to move from the initial Analysis of Alternatives, defining the requirements for a system, to an Initial Operational Capability.[205] In contrast, commercial companies typically execute an iterative development process for software systems (such as those involved in AI capabilities), delivering an initial

---

[200] Paul Scharre and Ainikki Riikonen, "The Defense Department Needs a Real Technology Strategy," *Defense One*, April 21, 2020, at https://www.defenseone.com/ ideas/2020/ 04/pentagon-needs-technology-strategy/164764/.

[201] M.L. Cummings, "Artificial Intelligence and the Future of Warfare," *Chatham House*, January 2017, p. 11, at https://www.chathamhouse.org/sites/default/files/publications/research/2017-01-26-artificial-intelligence-future-warfare-cummings-final.pdf.

[202] Amy Zegart and Kevin Childs, "The Divide between Silicon Valley and Washington Is a National-Security Threat," *The Atlantic*, December 13, 2018, at https://www.theatlantic. com/ideas/archive/2018/12/growing-gulf-between-silicon-valley-and-washington/577963/.

[203] See Defense Science Board, Applications of Quantum Technologies: Executive Summary; National Security Commission on Artificial Intelligence, First Quarter Recommendations, March 2020, pp. 21-43, at https://drive.google.com/file/d/1wkPh8Gb5dr BrKBg 6OhGu5oNaTEERbKss/view; and Amy Zegart and Kevin Childs, "The Divide between Silicon Valley and Washington." For example, DOD is establishing a university consortium for hypersonic research and workforce development, while the Defense Digital Service now offers one- to two-year assignments for commercial technology workers. Similarly, the National Security Innovation Network seeks to create models and pathways for recruiting technologists to the U.S. government.

[204] Andrew Ilachinski, *AI, Robots, and Swarms: Issues, Questions, and Recommended Studies*, Center for Naval Analysis, January 2017, pp. 190-191. For an overview of recent acquisition reform efforts, see CRS Report R45068, *Acquisition Reform in the FY2016-FY2018 National Defense Authorization Acts (NDAAs)*, by Heidi M. Peters.

[205] Andrew Ilachinski, *AI, Robots, and Swarms: Issues, Questions, and Recommended Studies*, p. 189.

product in six to nine months.[206] These findings prompted DOD to issue an interim software acquisition policy intended to "[simplify] the acquisition model to enable continuous integration and delivery of software capability on timelines relevant to the Warfighter/end user."[207] Similar efforts may be needed for other emerging military technologies.

Furthermore, the commercial companies that are often at the forefront of innovation in emerging technologies may be reluctant to partner with DOD due to the complexity of the defense acquisition process. A Government Accountability Office (GAO) study of this issue found that, of 12 U.S. commercial companies who choose not to do business with DOD, all 12 cited the complexity of the defense acquisition process as a rationale for their decision.[208] DOD has created a number of avenues for rapid acquisitions—including the Strategic Capabilities Office, the Defense Innovation Unit, and Project Maven—that are intended to streamline cumbersome processes and accelerate the acquisitions timeline.[209] Project Maven, for example, was established in April 2017; by December, the team was fielding a commercially acquired prototype AI system in combat.[210] Although some analysts argue that these are promising developments, critics point out that the department must replicate such results at scale and implement more comprehensive acquisitions reform.[211]

Intellectual Property

Commercial technology companies are often reluctant to partner with DOD due to concerns about intellectual property and data rights.[212] As an official interviewed for a 2017 GAO report on broader challenges in military acquisitions noted, intellectual property is the "life blood" of commercial technology companies, yet "DOD is putting increased pressure on companies to grant unlimited technical data and software rights or government purpose rights rather than limited or restricted rights."[213] In an effort to manage these concerns, DOD released an instruction that "establishes policy, assigns responsibilities, and prescribes procedures for the

---

206 Defense Science Board, "Design and Acquisition of Software for Defense Systems," February 2018, at https://apps.dtic.mil/dtic/tr/fulltext/u2/1048883.pdf. See also Defense Innovation Board, *Software is Never Done: Refactoring the Acquisition Code for Competitive Advantage*, May 3, 2019, at https://media.defense.gov/2019/Apr/30/2002124828/-1/-1/0/SOFTWAREISNEVERDONE_REFACTORING THEACQUISITIONCODEFORCOMPETITIVEADVANTAGE_FIN AL.SWAP.REPORT.PDF.

207 Office of the Under Secretary of Defense for Acquisition and Sustainment, "Software Acquisition Pathway Interim Policy and Procedures," January 3, 2020, at https://www. acq.osd.mil/ae/assets/docs/USA002825-19%20Signed%20Memo%20 (Software).pdf.

208 U.S. Government Accountability Office, *Military Acquisitions, DOD is Taking Step to Address Challenges Faced by Certain Companies*, GAO-17-644, July 20, 2017, p. 9. Other rationales cited include unstable budget environment, lengthy contracting timeline, government-specific contract terms and conditions, and inexperienced DOD contracting workforce.

209 In certain circumstances, DOD may also use other transaction authorities (OTAs) to accelerate research, prototyping, and production. For additional information about OTAs, see CRS Report R45521, *Department of Defense Use of Other Transaction Authority: Background, Analysis, and Issues for Congress*, by Heidi M. Peters.

210 Marcus Weisgerber, "The Pentagon's New Artificial Intelligence is Already Hunting Terrorists," *Defense One*, December 21, 2017, at http://www.defenseone.com/technology/ 2017/12/pentagons-new-artificial-intelligence-already-hunting-terrorists/144742/.

211 Andrew Ilachinski, *AI, Robots, and Swarms: Issues, Questions, and Recommended Studies*, Center for Naval Analysis, January 2017, p. 190.

212 U.S. Government Accountability Office, *Military Acquisitions, DOD is Taking Steps to Address Challenges Faced by Certain Companies.*

213 Ibid., p. 20.

acquisition, licensing, and management of IP."[214] The instruction additionally establishes a DOD IP Cadre to advise and assist the acquisition workforce on matters related to IP and calls for the development of an IP strategy to "identify and manage the full spectrum of IP and related matters" for each acquisition program.[215]

Supply Chain Security

A number of recent reports have raised concerns about the security of the U.S. supply chain for emerging military technologies. For example, one assessment found that China "may have opportunities to jeopardize the development of hypersonics through industrial espionage, transfers of technology, or providing unreliable components" due to its potential exposure to low- level U.S. suppliers.[216] Similarly the National Security Commission on Artificial Intelligence found that "the United States lacks domestic facilities capable of producing, integrating, assembling, and testing" the microelectronics needed to enable AI, forcing the U.S. "to rely on foreign fabrication and complex global supply chains for production."[217] In response to such concerns, Title XCIX of the FY2021 NDAA (P.L. 116-283), Creating Helpful Incentives to Produce Semiconductors (CHIPS) for America, authorized an incentive program for building and equipping semiconductor fabrication facilities in the United States. Congress subsequently provided funding for the program in the Chips and Science Act (P.L. 117-167).[218]

Technology Protection

Estimates indicate "that American industry loses more than $600 billion dollars [each year] to theft and expropriation," including the theft and expropriation of emerging military technologies and related intellectual property.[219] The United States has a number of programs devoted to addressing this issue. For example, pursuant to the Foreign Investment Risk Review Modernization Act of 2018 (FIRRMA), the Committee on Foreign Investment in the United States (CFIUS) now reviews certain foreign investments, including those involving "emerging and foundational technologies." In addition, FIRRMA authorized CFIUS to consider "whether a covered transaction involves a country of special concern that has a demonstrated or declared strategic goal of acquiring a type of critical technology or critical infrastructure that would affect United States leadership in areas related to national security."[220] Similarly, DOD's

---

[214] Office of the Under Secretary of Defense for Acquisition and Sustainment, "DOD Instruction 5010.44 Intellectual Property (IP) Acquisition and Licensing," October 16, 2019, at https://www.esd.whs.mil/Portals/54/Documents/DD/issuances/dodi/501044p.PDF?ver=2019-10-16-144448-070.

[215] Ibid., pp. 8-11.

[216] Govini, *The 2020 Federal Scorecard: High-Intensity Warfare Edition*, p. 67, at https://www.govini.com/wp-content/uploads/2020/06/Govini-2020-Federal-Scorecard. pdf.

[217] National Security Commission on Artificial Intelligence, *First Quarter Recommendations*, p. 46.

[218] For additional information about these provisions, see CRS Report R47558, *Semiconductors and the CHIPS Act: The Global Context*, by Karen M. Sutter, John F. Sargent Jr., and Manpreet Singh.

[219] Office of the Secretary of Defense, "Memorandum on the Establishment of the Protecting Critical Technology Task Force," October 24, 2018, at https://insidecybersecurity.com/sites/insidecybersecurity.com/files/documents/2018/nov/ cs2018_0459.pdf.

[220] The specific technologies that qualify as "emerging and foundational technologies" are to be identified by an interagency process led by the Department of Commerce. See P.L. 115-232, Title XVII, §1702(c). For more information on FIRRMA, see CRS In Focus IF10952, *CFIUS Reform Under FIRRMA*, by James K. Jackson and Cathleen D. Cimino-Isaacs. Some entities, including the National Security Commission on Artificial Intelligence, have argued that the U.S. government should consider additional measures of technology

Protecting Critical Technology Task Force helps protect universities, labs, and the U.S. defense industrial base against the theft of "classified information, controlled unclassified information, and key data."[221] As part of this effort, the task force intends to institute cybersecurity training programs for small businesses, enhance DOD's understanding of supply chain vulnerabilities, and develop a prioritized list of technologies that are critical to national security—as mandated by Section 1049 of the FY2019 NDAA—among other activities.[222] Some analysts have recommended expanding technology protection efforts to include U.S. allies and partners.[223]

Governance and Regulation

According to then-Director of National Intelligence Daniel Coats, "technology developments … are likely to outpace regulation, which could create international norms that are contrary to US interests and increase the likelihood of technology surprise."[224] To address this concern, some analysts have argued that "the United States should undertake broad, sustained diplomatic engagement to advance collaboration on emerging technologies, norms, and standards setting."[225]

Similarly, Section 9414 of the FY2021 NDAA directs the Director of the National Institute of Standards and Technology to oversee a study that assesses China's role in international standards setting organizations and provides recommendations for mitigating China's influence and strengthening U.S. participation in these organizations.

Oversight[226]

As Congress conducts oversight of emerging military technologies, it may be challenged in its ability to independently evaluate and assess complex, disparate technical disciplines. In 1972, Congress established the Office of Technology Assessment (OTA) to provide expert "assessments, background papers, technical memoranda, case studies, and workshop proceedings" that were to inform congressional decisionmaking and legislative activities.[227]

---

protection, such as "heavier scrutiny of the potential end use and end user of specific items." See National Security Commission on Artificial Intelligence, *Interim Report*, November 2019, p. 42, at https://drive.google.com/file/d/153OrxnuGEjsUvlxWsFYauslwNeCEkv Ub/view.

221 Office of the Secretary of Defense, "Memorandum on the Establishment of the Protecting Critical Technology Task Force."

222 C. Todd Lopez, "Task Force Curbs Technology Theft to Keep Joint Force Strong," DOD News, November 26, 2019, at https://www.defense.gov/Explore/News/ Article/Article/ 2027555/task-force-curbs-technology-theft-to-keep- joint-force-strong/.

223 See, for example, Daniel Kliman, Ben FitzGerald, Kristine Lee, and Joshua Fitt, *Forging an Alliance Innovation Base*, Center for a New American Security, March 2020, at https://s3.amazonaws.com/files.cnas.org/documents/CNAS-Report-Alliance-Innovation-Base-Final.pdf?mtime=20200329174909.

224 Daniel R. Coats, "Statement for the Record: Worldwide Threat Assessment of the US Intelligence Community," delivered before the U.S. Senate Committee on Armed Services, March 6, 2018.

225 Samuel J. Brannen, Christian S. Haig, Katherine Schmidt, and Kathleen H. Hicks, *Twin Pillars: Upholding National Security and National Innovation in Emerging Technologies Governance*, Center for Strategic and International Studies, January 2020, at https://csis-prod.s3.amazonaws.com/s3fs-public/publication/200123_Brannen_TwinPillars_WEB_FINAL.pdf?eljUpAKOjVauOujYfnvuSGDK0xvsQGZF.

226 For a full discussion of issues surrounding congressional oversight of technology, see CRS Report R46327, *The Office of Technology Assessment: History, Authorities, Issues, and Options*, by John F. Sargent Jr.

227 Ibid.

Congress eliminated funding for OTA in 1995 “amid broader efforts to reduce the size of government.[228] Since then, Congress has continued to debate the need for OTA or a similar technology assessment organization.[229]

[228] Ibid.

[229] For an overview of OTA/technology assessment-related legislation in the 107th-116th Congresses, see Appendix C in CRS Report R46327, *The Office of Technology Assessment: History, Authorities, Issues, and Options*, by John F. Sargent Jr.

Chapter 2

Defense Primer: Emerging Technologies*

Kelley M. Sayler

Both the 2022 National Defense Strategy and senior U.S. defense and intelligence officials have identified a number of emerging technologies that could have a disruptive impact on U.S. national security in the years to come. These technologies include

- artificial intelligence,
- lethal autonomous weapon systems,
- hypersonic weapons,
- directed energy weapons,
- biotechnology, and
- quantum technology.

As these technologies continue to mature, they could hold significant implications for congressional oversight, U.S. defense authorizations and appropriations, military concepts of operations, and the future of war.

Artificial Intelligence

Although the U.S. government has no official definition of artificial intelligence, policymakers generally use the term AI to refer to a computer system capable of human-level cognition. AI is further divided into three categories: narrow AI, general AI, and artificial superintelligence.

Narrow AI systems can perform only the specific task that they were trained to perform, while general AI systems would be capable of performing a broad range of tasks, including those for which they were not specifically trained. Artificial superintelligence refers to a system that could exceed human-level cognition across most tasks. General AI systems and artificial superintelligence do not yet—and may never—exist.

Narrow AI is currently being incorporated into a number of military applications by both the United States and its competitors. Such applications include but are not limited to intelligence, surveillance, and reconnaissance; logistics; cyber operations; command and control; and semi- autonomous and autonomous vehicles. These technologies are intended in

* This is an edited, reformatted and augmented version of Congressional Research Service Publication No. IF11105, updated November 4, 2024.

In: Emerging Military Technologies
Editor: Jerry Shields
ISBN: 979-8-89530-613-0

part to augment or replace human operators, freeing them to perform more complex and cognitively demanding work. In addition, AI-enabled systems could (1) react significantly faster than systems that rely on operator input, (2) cope with an exponential increase in the amount of data available for analysis, and (3) enable new concepts of operations, such as swarming (i.e., cooperative behavior in which unmanned vehicles autonomously coordinate to achieve a task) that could confer a warfighting advantage by overwhelming adversary defensive systems.

Narrow AI could, however, introduce a number of challenges. For example, such systems may be subject to algorithmic bias as a result of their training data or models. Researchers have repeatedly discovered instances of racial bias in AI facial recognition programs due to the lack of diversity in the images on which the systems were trained, while some natural language processing programs have developed gender bias. Such biases could hold significant implications for AI applications in a military context. A number of U.S. government documents, including DOD's *Responsible Artificial Intelligence Strategy and Implementation Pathway* and the *Framework to Advance AI Governance and Risk Management in National Security*, provide guidance on these applications.

Lethal Autonomous Weapon Systems

Lethal Autonomous Weapon Systems (LAWS) are a class of weapon systems capable of independently identifying a target and employing an onboard weapon system to engage and destroy the target without manual human control.

LAWS require computer algorithms and sensor suites to classify an object as hostile, make an engagement decision, and guide a weapon to the target. This capability would enable the system to operate in communications-degraded or -denied environments where traditional systems may not be able to operate.

LAWS are not yet in widespread development, and some senior military and defense leaders have expressed concerns about the ethics of ever fielding such systems. For example, in 2017 testimony before the Senate Armed Services Committee, then-Vice Chairman of the Joint Chiefs of Staff General Paul Selva stated, "I do not think it is reasonable for us to put robots in charge of whether or not we take a human life." Currently, there are no domestic or international legal prohibitions on the development of LAWS. However, approximately 30 countries have called for a preemptive ban on the systems due to ethical considerations, while others have called for formal regulation or guidelines for development and use. DOD Directive 3000.09 establishes department guidelines for the development and fielding of LAWS to ensure that they comply with "the law of war, applicable treaties, weapon system safety rules, and applicable rules of engagement."

Hypersonic Weapons

Hypersonic weapons—which fly at speeds of at least Mach 5—are in development in a number of countries, including the United States. There are two categories of hypersonic weapons:

- Hypersonic glide vehicles (HGV) are launched from a rocket before gliding to a target. (When HGVs are mated with their rocket booster, the resulting weapon system is often referred to as a hypersonic boost-glide weapon.)
- Hypersonic cruise missiles (HCM) are powered by high-speed engines throughout the duration of their flight.

In contrast to ballistic missiles, hypersonic weapons do not follow a ballistic trajectory and can maneuver en route to their destination, making defense against them difficult.

Currently, no such defense against hypersonic weapons exists, and experts disagree on the affordability and technological feasibility of hypersonic missile defense options.

According to open-source reporting, Russia fielded its first HGV—the Avangard—in December 2019 and its first HCM—the Tsirkon—in 2023. China has also reportedly fielded an operational HGV and may field an HCM in 2025. In contrast, the United States has not fielded an operational hypersonic weapon.

Analysts disagree about the strategic implications of hypersonic weapons. Some have identified two factors that could hold significant implications for strategic stability:

1) the weapon's short time-of-flight, which, in turn, compresses the timeline for response, and
2) its unpredictable flight path, which could generate uncertainty about the weapon's intended target and therefore heighten the risk of miscalculation or unintended escalation in the event of a conflict. Other analysts have argued that the strategic implications of hypersonic weapons are minimal because U.S. competitors such as China and Russia already possess the ability to strike the United States with intercontinental ballistic missiles, which, when launched in salvos, could overwhelm U.S. missile defenses.

Directed Energy Weapons

DOD defines directed energy (DE) weapons as those using concentrated electromagnetic energy, rather than kinetic energy, to "incapacitate, damage, disable, or destroy enemy equipment, facilities, and/or personnel." DE weapons— often colloquially referred to as "lasers"—could be used by ground forces in counter rocket, artillery, and mortar (C- RAM) or short-range air defense (SHORAD) missions.

They could offer low costs per shot and nearly limitless magazines that, in contrast to existing conventional systems, could enable an efficient and effective means of defending against missile salvos and swarms of uninhabited vehicles. Theoretically, DE weapons could also provide options for boost-phase missile intercept, given their speed- of-light travel time; however, as in the case of hypersonic missile defense, experts disagree on the affordability and technological feasibility of this application.

High-powered microwave (HPM) weapons, a subset of DE weapons, could be used as a nonkinetic means of disabling electronics, communications systems, and improvised explosive devices in the event of a conflict. In addition, the U.S. military has explored using HPM in a nonlethal "heat ray" system for crowd control; however, the system was recalled—likely due to ethical and operational considerations.

Biotechnology

Biotechnology leverages life sciences for technological applications. A number of developments in biotechnology hold potential implications for national security. As a 2018 Government Accountability Office (GAO) report notes, the Departments of Defense, State (State), and Homeland Security (DHS), and the Office of the Director of National Intelligence (ODNI), all assess that biotechnologies, such as the low-cost gene-editing tool CRISPR-Cas9, have the potential to "alter genes or create DNA to modify plants, animals, and humans. Such biotechnologies could be used to enhance [or degrade] the performance of military personnel. The proliferation of synthetic biology—used to create genetic code that does not exist in nature—may increase the number of actors that can create chemical and biological weapons" and could additionally enable the creation of adaptive camouflage, cloaking devices, or lighter, stronger, and—potentially—self-healing body and vehicle armor. U.S. adversaries may be less restrained in both researching and applying biotechnology, particularly as it relates to human performance modification and biological weapons.

Quantum Technology

Quantum technology, which employs the principles of quantum physics, has not yet reached maturity; however, it could hold significant implications for the future of encryption and stealth technologies. GAO reports that DOD, State, DHS, and the ODNI have assessed that "quantum communications could enable adversaries to develop secure communications that U.S. personnel would not be able to intercept or decrypt. Quantum computing may allow adversaries to decrypt [unclassified, classified, or sensitive] information, which could enable them to target U.S. personnel and military operations."

Quantum technology could have other military applications, such as quantum sensing, which could theoretically enable significant improvements in submarine detection, rendering the oceans "transparent." This could, in turn, hold implications for the survivability of the U.S. sea-based nuclear deterrent. Quantum sensing could also provide alternative positioning, navigation, and timing options that could in theory allow militaries to continue to operate at full performance in GPS-degraded or GPS-denied environments. Military application of quantum technologies could be constrained, however, by the fragility of quantum states, which can be disrupted by minute movements, changes in temperature, or other environmental factors.

Chapter 3

Defense Primer: Directed-Energy Weapons*

Kelley M. Sayler

Both the 2022 National Defense Strategy and the House Armed Services Committee's bipartisan Future of Defense Task Force Report have identified directed energy as a technology that could have a significant impact on U.S. national security in the years to come. As the Department of Defense (DOD) continues to invest in directed-energy (DE) weapons, Congress may consider implications for defense authorizations, appropriations, and oversight.

Overview

DOD defines DE weapons as those using concentrated electromagnetic energy, rather than kinetic energy, to "incapacitate, damage, disable, or destroy enemy equipment, facilities, and/or personnel." DE weapons include high-energy lasers (HEL) and high-powered microwave (HPM) weapons; other DE weapons, such as particle beam weapons, are outside the scope of this chapter.

HELs might be used by ground forces in short-range air defense (SHORAD), counter-unmanned aircraft systems (C-UAS), or counter-rocket, artillery, and mortar (C-RAM) missions. The weapons might be used to "dazzle" (i.e., temporarily disable) or damage satellites and sensors. This could in turn interfere with intelligence-gathering operations; military communications; and positioning, navigation, and timing systems used for weapons targeting. In addition, HELs could theoretically provide options for boost-phase missile intercept, given their speed-of-light travel time; however, experts disagree on the affordability, technological feasibility, and utility of this application.

In general, HELs might offer lower costs per shot and— assuming access to a sufficient power supply—deeper magazines compared with traditional munitions. (Although a number of different types of HELs exist, many of the United States' current programs are solid state lasers, which are fueled by electrical power. As a result, the cost per shot is equivalent to the cost of the electrical power required to fire the shot.) This could in turn produce a favorable cost-exchange ratio for the defender, whose marginal costs would be significantly lower than those of the aggressor.

* This is an edited, reformatted and augmented version of Congressional Research Service Publication No. IF11882, updated November 4, 2024.

In: Emerging Military Technologies
Editor: Jerry Shields
ISBN: 979-8-89530-613-0

Similarly, HPM weapons could provide a nonkinetic means of disabling adversary electronics and communications systems. These weapons could potentially generate effects over wider areas than HELs, which emit a narrower beam of energy. As a result, some analysts have noted that HPM weapons might provide more effective area defense against missile salvos and swarms of drones. HPM weapons have also been considered for use as nonlethal "heat ray" systems for crowd control.

Directed-Energy Weapons Programs

A number of countries are investing in directed-energy weapons programs. This chapter discusses a selection of unclassified DE weapons programs in three leading military powers: the United States, China, and Russia.

United States

DOD requested approximately $789.7 million for unclassified DE programs in FY2025—down from its $962.4 million request and $1.1 billion appropriation in FY2024. DOD does not currently have any DE programs of record.

DOD directed energy programs are coordinated by the Principal Director for Directed Energy within the Office of the Under Secretary of Defense for Research and Engineering (OUSD[R&E]). The Principal Director for Directed Energy is additionally responsible for development and oversight of the Directed Energy Roadmap. According to OUSD(R&E), DOD's DE roadmap outlines DOD's plan to increase power levels of DE weapons from around 150 kilowatts (kW—a unit of power), as is currently feasible, to 500 kW—with reduced size and weight—by FY2025. DOD seeks "to further reduce size and weight and increase power to MW [megawatt] levels by FY2026." For reference, although there is no consensus regarding the precise power level that would be needed to neutralize different target sets, some analysts believe that lasers of around 100 kW could engage unmanned aircraft systems, small boats, rockets, artillery, and mortars, whereas lasers of around 300 kW could additionally engage cruise missiles flying in certain profiles (i.e., flying across—rather than at—the laser). Lasers of 1 MW could potentially neutralize ballistic missiles and hypersonic weapons.

In addition to the DE roadmap, OUSD(R&E) manages the High Energy Laser Scaling Initiative (HELSI), which former Principal Director for Directed Energy Dr. Jim Trebes notes is "to demonstrate laser output power scaling while maintaining or improving beam quality and efficiency." HELSI is intended to strengthen the defense industrial base for potential future directed-energy weapons by providing near-term prototyping opportunities for industry partners. Dr. Trebes additionally notes that OUSD(R&E) has completed a DOD-wide Laser Lethality Analysis Process Review to identify future needs for the Department and best practices for DE development and use.

According to OUSD(R&E), DOD is also testing a Directed Energy Lethality Database that is to serve as a searchable repository for the department's DE analyses.

China

According to the U.S.-China Economic and Security Review Commission, China has been developing DE weapons since at least the 1980s and has made steady progress in developing HPM and increasingly powerful HELs. China has reportedly developed a 30-kilowatt road-mobile HEL, LW-30, designed to engage unmanned aircraft systems and precision-guided weapons. Reports indicate that China is also developing an airborne HEL pod.

According to the Defense Intelligence Agency, China is additionally pursuing DE weapons to disrupt, degrade, or damage satellites and their sensors and possibly already has a limited capability to employ laser systems against satellite sensors. China [has likely fielded] a ground-based laser weapon that can counter low-orbit space- based sensors ... and by the mid-to-late 2020s, it may field higher power systems that extend the threat to the structures of non-optical satellites.

Russia

Russia has been conducting DE weapons research since the 1960s, with a particular emphasis on HELs. Russia has reportedly deployed the Peresvet ground-based HEL with several mobile intercontinental ballistic missile units.

Although little is publicly known about Peresvet, including its power level, some analysts assert it is to dazzle satellites and provide point defense against unmanned aircraft systems. Russia's deputy defense minister Alexei Krivoruchko has stated that efforts are underway to increase Peresvet's power level and to deploy it on military aircraft. Reports suggest that Russia may also be developing HPMs as well as additional HELs capable of performing anti- satellite missions.

Potential Issues and Questions for Congress

Technological Maturity

Directed-energy weapons programs continue to face questions about their technological maturity, including the ability to improve beam quality and control to militarily useful levels and the ability to meet size, weight, and power (SWaP) and cooling requirements for integration into current platforms. Some DE systems are small enough to fit on military vehicles, but many require larger and/or fixed platforms that could potentially limit deployment options and operational utility. In what ways, if any, are DOD technology maturation efforts reducing the SWaP and cooling requirements of DE systems?

Weapons Characteristics

Although HELs may offer a lower cost per shot than traditional weapons such as missiles, they are also subject to a number of limitations. For example, atmospheric conditions (e.g., rain, fog, obscurants) and SWaP and cooling requirements can limit the range and beam quality of HELs, in turn reducing their effectiveness. Traditional weapons, in contrast, are not affected by these factors.

How, if at all, might HEL limitations be mitigated by technological developments, concepts of operation, or other methods? What impact might a failure to mitigate these limitations have on future military operations?

Mission Utility

Given the strengths and weaknesses of DE weapons, DOD is continuing to examine their role within the military. DOD is additionally conducting multiple utility studies to analyze potential concepts of operation for DE weapons and to assess the scenarios in which they might be militarily useful. How might Congress draw upon the conclusions of these analyses as it conducts oversight of DE weapons programs? What is the appropriate balance between DE weapons and traditional munitions within the military's portfolio of capabilities?

Defense Industrial Base

Some analysts have expressed concerns that, in the past, DOD did not provide stable funding for DE weapons programs or sufficient opportunities for the DE workforce. According to OUSD(R&E), HELSI is intended to address these concerns by providing industry with assured prototyping opportunities. In what ways, if any, has HELSI strengthened the defense industrial base for DE weapons? What, if any, challenges does the base continue to face and how might they be mitigated?

Intelligence Requirements

Some analysts have questioned whether DOD has sufficient knowledge of adversary DE weapons systems and materials to develop its weapons requirements. DOD is currently attempting to further define its DE collection requirements for the intelligence community (IC) through the Directed Energy Lethality Intelligence initiative. To what extent, if at all, is this initiative improving connectivity between DOD's DE community and the IC? What collection requirements, if any, remain?

Coordination within DOD

Pursuant to Section 219 of the FY2017 National Defense Authorization Act (P.L. 114-328), OUSD(R&E)'s Principal Director for Directed Energy is tasked with coordinating DE efforts across DOD and with developing DOD's Directed Energy Roadmap, which is to guide development efforts. To what extent are the military departments and defense agencies adhering to this roadmap? What, if any, additional authorities or structural changes would be required to ensure proper coordination throughout DOD?

This report was originally co-authored by John Hoehn.

Chapter 4

Department of Defense Directed Energy Weapons: Background and Issues for Congress*

Kelley M. Sayler
Jennifer DiMascio
Andrew Feickert
and Ronald O'Rourke

Summary

Directed energy (DE) weapons use concentrated electromagnetic energy, rather than kinetic energy, to combat enemy forces. Although the United States has been researching directed energy since the 1960s, some experts have observed that the Department of Defense (DOD) has invested billions of dollars in DE programs that failed to reach maturity and were ultimately cancelled. In recent years, however, DOD has made progress on DE weapons development, deploying the first operational U.S. DE weapon in 2014 aboard the USS *Ponce*. Since then, DE weapons development has continued, with DOD issuing a Directed Energy Roadmap to coordinate the department's efforts. DOD has also introduced a High Energy Laser Scaling Initiative, which seeks to strengthen the defense industrial base for DE weapons and improve laser beam quality and efficiency.

This report provides background information and issues for Congress on DE weapons, including high-energy lasers (HELs) and high-powered microwave (HPM) weapons, and outlines selected unclassified DOD, Air Force, Army, and Navy DE programs. If successfully fielded, HELs could be used by ground forces in a range of missions, including short-range air defense (SHORAD); counter-unmanned aircraft systems (C-UAS); and counter-rocket, artillery, and mortar (C-RAM) missions. HPM weapons could provide a nonkinetic means of disabling adversary electronics and communications systems. Compared with traditional munitions, DE weapons could offer lower logistical requirements, lower costs per shot, and—assuming access to a sufficient power supply—deeper magazines. These weapons could, however, face limitations not faced by their kinetic counterparts. For example, atmospheric conditions (e.g., rain, fog, obscurants) could potentially limit the range and beam quality of DE weapons, in turn reducing their effectiveness.

As DOD continues to invest in DE weapons, Congress may consider the weapons' technological maturity, lifecycle cost, characteristics, mission utility, industrial base,

---

* This is an edited, reformatted and augmented version of Congressional Research Service Publication No. R46925, updated July 11, 2024.

In: Emerging Military Technologies
Editor: Jerry Shields
ISBN: 979-8-89530-613-0

intelligence requirements, and oversight structure. Congress may also consider the implications of DE weapons for future arms control agreements.

Introduction

This report provides background information and issues for Congress on Department of Defense (DOD) efforts to develop and procure directed energy (DE) weapons. The report provides an overview of certain DOD, Air Force, Army, and Navy DE programs. Two other CRS reports provide additional discussion of Army and Navy DE programs.[1] Some types of DE weapons, such as particle-beam weapons, are outside the scope of this report.

DOD's efforts on DE weapons pose a number of potential issues for Congress. Decisions that Congress makes on these issues could have substantial implications for future DOD capabilities and funding requirements and the U.S. defense industrial base.

Overview of Directed Energy Weapons[2]

DOD defines directed energy weapons as those using concentrated electromagnetic energy, rather than kinetic energy, to "incapacitate, damage, disable, or destroy enemy equipment, facilities, and/or personnel."[3] DE weapons include high-energy laser (HEL) and high-powered microwave (HPM) weapons.

HEL weapons might be used by ground forces in various missions, including short-range air defense (SHORAD); counter-unmanned aircraft systems (C-UAS); and counter-rocket, artillery, and mortar (C-RAM) missions.[4] The weapons might be used to "dazzle" (i.e., temporarily disable) or damage satellites and sensors. This could in turn interfere with intelligence-gathering operations; military communications; and positioning, navigation, and timing systems used for weapons targeting. In addition, HEL weapons could theoretically provide options for boost-phase missile intercept, given their speed-of-light travel time; however, experts disagree on the affordability, technological feasibility, and utility of this application.[5]

In general, HEL weapons might offer lower logistical requirements, lower costs per shot, and— assuming access to a sufficient power supply—deeper magazines compared with traditional munitions. (Although a number of different types of HELs exist, many of the United

---

[1] See CRS Report R45098, U.S. Army Weapons-Related Directed Energy (DE) Programs: Background and Potential Issues for Congress, by Andrew Feickert, and CRS Report R44175, Navy Shipboard Lasers: Background and Issues for Congress, by Ronald O'Rourke.

[2] This section was written by Kelley M. Sayler, CRS Specialist in Advanced Technology and Global Security. For more information—including information about DE weapons programs in China and Russia—see CRS Report R46458, Emerging Military Technologies: Background and Issues for Congress, by Kelley M. Sayler.

[3] Joint Chiefs of Staff, Joint Electromagnetic Spectrum Operations, Joint Publication 3-85, May 22, 2020, p. GL-6.

[4] For more information about the role of DE weapons in C-UAS missions, see CRS In Focus IF11426, Department of Defense Counter-Unmanned Aircraft Systems, by John R. Hoehn and Kelley M. Sayler.

[5] See, for example, James N. Miller and Frank A. Rose, "Bad Idea: Space-Based Interceptors and Space-Based Directed Energy Systems," Center for Strategic and International Studies, December 13, 2018, at https://defense360.csis.org/bad-idea-space-based-interceptors-and-space-based-directed energy-systems/; and Justin Doubleday, "Pentagon punts MDA's laser ambitions, shifts funding toward OSD-led 'laser scaling,'" Inside Defense, February 19, 2020, at https://insidedefense.com/daily-news/pentagon-punts-mdas-laser-ambitions-shifts-funding- toward-osd-led-laser-scaling.

States' current programs are solid state lasers, which are fueled by electrical power. As a result, the cost per shot would be equivalent to the cost of the electrical power required to fire the shot.[6]) These characteristics could in turn produce a favorable cost-exchange ratio for a defender, whose marginal costs would be significantly lower than those of an aggressor.

Similarly, HPM weapons could provide a nonkinetic means of disabling adversary electronics and communications systems. These weapons could potentially generate effects over wider areas— disabling any electronics within their electromagnetic cone—than HEL weapons, which emit a narrower beam of energy (see Figure 1). Some analysts have noted that HPM weapons might provide more effective area defense against missile salvos and swarms of unmanned aircraft systems. HPM weapons in an anti-personnel configuration might provide a means of nonlethal crowd control, perimeter defense, or patrol or convoy protection.[7] Potential advantages and limitations of both HEL and HPM weapons are discussed in greater detail in Appendix A.

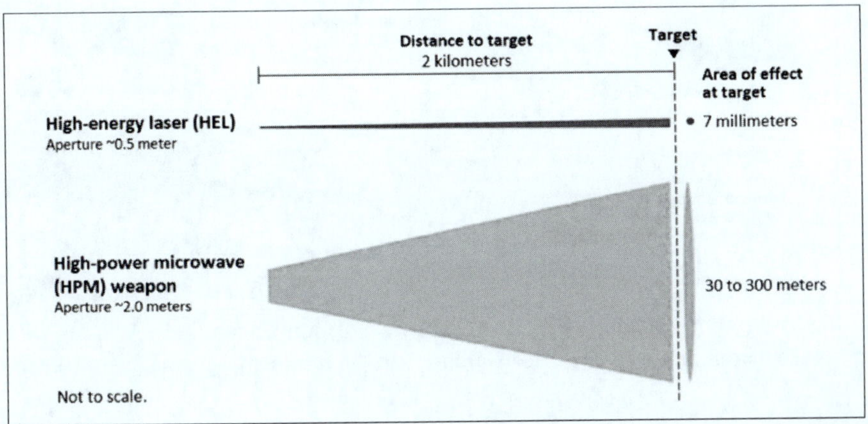

Source: CRS image based on an image in Mark Gunzinger and Chris Dougherty, Changing the Game: The Promise of Directed-Energy Weapons, Center for Strategic and Budgetary Assessments, April 19, 2021, p. 40, at https://csbaonline. org/uploads/documents/CSBA_ChangingTheGame_e reader.pdf.

Note: Units of measurement are illustrative.

Figure 1. Illustrative Effects of HELs Versus HPM Weapons.

Selected Defense-Wide Directed Energy Programs[8]

DOD directed energy programs are coordinated by the Principal Director for Directed Energy within the Office of the Under Secretary of Defense for Research and Engineering (OUSD[R&E]). The Principal Director for Directed Energy is responsible for development and oversight of the Directed Energy Roadmap, which articulates DOD's objective of "[achieving]

---

[6] Ariel Robinson, "Directed Energy Weapons: Will They Ever Be Ready?," National Defense, July 1, 2015, at https://www.nationaldefensemagazine.org/articles/2015/7/1/2015july-directed-energy-weapons-will-they-ever-be- ready.

[7] See, for example, Joint Intermediate Force Capabilities Office, "Active Denial System FAQs," https://jnlwp.defense.gov/About/Frequently-Asked-Questions/Active-Denial-System-FAQs/.

[8] This section was written by Kelley M. Sayler, CRS Analyst in Advanced Technology and Global Security.

dominance in DE military applications in every mission and domain where they give advantage."[9] According to OUSD(R&E), the current roadmap outlines DOD's plans to increase power levels of HEL weapons from around 150 kilowatt (kW), as is currently feasible, to 500 kW class—with reduced size and weight—by FY2025.[10] DOD seeks "to further reduce size and weight and increase power to MW [megawatt] levels by FY2026."[11] For reference, although no consensus exists regarding the precise power level that would be needed to neutralize different target sets, DOD briefing documents (see *Figure 2*) suggest that a laser of approximately 100 kW could engage UASs, rockets, artillery, and mortars, whereas a laser of around 300 kW could additionally engage small boats and cruise missiles flying in certain profiles (i.e., flying across— rather than at—the laser).[12] Lasers of 1 MW could potentially neutralize ballistic missiles and hypersonic weapons.[13]

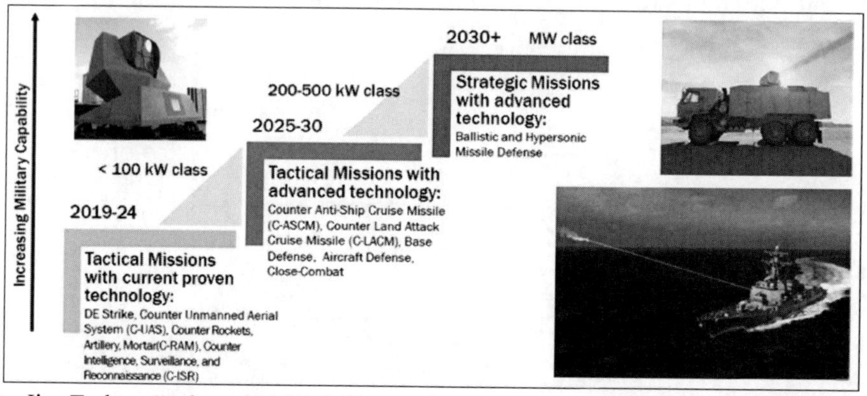

Source: Dr. Jim Trebes, "Advancing High Energy Laser Weapon Capabilities: What is OUSD (R&E) Doing?" presentation at the Institute for Defense and Government Advancement (IDGA), October 21, 2020.

Figure 2. Summary of DOD Directed Energy Roadmap.

In addition to managing the DE roadmap, OUSD(R&E) manages the High Energy Laser Scaling Initiative (HELSI), which seeks "to demonstrate laser output power scaling while maintaining or improving beam quality and efficiency."[14] HELSI is intended to strengthen the defense industrial base for potential future DE weapons by providing near-term prototyping opportunities for industry partners.[15] OUSD(R&E) has completed a DOD-wide Laser Lethality

---

[9] Dr. Jim Trebes, "Advancing High Energy Laser Weapon Capabilities: What is OUSD (R&E) Doing?" presentation at the Institute for Defense and Government Advancement (IDGA), October 21, 2020.

[10] CRS correspondence with the Office of the Under Secretary of Defense for Research and Engineering, September 8, 2022. Kilowatts and megawatts are units of power. For example, 1 kilowatt is equal to 1,000 watts, and 1 megawatt is equal to 1 million watts.

[11] Ibid.

[12] Dr. Jim Trebes, "Advancing High Energy Laser Weapon Capabilities: What is OUSD (R&E) Doing?" presentation at IDGA, October 21, 2020; and CRS conversation with Principal Director for Directed Energy Modernization Dr. Jim Trebes, November 17, 2020. Required power levels could be affected by additional factors such as adversary countermeasures and atmospheric conditions and effects.

[13] Dr. Jim Trebes, "Advancing High Energy Laser Weapon Capabilities: What is OUSD (R&E) Doing?" presentation at IDGA, October 21, 2020.

[14] Dr. Jim Trebes, "Advancing High Energy Laser Weapon Capabilities: What is OUSD (R&E) Doing?" presentation at IDGA, October 21, 2020.

[15] Industry participants in HELSI include nLight-Nutronics (sponsored by the Navy), Lockheed Martin (sponsored by the Army), General Atomics (sponsored by the Air Force), and Northrop Grumman. Nancy Jones-Bonbrest,

Analysis Process Review to identify future needs for the department and best practices for DE development and use. In addition, OUSD(R&E) is developing laser and HPM lethality databases, searchable repositories for DOD's DE analyses.[16]

In support of these initiatives, DOD maintains a number of research programs, including programs at the Missile Defense Agency (MDA), the Office of the Secretary of Defense (OSD), and the Defense Advanced Projects Research Agency (DARPA). For example, MDA's Directed Energy Demonstrator Development program addresses "technology risk reduction and maturation for high powered strategic lasers, beam control, lethality, and related technologies" in support of OUSD(R&E)'s Directed Energy Roadmap.[17] The program received $42 million in FY2021.

Although MDA did not request funding for the program in FY2022 or FY2023 "due to a shift in Department of Defense priorities," Congress appropriated $39 million and $16 million, respectively, to continue development efforts.[18] MDA did not request funds for the Directed Energy Demonstrator Development program in FY2024 or FY2025.[19]

In FY2025, OSD requested $16.5 million for High Energy Laser Research Initiatives, including basic research and educational grants, and $48.6 million for High Energy Laser Development, which funds applied research.[20] OSD additionally requested $110.4 million in FY2025 for High Energy Laser Advanced Development, which is focused on "scaling the

---

"Scaling Up: Army Advances 300kW-class Laser Prototype," Army Rapid Capabilities and Critical Technologies Office, March 3, 2020, at https://www.army.mil/article/233346/scaling_up_army_advances_300kw_class_laser_prototype; and CRS conversation with Principal Director for Directed Energy Dr. Frank Peterkin, May 17, 2023.

[16] The laser lethality database has been populated with data and is being updated based on user feedback, while the HPM laser database is under development. OUSD(R&E) plans to transition the laser database to the Director, Operational Test & Evaluation (DOT&E) in early FY2025. The HPM database is expected to transition to DOT&E later in FY2025. Both databases are intended to serve as "analysts' tools to support more rapid target vulnerability package development for the Services (engagement/mission level studies, DE weapon system performance assessments, etc.)." CRS correspondence with Distinguished Scientist for Laser Weapon Systems Lethality Dr. Christopher Lloyd, June 27, 2024.

[17] DOD, Department of Defense Fiscal Year (FY) 2023 Budget Estimates, Missile Defense Agency, Defense-Wide Justification Book Volume 2a of 5 Research, Development, Test & Evaluation, Defense-Wide, pp. 601-603, at https://comptroller.defense.gov/Portals/45/Documents/defbudget/fy2023/budget_justification/pdfs/03_RDT_and_E/RDTE_Vol2_MDA_RDTE_PB23_Justification_Book.pdf.

[18] DOD, Department of Defense Fiscal Year (FY) 2024 Budget Estimates, Missile Defense Agency, Defense-Wide Justification Book Volume 2a of 5 Research, Development, Test & Evaluation, Defense-Wide, p. 605, at https://comptroller.defense.gov/Portals/45/Documents/defbudget/fy2024/budget_justification/pdfs/03_RDT_and_E/RDTE_Vol2_MDA_RDTE_PB24_Justification_Book.pdf.

[19] DOD, Department of Defense Fiscal Year (FY) 2025 Budget Estimates, Missile Defense Agency, Defense-Wide Justification Book Volume 2a of 5 Research, Development, Test & Evaluation, Defense-Wide, p. xliii, at https://comptroller.defense.gov/Portals/45/Documents/defbudget/FY2025/budget_justification/pdfs/03_RDT_and_E/RDTE_Vol2_MDA_RDTE_PB25_Justification_Book.pdf.

[20] DOD, Department of Defense Fiscal Year (FY) 2025 Budget Estimates, Office of the Secretary of Defense, Defense- Wide Justification Book Volume 3 of 5 Research, Development, Test & Evaluation, pp. 1 and 93, at https://comptroller.defense.gov/Portals/45/Documents/defbudget/FY2025/budget_justification/pdfs/03_RDT_and_E/R DTE_OSD_PB_2025.pdf. These programs were transferred to OSD from the Air Force to "better align [the] research area to Department of Defense Science and Technology strategy and priorities for Directed Energy." This transfer could reflect greater coordination across DOD DE programs. DOD, Department of Defense Fiscal Year (FY) 2022 Budget Estimates, Office of the Secretary of Defense, Defense-Wide Justification Book Volume 3 of 5 Research, Development, Test & Evaluation, p. 79, at https://comptroller.defense.gov/Portals/45/Documents/defbudget/fy2022/budget_justification/pdfs/03_RDT_and_E/RDTE_Vol3_OSD_RDTE_PB22_Justification_Book.pdf.

output power of directed energy weapon systems to reach operationally effective power levels applicable to broad mission areas across the [DOD].”[21] OSD requested $10 million in FY2025 to continue assessments of directed energy weapons, including assessments of the weapons’ effects, effectiveness, and limitations.[22] Finally, DARPA’s Waveform Agile Radio-frequency Directed Energy (WARDEN) program seeks to “extend the range and lethality of high power microwave weapons ... [for] counter-unmanned aerial systems, vehicle and vessel disruption, electronic strike, and guided missile defense.”[23] DARPA received $20 million for WARDEN in FY2024 and requested $8 million for the program in FY2025.[24]

Overall, DOD requested $789.7 million for directed energy weapons programs in FY2025, down from its $962.4 million request and $1.1 billion appropriation in FY2024.[25] For a breakout of funding by military department, see *Appendix A*.

Selected Air Force Directed Energy Weapons Programs[26]

The Air Force is developing and testing a number of DE technologies through the Directed Energy Directorate of the Air Force Research Laboratory (AFRL). The following section provides a brief description of selected unclassified efforts.

Tactical High-Power Operational Responder (THOR)

The Tactical High-Power Microwave Operational Responder (THOR) technology demonstrator (see Figure 3), designed by AFRL in collaboration with industry partners, is intended to provide a viable DE C-UAS weapon system focused on short-range air base defense.[27] THOR is housed in a standardized 20-foot transport container that enables it to fit inside a C-130 transport aircraft.

---

[21] DOD, Department of Defense Fiscal Year (FY) 2025 Budget Estimates, Office of the Secretary of Defense, Defense- Wide Justification Book Volume 3 of 5 Research, Development, Test & Evaluation, p. 355, at https://comptroller.defense.gov/Portals/45/ Documents/defbudget/FY2025/budget_justification/pdfs/03_RDT_and_E/R DTE_OSD_PB_2025.pdf.

[22] DOD, Department of Defense Fiscal Year (FY) 2025 Budget Estimates, Office of the Secretary of Defense, Defense- Wide Justification Book Volume 3 of 5 Research, Development, Test & Evaluation, p. 376, at https://comptroller.defense.gov/Portals/45/Documents/defbudget/ FY2025/budget_justification/pdfs/03_RDT_and_E/R DTE_OSD_PB_2025.pdf.

[23] DOD, Department of Defense Fiscal Year (FY) 2023 Budget Estimates, Defense Advanced Research Projects Agency, Defense-Wide Justification Book Volume 1 of 5 Research, Development, Test & Evaluation, p. 145, at https://comptroller.defense.gov/Portals/45/ Documents/defbudget/fy2023/budget_justification/pdfs/03_RDT _and_E/RDTE_Vol1_DARPA_MasterJustificationBook_PB_2023.pdf.

24 DOD, Department of Defense Fiscal Year (FY) 2025 Budget Estimates, Defense Advanced Research Projects Agency, Defense-Wide Justification Book Volume 1 of 5 Research, Development, Test & Evaluation, p. 128, at https://comptroller.defense.gov/Portals/ 45/Documents/defbudget/FY2025/budget_justification/pdfs/03_ RDT _and_E/R DTE_Vol1_DARPA_MasterJustificationBook_PB_2025.pdf.

[25] CRS correspondence with the Office of the Under Secretary of Defense for Research and Engineering, April 30, 2024.

[26] This section was written by former CRS Research Assistant Samuel D. Ryder and former CRS Analyst in Military Capabilities and Programs John R. Hoehn. It has been updated by CRS Analyst in U.S. Defense Policy Jennifer DiMascio.

[27] Industry partners include BAE Systems, Leidos, and Verus Research. THOR also features a proprietary radar system developed by Black Sage.

Users reportedly can deploy the system in three hours and operate its user interface with only rudimentary training.[28] According to Air Force press releases, THOR has successfully completed a two-year test period and, in 2023, engaged a swarm of multiple targets.[29] THOR has informed follow-on prototype efforts such as Mjölnir.[30]

Source: U.S. Air Force, AFRL Directed Energy Directorate, press release, September 24, 2019.

Figure 3.THOR Demonstrator.

Phaser High-Powered Microwave

The Phaser High-Powered Microwave system (see Figure 4), developed by Raytheon, is intended to provide a short-range C-UAS capability similar to that of THOR. The Air Force reportedly procured a $16.3 million prototype Phaser for testing and overseas field assessments; however, it is unclear whether the system has been deployed outside the United States.[31]

---

[28] Bryan Ripple, "Enemy drone operators may soon face the power of THOR," 88th Air Base Wing Public Affairs, September 24, 2019, at https://www.af.mil/News/Article-Display/ Article/1836495/air-force-research-laboratory-completes-successful-shoot-down-of-air-launched-m/.

[29] 1st Lt. James Wymer, "AFRL's drone killer, THOR will welcome new drone 'hammer,'" *U.S. Air Force*, August 2, 2021, at https://www.af.mil/News/Article-Display/Article/ 2713908/ afrls-drone-killer-thor-will-welcome-new-drone-hammer/. 1st Lt. Nina Rogers, "AFRL conducts swarm technology demonstration," *Air Force Research Laboratory Public Affairs*, May 16, 2023, at https://www.afrl.af.mil/News/Article-Display/Article/ 3396995/ afrl-conducts-swarm-technology-demonstration/.

[30] Steve Trimble, "Pentagon Seeks Jump to Long-Range Targets For Microwave Weapons," *Aviation Week Intelligence Network*, February 2, 2024.

[31] Joe Pappalardo, "The Air Force Is Deploying Its First Drone-Killing Microwave Weapon," *Popular Mechanics*, September 24, 2019, at https://www.popularmechanics.com/ military/weapons/a29198555/phaser-weapon-air-

Source: Raytheon Missiles and Defense, Phaser product page, February 2020.

Figure 4. Phaser Demonstrator.

Counter-Electronic High-Power Microwave Extended-Range Air Base Defense (CHIMERA)

AFRL awarded Raytheon Missiles and Defense a contract for testing of the Counter-Electronic High-Power Microwave Extended-Range Air Base Defense (CHIMERA) system in October 2020. In contrast to THOR and Phaser, which are designed for a short-range C-UAS mission, the CHIMERA system is intended to be able to engage UAS at greater distances.[32] In January 2024, AFRL and Raytheon acquired and tracked aerial targets for an entire flight path during a three- week field test at White Sands Missile Range in New Mexico.[33] Unclassified information about the CHIMERA system is limited.

force/; and Theresa Hitchens, "AF Says Lasers Are Being Field Tested, but NOT THOR or Other Microwave Weapons," *Breaking Defense*, December 22, 2020, at https://breakingdefense.com/2020/12/af-says-lasers-are-being-field-tested- but-not-thor-or-other-microwave-weapon/.

[32] Sara Sirota, "AFRL to award Raytheon sole-sourced contract for directed energy weapon," *Inside Defense*, October 29, 2020, at https://insidedefense.com/insider/afrl-award-raytheon-sole-sourced-contract-directed-energy-weapon.

[33] RTX press release, "RTX's CHIMERA high-power microwave system excels during three-week field test," January 29, 2024, at https://www.rtx.com/news/news-center/2024/01/29/ rtxs-chimera-high-power-microwave-system-excels- during-three-week-field-test.

High-Energy Laser Weapon System (HELWS)

The High-Energy Laser Weapon System (HELWS) is to serve as a mobile C-UAS capability for air base defense (see Figure 5). The system comprises a laser weapon and multispectral targeting system mounted on the back of a Polaris MRZR all-terrain vehicle and can reportedly operate at distances of up to 3 km.[34] HELWS developer Raytheon claims the laser can fire dozens of shots using a single charge from a standard 220-volt outlet, and an indefinite number of shots if connected to an external power source such as a generator.[35] The Air Force acquired the first HELWS in October 2019 and reportedly deployed HELWS overseas for field assessments in April 2020.[36] The Air Force awarded Raytheon a $15.5 million contract for an upgraded version of HELWS in April 2021.[37] This version is to be "delivered unmounted on pallets for potential use with different platforms."[38]

Source: Raytheon Missiles and Defense, HELWS product page, April 2020.

Figure 5. HELWS Prototype.

[34] Raytheon, "Raytheon Intelligence & Space delivers another Air Force laser system ready for operational use," September 14, 2020, https://www.raytheonintelligenceandspace.com/ news/advisories/raytheon-intelligence-space-delivers-another-air-force-laser-system-ready; and Nathan Strout, "Raytheon awarded $15.5 million to upgrade laser weapon," *C4ISRNET*, April 7, 2021, at https://www.c4isrnet.com/unmanned/2021/04/07/raytheon-awarded-155-to- upgrade-laser-weapon/.

[35] Kyle Mizokami, "The Air Force Mobilizes Its Laser and Microwave Weapons Abroad," *Popular Mechanics*, April 9, 2020, at https://www.popularmechanics.com/ military/wea pons/a32083799/laser-microwave-weapons/; and Raytheon, "Raytheon Intelligence & Space delivers another Air Force laser system ready for operational use," September 14, 2020, at https://www.raytheonintelligenceandspace.com/news/advisories/raytheon-intelli gence-space-delivers-another- air-force-laser-system-ready.

[36] Raytheon, "Raytheon Delivers First Laser Counter-UAS System to U.S. Air Force," October 22, 2019, at https://raytheon.mediaroom.com/2019-10-22-Raytheon-delivers-first-laser-counter-UAS-System-to-U-S-Air-Force#:~:text=Laser%20dune%20buggy%20set% 20for, Air%20Force%20earlier%20this%20month; and 88th Air Base Wing Public Affairs, "AFRL gives warfighters new weapons system," April 6, 2020, at https://www.whs. mil/News/News-Display/Article/2138161/afrl-gives-warfighters-new-weapons-system/.

[37] Nathan Strout, "Raytheon awarded $15.5 million to upgrade laser weapon," *C4ISRNET*, April 7, 2021, at https://www.c4isrnet.com/unmanned/2021/04/07/raytheon-awarded-155-to-upgrade-laser-weapon/.

[38] Ibid.

Self-Protect High-Energy Laser Demonstrator (SHiELD)

Source: Lockheed Martin, Tactical Airborne Laser Weapon System, September 14, 2020.

Figure 6. SHiELD Prototype Rendering.

The Self-Protect High-Energy Laser Demonstrator (SHiELD) is a prototype system in development by AFRL, Boeing, Lockheed Martin, and Northrop Grumman (see Figure 6). It is intended to mount as an external pod on Air Force aircraft—from fourth-generation F-15 fighters to sixth-generation aircraft currently in development—and target incoming air-to-air and surface- to-air missiles.[39] The Air Force conducted a series of tests of the Demonstrator Laser Weapon System, a ground-based test surrogate for SHiELD, in April 2019. The demonstrator successfully engaged incoming missiles and helped validate SHiELD's technology; however, technical issues and challenges related to the COVID-19 pandemic have reportedly pushed SHiELD's first flight demonstration from FY2021 to FY2024.[40] Furthermore, at a June 2020 Mitchell Institute event, then-Assistant Secretary of the Air Force Will Roper stated that the Air Force is reassessing the technological maturity of and use cases for SHiELD, as well as its potential role in missile defense missions.[41] Former Under Secretary of Defense for Research and Engineering Mike Griffin has noted that he is "extremely skeptical that we can put a large laser on an aircraft and use it to shoot down an adversary missile, even from fairly close."[42] As part of SHiELD, Lockheed Martin in 2022 delivered a subsystem to AFRL, a compact directed energy weapon called Laser Advancements for Next-Generation

---

[39] See Joanne Perkins, "AFRL's SHiELD set to receive critical assembly," *Air Force Research Laboratory*, February 23, 2021, at https://www.afrl.af.mil/News/Article-Display/Article/ 2511692/afrls-shield-set-to-receive-critical-assembly/.

[40] "Air Force Research Laboratory completes successful shoot down of air-launched missiles," 88th Air Base Wing Public Affairs, May 3, 2019, at https://www.af.mil/News/Article-Display/Article/1836495/air-force-research-laboratory-completes-successful-shoot-down-of-air-launched-m/; Valerie Insinna, "US Air Force delays timeline for testing a laser on a fighter jet," *Defense News*, June 30, 2020, at https://www.defensenews.com/ air/2020/ 06/30/us-air-force-delays-timeline-for-testing-a-laser-on-a-fighter-jet/; and Nathan Strout, "Air Force to begin assembly of airborne laser," *C4ISRNET*, February 23, 2021, at https://www.c4isrnet.com/battlefield-tech/2021/02/23/air-force-to-begin-assembly-of-airborne-laser/.

[41] Valerie Insinna, "US Air Force delays timeline for testing a laser on a fighter jet," *Defense News*, June 30, 2020, at https://www.defensenews.com/air/2020/06/30/us-air-force-delays-timeline-for-testing-a-laser-on-a-fighter-jet/.

[42] Aaron Mehta, "Griffin 'extremely skeptical' of airborne lasers for missile defense," *Defense News*, May 20, 2020, at https://www.defensenews.com/2020/05/20/griffin-extremely-skeptical-of-airborne-lasers-for-missile-defense/.

Compact Environments (LANCE).[43] In May 2024, SHiELD program manager Dr. Ted Ortiz noted that "the SHiELD program has concluded, and there are no plans for further testing and evaluation."[44] SHiELD was not installed on a fighter jet test bed.[45]

Selected Army Directed Energy Weapons Programs[46]

The Army Rapid Capabilities and Critical Technologies Office (RCCTO) is currently managing three major Army DE weapons programs:

- Directed Energy Maneuver-Short Range Air Defense (DE M-SHORAD),
- Indirect Fire Protection Capability-High Energy Laser (IFPC-HEL), and
- Indirect Fire Protection Capability-High Power Microwave (IFPC-HPM).[47]

The Army is developing DE weapons to counter UAS and rockets, artillery, and mortars (RAM), in turn increasing Army air and missile defense capability and reducing total system lifecycle costs by means of reduced logistical demands.[48]

Directed Energy Maneuver-Short-Range Air Defense (DE M-SHORAD)[49]

DE M-SHORAD, also known as Guardian (see *Figure 7*), seeks to integrate a 50 kW laser on a Stryker combat vehicle to provide short-range air defense support to the Army's combat brigades. The Army reportedly tested two DE M-SHORAD prototypes—one from Raytheon/Kord and one from Northrop Grumman—in a "shoot-off" at Ft. Sill, Oklahoma, in July 2021.[50] According to Director of the RCCTO Lieutenant General Neil Thurgood, DE M-SHORAD successfully defeated UAS but failed to defeat mortar rounds during this test; Northrop Grumman subsequently withdrew from the program.[51] The Army tested the Raytheon/Kord prototype again in 2022 during a four-week exercise at White Sands Missile

43 Justin Katz, "Lockheed delivers airborne laser LANCE to Air Force Research Lab," Breaking Defense, July 11, 2020, at https://breakingdefense.com/2022/07/lockheed-delivers-airborne-laser-lance-to-air-force-research-lab/.

44 Jared Keller, "Another Dead End for Airborne Lasers: Air Force Scraps Effort to Mount Directed-Energy Weapon on Fighter Jet," *Military.com*, May 17, 2024, https://www. military.com/daily-news/2024/05/17/air-force-abandons-plan-mount-laser-weapon-fighter-jet-after-scrapping-similar-gunship-project.html.

45 Ibid.

46 This section was written by Andrew Feickert, CRS Specialist in Military Ground Forces.

47 Lieutenant General (LTG) L. Neil Thurgood, "Space and Missile Defense Symposium," Army Rapid Capabilities and Critical Technologies Office, August 11, 2021.

48 Nancy Jones-Bonbrest, "Army Advances First Laser Weapon Through Combat Shoot-Off," Army Rapid Capabilities and Critical Technologies Office, August 10, 2021, at https://www.army.mil/article/249239/army_advances_first_laser_weapon_through_combat_shoot_off.

49 For additional information, see CRS In Focus IF12397, *U.S. Army's Maneuver Short-Range Air Defense (M-SHORAD) System*, by Andrew Feickert.

50 Jared Keller, "The Army's First Laser Weapon Almost Ready for a Fight," *Task and Purpose*, August 12, 2021; and Ethan Sterenfeld, "Laser M-SHORAD works against mortars in Army test," *Inside Defense*, May 17, 2022.

51 Evan Oschner, "Army set to deliver first 50-kilowatt lasers," *Inside Defense*, August 10, 2022; and Jen Judson, "Northrop bows out of competition to build laser weapon for Strykers," Defense News, August 18, 2021.

Range, New Mexico and in 2023 during a live-fire test at Yuma Proving Ground.[52] According to a Raytheon press release, the prototype "acquired, tracked, targeted, and defeated multiple mortars and successfully accomplished multiple tests simulating real-world scenarios."[53] Army plans called for DE M- SHORAD to start a user assessment beginning in the fourth quarter of FY2023, running through the first quarter of FY2024. The Army also plans for an additional final contract competition prior to FY2025.

Reportedly, in February 2024, the Army deployed four Stryker-mounted DE M-SHORAD prototypes to the Middle East for soldiers to test the system's performance against aerial threats. Initial soldier feedback was characterized as "not overwhelmingly positive," with officials suggesting that "results from the lab environment and test ranges are very different from the tactical environment."[54]

Source: Kristen Burroughs, "The Army Rapid Capabilities and Critical Technologies Office's Directed Energy Maneuver-Short Range Air Defense (DE M-SHORAD) Rapid Prototyping Effort is On-Track to Deliver," Army News, August 18, 2021.

Figure 7. Guardian DE M-SHORAD.

Indirect Fire Protection Capability-High Energy Laser (IFPC-HEL)[55]

IFPC-HEL, also known as Valkyrie (see Figure 8), is to protect fixed and semi-fixed sites from cruise missiles, UAS, and RAM threats.[56] According to Army budget documents, the system is to include "a vehicle, 300 kW class [>250 kW] laser subsystem, power and thermal subsystem,

[52] Ethan Sterenfeld, "Laser M-SHORAD works against mortars in Army test," *Inside Defense*, May 17, 2022; and Jen Judson, "Army short-range air defense laser prototypes take down drones at Yuma," *Defense News*, April 13, 2023.

[53] Ethan Sterenfeld, "Laser M-SHORAD works against mortars in Army test," *Inside Defense*, May 17, 2022. See also Jen Judson, "Army short-range air defense laser prototypes take down drones at Yuma," *Defense News*, April 13, 2023.

[54] Ashley Roque, "Army Soldiers not Impressed with Strykers Outfitted with 50-kilowatt Lasers, Service Official Says," *Breaking Defense,* May 16, 2024.

[55] For additional information, see CRS In Focus IF12421, *The U.S. Army's Indirect Fire Protection Capability (IFPC) System*, by Andrew Feickert.

[56] DOD, *Department of Defense Fiscal Year (FY) 2023 Budget Estimates, Army Justification Book Volume II Budget Activity 4, Research, Development, Test & Evaluation*, p. 405, at https://www.asafm.army.mil/Portals/72/Documents/BudgetMaterial/2023/Base%20Budget/rdte/vol_2-Budget_Activity_4.pdf.

and a beam control subsystem integrated with a battle management command, control, and communication software."[57] Army RCCTO issued a request for white papers for IFPC-HEL in January 2022, "with the intent of awarding one or more Other Transaction Agreement for Prototype."[58] Reports indicate that the Army subsequently selected Dynetics to serve as systems integrator for IFPC-HEL.[59] Four IFPC-HEL prototypes are to be delivered in the third quarter of FY2025, when IFPC-HEL is to transition to a program of record.[60]

Source: "Dynetics to Build and Increase Power of U.S. Army Laser Weapons," May 7, 2020, https://www.dynetics.com/newsroom/news/2020/dynetics-to-build-and-increase-power-of-us-army-laser- weapons, accessed August 12, 2022.

Figure 8. Valkyrie IFPC-HEL.

IFPC-High Power Microwave (IFPC-HPM)[61]

The Army is developing IFPC-HPM (see Figure 9)—a transportable, containerized system—to counter swarms of Group 1 and Group 2 UAS.[62] IFPC-HPM is to be "paired with IFPC-HEL

---

[57] DOD, Department of Defense Fiscal Year (FY) 2023 Budget Estimates, Army Justification Book Volume II Budget Activity 4, Research, Development, Test & Evaluation, p. 403, at https://www.asafm.army.mil/Portals/72/Documents/BudgetMaterial/2023/Base%20Budget/rdte/vol_2-Budget_Activity_4.pdf.

[58] SAM.gov, "Request for White Papers (RFWP) Indirect Fire Protection Capability-High Energy Laser (IFPC-HEL) Prototypes Prime Contractor," January 20, 2022, at https://sam.gov/opp/fe1cce00fde64c328b5234be24c795b1/view. For additional information about Other Transaction Agreements, see CRS Report R45521, *Department of Defense Use of Other Transaction Authority: Background, Analysis, and Issues for Congress*, by Heidi M. Peters.

[59] Andrew Eversden, "US Army successfully tests Iron Dome at White Sands Missile Range," *Breaking Defense*, August 2, 2022, at https://breakingdefense.com/2022/08/us-army-successfully-tests-iron-dome-at-white-sands-missile- range/.

[60] DOD, Department of Defense Fiscal Year (FY) 2024 Budget Estimates, Army Justification Book Volume II Budget Activity 4, Research, Development, Test & Evaluation, p. 344, at https://www.asafm.army.mil/Portals/72/Documents/BudgetMaterial/2024/Base%20Budget/rdte/RDTE-Vol%202- Budget%20Activity%204A.pdf.

[61] For additional information, see CRS In Focus IF12421, *The U.S. Army's Indirect Fire Protection Capability (IFPC) System*, by Andrew Feickert.

[62] DOD, *Department of Defense Fiscal Year (FY) 2023 Budget Estimates, Army Justification Book Volume II Budget Activity 4, Research, Development, Test & Evaluation*, p. 411, at https:/www.asafm.army.mil/Portals/72/Documents/ BudgetMaterial/2023/Base%20Budget/rdte/vol_2-Budget_Activity_4.pdf. Group 1 UAS are "typically hand-launched, portable systems," while Group 2 UAS are "typically medium-sized,

as part of a layered defense to protect fixed and semi-fixed sites."[63] According to Army budget documents, the program "leverages previous HPM technology demonstrations and experimentation campaigns such as the [the Air Force's THOR program]."[64] The Army intends to develop, test, and deliver four IFPC-HPM prototypes by the fourth quarter of FY2024 and to conduct planning for a potential transition to a program of record in FY2025.[65] On May 15, 2024, Epirus reportedly delivered the last two IFPC-HPM systems to the Army, with Epirus further noting that the systems had completed new equipment training and engineering developmental testing.[66]

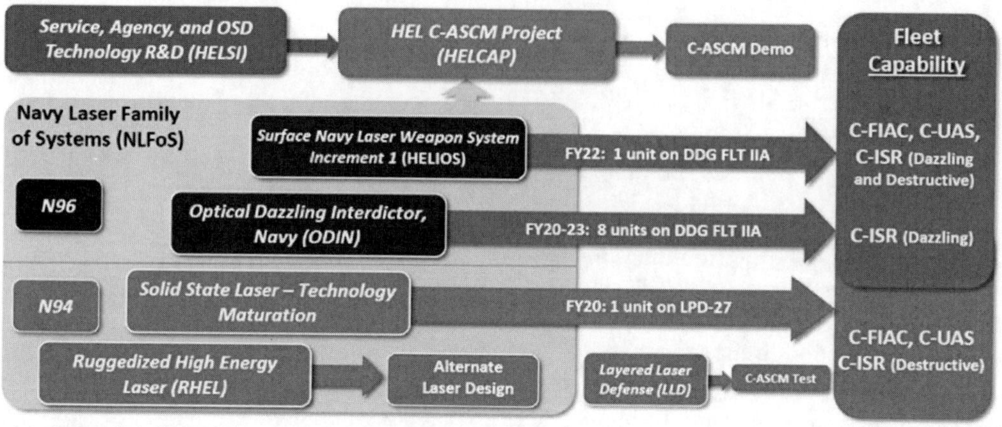

Source: Navy briefing slide provided by Navy Office of Legislative Affairs to CRS on May 16, 2024. N96 is the Surface Warfare Division of the Office of Chief of Naval Operations. N94 is the Innovation, Technology Missions, and Test and Evaluation Division.

Figure 9. Navy Laser Development Roadmap.

Lasers on Next-Generation Army Combat Vehicles

Army officials have suggested that next-generation combat vehicles could feature an active protection system employing directed energy to protect the vehicle and to replace traditional mounted weapons.[67] The Army asserts that active protection systems featuring lasers could provide 360-degree protection from incoming rounds or UASs, and that laser weapons might also be used to disable or possibly destroy enemy vehicles. Officials note that to begin fielding Army units with a next-generation combat vehicle in 2035, major decisions would need to be made by 2025. This time frame suggests that the Army has fewer than two years to advance

catapult-launched, mobile systems." For additional information about UAS groups, see U.S. Army, *US. Army Unmanned Aircraft Systems Roadmap 2010-2035*, pp. 12-13, at https://irp.fas.org/program/collect/uas-army.pdf.

[63] "Army Directed Energy Strategy," Army Rapid Capabilities and Critical Technologies Office, August 20, 2021.

[64] DOD, Department of Defense Fiscal Year (FY) 2023 Budget Estimates, Army Justification Book Volume II Budget Activity 4, Research, Development, Test & Evaluation, p. 411, at https://www.asafm.army.mil/Portals/72/Documents/BudgetMaterial/2023/Base%20Budget/rdte/vol_2-Budget_Activity_4.pdf.

[65] DOD, Department of Defense Fiscal Year (FY) 2024 Budget Estimates, Army Justification Book Volume II Budget Activity 4, Research, Development, Test & Evaluation, p. 350, at https://www.asafm.army.mil/Portals/72/Documents/BudgetMaterial/2024/Base%20Budget/rdte/RDTE-Vol%202- Budget%20Activity%204A.pdf.

[66] Dan Schere, "Epirus Delivers Last Two IFPC High-Power Microwave Prototypes to Army," *Inside Defense,* May 15, 2024.

[67] CRS Report R44598, Army and Marine Corps Active Protection System (APS) Efforts, by Andrew Feickert.

laser weapons technology to a point where it can be considered a viable option, if it is to be incorporated into next-generation combat vehicles.[68]

Selected Navy Directed Energy Programs[69]

The Navy installed its first prototype DE weapon, a 30 kW laser capable of countering small surface craft and UAS, on the USS Ponce in 2014.[70] Since then, the Navy has been developing lasers with improved capability for countering surface craft and UAS and is in the process of developing a capability for countering anti-ship cruise missiles (ASCMs). Current Navy DE programs include the following:

- Solid State Laser Technology Maturation (SSL-TM);
- Optical Dazzling Interdictor, Navy (ODIN);
- Surface Navy Laser Weapon System (SNLWS) Increment 1, also known as the High-Energy Laser with Integrated Optical-dazzler and Surveillance (HELIOS); and
- High Energy Laser Counter-ASCM Program (HELCAP). The Navy's laser development roadmap is illustrated in Figure 9.

As shown in *Figure 9*, SSL-TM, ODIN, and SNLWS Increment 1/HELIOS are included in the Navy Laser Family of Systems (NLFoS). (The Navy has since completed work on the fourth NLFoS effort shown in *Figure 9*, the Ruggedized High Energy Laser [RHEL].) As also shown in *Figure 9*, the Navy intends for both NLFoS and HELCAP efforts, along with DOD laser technologies, to provide fleet capability for counter-fast inshore attack craft (C-FIAC); C-UAS; and counter-intelligence, surveillance, and reconnaissance (C-ISR).

Solid State Laser Technology Maturation (SSL-TM)

The SSL-TM program (see Figure 10) is to develop a prototype shipboard laser called the Laser Weapons System Demonstrator (LWSD) "to address known capability gaps against asymmetric threats (UAS, small boats, and ISR sensors)."[71] The program is to additionally "inform future acquisition strategies, system designs, integration architectures, and fielding plans for laser weapon systems."[72] The Navy reportedly installed a 150 kW LWSD on the USS

---

[68] See Gary Sheftick, "The Next-Generation Combat Vehicle Could Have Lasers, Run on Hybrid Power," Army News Service, November 3, 2016, and Hope Hodge Seck, "Next Army Combat Vehicle May Feature Active Protection, Laser Weapons," Defense Tech, October 30, 2017.

[69] This section was written by Ronald O'Rourke, CRS Specialist in Naval Affairs. For more information about U.S. Navy DE programs, including information about the Navy's past DE development programs, see CRS Report R44175, Navy Shipboard Lasers: Background and Issues for Congress, by Ronald O'Rourke.

[70] Sam LaGrone, "U.S. Navy Allowed to Use Persian Gulf Laser for Defense," USNI News, December 10, 2014.

[71] DOD, *Department of Defense, Fiscal Year (FY) 2023 Budget Estimates, Navy Justification Book Volume 2 of 5*, Research, Development, Test & Evaluation, Navy, April 2022, p. 184, at https://www.secnav.navy.mil/fmc/fmb/Documents/23pres/RDTEN_BA4_Book.pdf.

[72] DOD, *Department of Defense, Fiscal Year (FY) 2023 Budget Estimates, Navy Justification Book Volume 2 of 5*, Research, Development, Test & Evaluation, Navy, April 2022, p. 184, at https://www.secnav.navy.mil/fmc/fmb/Documents/23pres/RDTEN_BA4_Book.pdf.

Portland in the fall of 2019 and has since completed onboard testing.[73] According to Navy FY2025 budget documents, SSL-TM deinstallation, final report, program closeout, and hardware disposition began in the second quarter of FY2023 and are scheduled to be completed in the fourth quarter of FY2024.[74] The Navy did not request funding for SSL-TM in FY2025.[75]

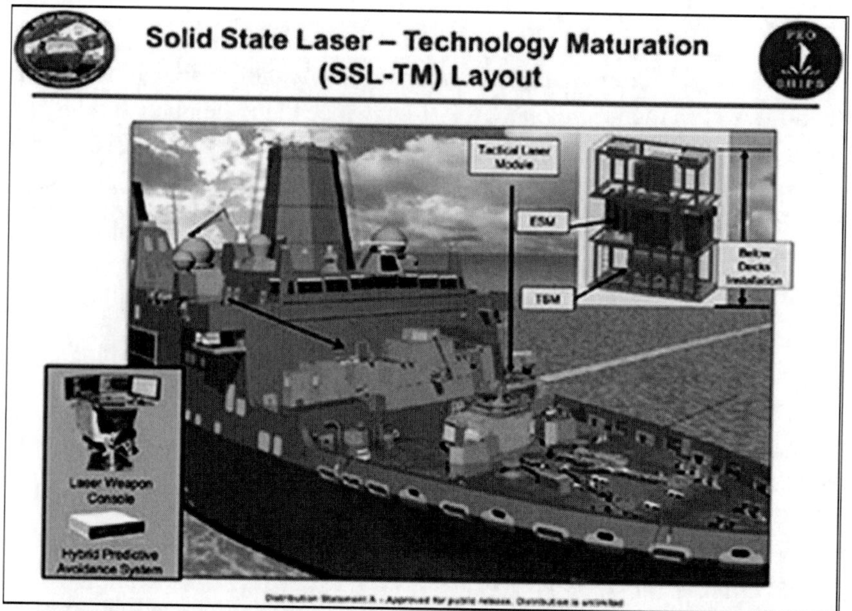

Source: Navy briefing slide accompanying Tyler Rogoway, "Mysterious Object Northrop Is Barging From Redondo Beach Is A High-Power Naval Laser," The Drive, October 18, 2019. The blog post credits the slide to the Navy and describes it as a "recent slide."

Figure 10. Navy Graphic of SSL-TM Laser System.

Optical Dazzling Interceptor, Navy (ODIN)

According to the Navy's FY2024 budget submission, the Optical Dazzling Interceptor, Navy (ODIN) effort is designed to provide "near-term, directed energy, shipboard C-ISR capabilities to dazzle UAS and other platforms that address urgent operational needs of the Fleet."[76] The Navy has deployed eight ODIN units on Arleigh Burke Flight IIA destroyers; the service did not request RDT&E funds for ODIN in FY2025.[77]

---

[73] Christopher P. Cavas, "Lasers Sprout in San Diego," *Defense & Aerospace Report*, March 1, 2020.

[74] DOD, *Department of Defense, Fiscal Year (FY) 2025 Budget Estimates, Navy Justification Book Volume 2 of 5*, Research, Development, Test & Evaluation, Navy, March 2024, pp. 180, 185, 186, https://www.secnav.navy.mil/fmc/fmb/Documents/25pres/RDTEN_ BA4_ Book.pdf.

[75] DOD, *Department of Defense, Fiscal Year (FY) 2025 Budget Estimates, Navy Justification Book Volume 2 of 5*, Research, Development, Test & Evaluation, Navy, March 2024, p. 180, https://www.secnav.navy.mil/fmc/fmb/Documents/25pres/RDTEN_BA4_Book.pdf.

[76] DOD, *Department of Defense, Fiscal Year (FY) 2024 Budget Estimates, Navy Justification Book Volume 2 of 5*, Research, Development, Test & Evaluation, Navy, March 2023, p. 958, at https://www.secnav.navy.mil/fmc/fmb/Documents/24pres/RDTEN_BA4_Book.pdf.

[77] DOD, *Department of Defense, Fiscal Year (FY) 2025 Budget Estimates, Navy Justification Book Volume 2 of 5*, Research, Development, Test & Evaluation, Navy, March 2024, p. 951, https://www.secnav.navy.mil/fmc/fmb/Documents/25pres/RDTEN_BA4_Book.pdf.

SNLWS Increment 1 (HELIOS)

SNLWS Increment 1 is also known as the High-Energy Laser with Integrated Optical-dazzler and Surveillance (HELIOS). The HELIOS effort is focused on rapid development and rapid fielding of a 60 kW-class high-energy laser (with growth potential to 120 kW) and dazzler in an integrated weapon system, for use in countering UAS, fast inshore attack craft, and ISR sensors, and for combat identification and battle damage assessment.[78] According to Navy budget documents, HELIOS was installed on an Arleigh Burke-class destroyer, the USS *Preble*, in FY2022 and was to conduct sea trials in FY2023.[79] The system is to remain on the ship for fleet testing and sustainment through at least the end of FY2029.[80]

High Energy Laser Counter ASCM Project (HELCAP)

The Navy's FY2025 budget submission states that the HELCAP effort

> will expedite the development, experimentation, integration and demonstration of critical technologies to defeat crossing Anti-Ship Cruise Missiles (ASCM) by addressing the remaining technical challenges, e.g.: atmospheric turbulence, automatic target identification and aim point selection, precision target tracking with low jitter in high clutter conditions, advanced beam control, and higher power HEL development. HELCAP will assess, develop, experiment, and demonstrate the various laser weapon system technologies and methods of implementation required to defeat ASCMs in a crossing engagement.[81]

The HELCAP prototype system is to include a beam control testbed, 300 kW+ class laser source—selected and adapted from a laser source developed under OSD's laser scaling initiative, prototype control system, and auxiliary prime power and cooling[82] The Navy planned to conduct system experimentation in FY2023 and FY2024, focusing on "ASCM detect to engage experimentation against targets of increasing complexity up to and including static and

---

[78] DOD, Department of Defense Fiscal Year (FY) 2021 Budget Estimates, Navy, Justification Book Volume 2 of 5, Research, Development, Test & Evaluation, Navy, February 2020, p. 1021. Although the Navy previously identified HELIOS as being scalable to 150 kW, recent reports indicate that the system is to be scalable to only 120 kW. See, for example, Richard R. Burgess, "HELIOS Laser Weapon System Delivered for Installation on USS Preble," Seapower Magazine, March 31, 2022, at https://seapowermagazine.org/helios-laser-weapon-system-delivered-for-installation-on- uss-preble/.

[79] DOD, *Department of Defense, Fiscal Year (FY) 2023 Budget Estimates, Navy Justification Book Volume 2 of 5,* Research, Development, Test & Evaluation, Navy, April 2022, p. 1011, at https://www.secnav.navy.mil/fmc/fmb/ Documents/23pres/RDTEN_BA4_Book.pdf.

[80] DOD, *Department of Defense, Fiscal Year (FY) 2025 Budget Estimates, Navy Justification Book Volume 2 of 5,* Research, Development, Test & Evaluation, Navy, March 2024, p. 944, at https://www.secnav.navy.mil/fmc /fmb/Documents/25pres/RDTEN_BA4_Book.pdf.

[81] DOD, *Department of Defense, Fiscal Year (FY) 2025 Budget Estimates, Navy Justification Book Volume 2 of 5,* Research, Development, Test & Evaluation, Navy, March 2024, p. 926, at https://www.secnav.navy.mil/fmc/fmb/Documents/25pres/RDTEN_BA4_Book.pdf.

[82] DOD, *Department of Defense, Fiscal Year (FY) 2024 Budget Estimates, Navy Justification Book Volume 2 of 5,* Research, Development, Test & Evaluation, Navy, March 2023, p. 962, at https://www.secnav.navy.mil/fmc/fmb/Documents/24pres/RDTEN_BA4_Book.pdf.

dynamic ground targets and low-cost unmanned aerial targets."[83] According to Navy budget documents, the service is to complete "primary demonstration activities" in FY2025.[84] HELCAP experimentation is to continue through at least FY2026.[85]

Layered Laser Defense (LLD) System

An additional Navy laser development effort (not shown in Figure 9) is called the Layered Laser Defense (LLD) system. A March 9, 2020, DOD contract award announcement stated that the Navy awarded Lockheed Martin a $22 million contract for

> the integration, demonstration, testing and operation of the Layered Laser Defense (LLD) weapon system prototype onboard a Navy littoral combat ship [LCS] while that vessel is underway.... Key areas of work to be performed include development of a prototype structure and enclosure to protect the LLD from ships motion and maritime environment in a mission module format; system integration and test with government-furnished equipment; platform integration and system operational verification and test; systems engineering; test planning; data collection and analysis support; and operational demonstration.[86]

Press reports indicate that the Office of Naval Research—in partnership with Lockheed Martin and the Office of the Under Secretary of Defense for Research and Engineering—demonstrated the system in February 2022 against a "target representing a subsonic cruise missile in flight."[87] The Navy reportedly does not plan to field the LLD.[88]

---

[83] DOD, *Department of Defense, Fiscal Year (FY) 2024 Budget Estimates, Navy Justification Book Volume 2 of 5*, Research, Development, Test & Evaluation, Navy, March 2023, p. 961, at https://www.secnav.navy.mil/fmc/fmb/Documents/24pres/RDTEN_BA4_Book.pdf.

[84] DOD, *Department of Defense, Fiscal Year (FY) 2025 Budget Estimates, Navy Justification Book Volume 2 of 5*, Research, Development, Test & Evaluation, Navy, March 2024, p. 927, at https://www.secnav.navy.mil/fmc/fmb/Documents/25pres/RDTEN_BA4_Book.pdf.

[85] DOD, *Department of Defense, Fiscal Year (FY) 2025 Budget Estimates, Navy Justification Book Volume 2 of 5*, Research, Development, Test & Evaluation, Navy, March 2024, p. 935, at https://www.secnav.navy.mil/fmc/fmb/Documents/25pres/RDTEN_BA4_Book.pdf.

[86] Department of Defense, "Contracts for March 9, 2020." See also Rich Abott, "Lockheed Martin Nabs $22 Million Contract for Layered Laser Defense Prototype On LCS," *Defense Daily*, March 16, 2020.

[87] Warren Duffie Jr., "Laser Trailblazer: Navy Conducts Historic Test of New Laser Weapon System," Office of Naval Research, April, 13, 2022, at https://www.navy.mil/Press-Office/News-Stories/Article/2998829/laser-trailblazer-navy-conducts-historic-test-of-new-laser-weapon-system/.

[88] Warren Duffie Jr., "Laser Trailblazer: Navy Conducts Historic Test of New Laser Weapon System," Office of Naval Research, April, 13, 2022, at https://www.navy.mil/Press-Office/News-Stories/Article/2998829/laser-trailblazer-navy- conducts-historic-test-of-new-laser-weapon-system/.

Potential Issues and Questions for Congress[89]

Technological Maturity

One question regarding directed energy weapons programs involves their technological maturity, including the ability to improve beam quality and control to militarily useful levels, and to meet size, weight, and power (SWaP) and cooling requirements for integration into current platforms.[90] Some DE systems are small enough to fit on military vehicles, but many require larger and/or fixed platforms that could potentially limit deployment options and operational utility. Congress may consider directing DOD to establish metrics for assessing the pace of technological advancement. In what ways, if any, are DOD technology maturation efforts reducing the SWaP and cooling requirements of DE systems?

Cost

The United States has been researching directed energy since the 1960s, yet some experts have observed that "actual directed-energy programs … have frequently fallen short of expectations," with DOD investing billions of dollars in programs that failed to reach maturity and were ultimately cancelled.[91] Directed energy weapons may therefore require greater up-front investment than traditional kinetic weapons in order to field a successful weapons system.

Congress may consider requesting an independent assessment of the technological maturity and life cycle cost estimates for various DE weapons, as well as a comparative assessment of costs of DE weapons versus comparable kinetic weapons. How do estimates of the total lifecycle costs of DE weapons compare with those of their kinetic counterparts? Does the technological maturity of DE weapons warrant current funding levels?

Weapons Characteristics

Although DE weapons may offer a lower cost per shot than traditional weapons such as missiles, DE weapons are subject to limitations. For example, atmospheric conditions (e.g., rain, fog, obscurants) and SWaP and cooling requirements can limit the range and beam quality of DE weapons, in turn reducing their effectiveness. Traditional weapons, in contrast, are less affected by these factors.[92] Furthermore, DE weapons may be more difficult to maintain than traditional weapons. As the Government Accountability Office notes, "the internal mechanisms

---

[89] This section was written by Kelley M. Sayler, CRS Analyst in Advanced Technology and Global Security, and John R. Hoehn, CRS Analyst in Military Capabilities and Programs.

[90] Ariel Robinson, "Directed Energy Weapons: Will They Ever Be Ready?," *National Defense*, July 1, 2015, at https://www.nationaldefensemagazine.org/articles/2015/7/ 1/2015july-directed-energy-weapons-will-they-ever-be- ready.

[91] Paul Scharre, *Directed-Energy Weapons: Promise and Prospects*, Center for a New American Security, April 2015, p. 4.

[92] Ariel Robinson, "Directed Energy Weapons: Will They Ever Be Ready?," *National Defense*, July 1, 2015, at https://www.nationaldefensemagazine.org/articles/2015/7/1/ 2015july-directed-energy-weapons-will-they-ever-be- ready; and David Vergun, "Army developing lasers that pierce fog, dust to destroy targets," *Army News Service*, October 23, 2017, at https://www.army.mil/article/195650/army_developing_lasers_that_pierce _fog_dust_to_destroy_targets.

for DE weapons are sensitive, and typically require a specialized clean room for repairs."[93] Such challenges could impact their sustainability in the field.[94] How, if at all, might the limitations of DE weapons be mitigated by technological developments such as adaptive optics, concepts of operation, or other methods? What impact might a failure to mitigate these limitations have on future military operations?

Mission Utility

Given the strengths and weaknesses of DE weapons, DOD is conducting multiple utility studies to analyze potential concepts of operation for DE weapons and to assess the scenarios in which they might be militarily useful.[95] How might Congress draw upon the conclusions of these analyses as it conducts oversight of DE weapons programs? What is the appropriate balance between DE weapons and traditional munitions within the military's portfolio of capabilities?

Defense Industrial Base

Some analysts have expressed concerns that, in the past, DOD did not provide stable funding for DE weapons programs or sufficient opportunities for the DE workforce. Acknowledging these concerns, DOD former Principal Director for Directed Energy Dr. Jim Trebes has stated that, although he believes the DE industrial base is currently healthy, its capacity could be strained in the future if DOD begins to buy larger numbers of DE systems. Dr. Trebes has noted that although today's DE workforce is sufficient, it may face a demographic problem in the future due to retirement.[96] According to OUSD(R&E), HELSI is intended to address such concerns about the future of the DE industrial base by providing industry with assured prototyping opportunities. In what ways, if any, has HELSI strengthened the defense industrial base for DE weapons? What, if any, challenges does the base continue to face, and how might they be mitigated?

Intelligence Requirements

Some analysts have questioned whether DOD has sufficient knowledge of adversary DE weapons systems and materials to develop its own weapons requirements. DOD is currently attempting to further define its DE collection requirements for the intelligence community (IC)

---

[93] Government Accountability Office, Directed Energy Weapons: DOD Should Focus Transition on Planning, April 2023, p. 22.

[94] See Jen Judson, "US Army working through challenges with laser weapons," *Defense News*, August 11, 2023, https://www.defensenews.com/land/2023/08/11/us-army-working-through-challenges-with-laser-weapons/.

[95] Dr. Jim Trebes, "Advancing High Energy Laser Weapon Capabilities: What is OUSD (R&E) Doing?" presentation at IDGA, October 21, 2020.

[96] CRS conversation with then-Principal Director for Directed Energy Dr. Jim Trebes, November 17, 2020. See also Dr. Jim Trebes, "Advancing High Energy Laser Weapon Capabilities: What is OUSD (R&E) Doing?" presentation at IDGA, October 21, 2020.

through the Directed Energy Lethality Intelligence initiative.[97] To what extent, if at all, is this initiative improving connectivity between DOD's DE community and the IC? What collection requirements, if any, remain?

Coordination Within DOD

Pursuant to Section 219 of the FY2017 National Defense Authorization Act (NDAA) (P.L. 114- 328), OUSD(R&E)'s Principal Director for Directed Energy is tasked with coordinating DE efforts across DOD and with developing DOD's Directed Energy Roadmap, which is to guide development efforts. Section 215 of the FY2020 NDAA (P.L. 116-283) established a Directed Energy Working Group to "analyze and evaluate the current and planned directed energy programs of each of the military departments ... [and] make recommendations to the Secretary of Defense." These recommendations are intended to improve DOD DE coordination activities and accelerate the fielding of DE capabilities. To what extent are the military departments and defense agencies adhering to OUSD(R&E)'s roadmap? What, if any, additional authorities or structural changes would be required to ensure proper implementation of the roadmap and execution of the working group's recommendations?

Arms Control

DE weapons "are not authoritatively defined under international law, nor are they currently on the agenda of any existing multilateral mechanism."[98] However, some applications of DE weapons are prohibited. Article 1 of the Protocol on Blinding Lasers prohibits the employment of "laser weapons specifically designed, as their sole combat function or as one of their combat functions, to cause permanent blindness to unenhanced vision."[99]

Some analysts have suggested that additional multilateral agreements should be considered. For example, Congress may consider prohibitions on nonlethal anti-personnel uses of DE weapons— such as "heat rays"[100] or lasers intended to cause temporary visual impairment—or on certain military applications of DE weapons—such as aircraft interference—in peacetime.[101] Other analysts have argued that DE weapons could be

[97] Dr. Jim Trebes, "Advancing High Energy Laser Weapon Capabilities: What is OUSD (R&E) Doing?," presentation at IDGA, October 21, 2020.

[98] "Directed Energy Weapons: Discussion paper for the Convention on Certain Conventional Weapons (CCW)," Article 36, November 2017.

[99] The protocol does not cover the development, procurement, or possession of such weapons, nor does it prohibit the employment of laser weapons that may cause blindness "as an incidental or collateral effect." *Additional Protocol to the Convention on Prohibitions or Restrictions on the Use of Certain Conventional Weapons Which May Be Deemed to Be Excessively Injurious or to Have Indiscriminate Effects*, Vienna, October 13, 1995, United Nations, Treaty Series, vol. 1380, p. 370, at https://treaties.un.org/doc/Treaties/ 1995/10/19951013%2001-30%20AM/Ch_XXVI_02_ap.pdf. For additional information about the protocol and its relationship to DE weapons programs, see Appendix I of CRS Report R41526, *Navy Shipboard Lasers for Surface, Air, and Missile Defense: Background and Issues for Congress*, by Ronald O'Rourke.

[100] See "Active Denial Technology: Fact Sheet," Joint Intermediate Force Capabilities Office, May 11, 2020, at https://jnlwp.defense.gov/Press-Room/Fact-Sheets/Article-View-Fact-sheets/Article/577989/active-denial-technology/.

[101] Patrick M. Cronin and Ryan D. Neuhard, "Countering China's Laser Offensive," *The Diplomat*, April 2, 2020, at https://thediplomat.com/2020/04/countering-chinas-laser-offensive/.

considered more humane than conventional weapons because their accuracy could reduce collateral damage and because they could provide a nonlethal anti-personnel capability in circumstances when lethal force might otherwise be used.[102] In what circumstances and for what purposes should the U.S. military's use of DE weapons be permissible? What, if any, regulations, treaties, or other measures should the United States consider regarding the use of DE weapons in both war and peacetime?

Appendix A. Potential Advantages and Limitations of Directed Energy Weapons[103]

This appendix provides additional information on potential advantages and limitations of High-Energy Laser (HEL) and High-Powered Microwave (HPM) weapons. The advantages and limitations of any HEL or HPM weapons would be specific to the system; as such, all advantages and limitations might not equally apply to each system.

Potential Advantages of HEL Weapons

In addition to deeper magazines, lower logistics requirements, and lower costs per shot, potential advantages of HEL weapons include the following:

Potential Limitations of HEL Weapons

Potential limitations of HEL weapons include the following:

- *Line of sight.* Since laser light passes through the atmosphere on an essentially straight path, HEL weapons would be limited to line-of-sight engagements, and consequently could not counter over-the-horizon targets or targets obscured by intervening objects. As a result, potential engagement ranges against certain targets (e.g., low-flying targets) would be limited.
- Atmospheric absorption, scattering, and turbulence. Substances in the atmosphere—particularly water vapor, but also sand, dust, salt particles, smoke, and other air pollution—absorb and scatter light, and atmospheric turbulence can defocus a laser beam. These effects can reduce the effective range of an HEL weapon. Absorption by water vapor is a particular consideration for shipboard lasers because marine environments feature substantial amounts of water vapor in the air. There are certain wavelengths of light (i.e., "sweet spots" in the electromagnetic spectrum) where atmospheric absorption by water vapor is markedly reduced. Lasers can be designed

---

[102] See, for example, Mark Gunzinger and Chris Dougherty, *Changing the Game: The Promise of Directed-Energy Weapons*, Center for Strategic and Budgetary Assessments, April 19, 2021, at https://csbaonline.org/uploads/documents/CSBA_ChangingThe Game_ereader.pdf.

[103] This appendix was written by Ronald O'Rourke (HEL weapons) and Andrew Feickert (HPM weapons), CRS Specialist in Naval Affairs and CRS Specialist in Military Ground Forces, respectively.

to emit light at or near those sweet spots, so as to maximize their potential effectiveness. Absorption generally grows with distance to target, making it in general less of a potential problem for short- range operations than for longer-range operations. Adaptive optics, which make rapid, fine adjustments to a laser beam on a continuous basis in response to observed turbulence, can counteract the effects of atmospheric turbulence. Even so, lasers might not work well, or at all, in rain or fog, preventing lasers from being an all-weather solution.

- *Thermal blooming.* A laser that continues firing in the same exact direction for a certain amount of time can heat up the air it is passing through, which in turn can defocus the laser beam, reducing its ability to disable the intended target. This effect, called *thermal blooming*, can make lasers less effective for countering targets that are coming straight at them, on a constant bearing (i.e., "down-the- throat" shots). Most tests of laser systems have been against crossing targets rather than "down-the-throat" shots. In general, thermal blooming becomes more of a concern as the power of the laser beam increases.
- *Saturation attacks.* Since a HEL weapon can attack only one target at a time, requires several seconds to disable the target, and requires several more to be redirected to the next one, a HEL weapon can disable only so many targets within a given period of time. This places an upper limit on the ability of an individual laser to deal with saturation attacks—attacks by multiple weapons that approach the platform simultaneously or within a few seconds of one another. This limitation can be mitigated by installing more than one laser on the platform, up to space and energy availability.
- *Hardened targets and countermeasures.* Less powerful lasers—that is, lasers with beam powers measured in kilowatts (kW) rather than megawatts (MW)— can be less effective against targets that incorporate shielding, ablative material, or highly reflective surfaces, or that tumble or rotate rapidly (so that the laser spot does not remain continuously on a single location on the target's surface). Smoke or other obscurants can reduce the susceptibility of a target platform to laser attack. Such measures, however, can increase the cost and/or weight of the target platform.

Potential Limitations of HPM Weapons

Potential limitations of HPM weapons include the following:

- Range constraints. Because HPM beams are more diffuse than lasers and cannot be as tightly focused, the "energy per unit area in HPM beams decreases significantly over distance."[104] This characteristic could limit the range at which HPM weapons are operationally effective.

---

[104] Mark Gunzinger and Chris Dougherty, *Changing the Game: The Promise of Directed-Energy Weapons*, Center for Strategic and Budgetary Assessments, April 19, 2021, p. 39, at https://csbaonline.org/uploads/documents/CSBA_ChangingTheGame_ereader.pdf.

- Potential for fratricide. Because HPM weapons could affect all unshielded electronic systems within range, measures must be taken to ensure that friendly systems are properly shielded or kept outside of the weapon's range when the weapon is in use.
- Effectiveness of countermeasures. Because electromagnetic radiation can be absorbed by shielding, HPM weapons may not be effective against shielded targets.

Appendix B. Funding for Directed Energy Programs

Table B-1. Funding for directed energy programs (FY2023-FY2025) (in millions of dollars)

FY2023			FY2024		FY2025
Military Department	President's Budget	Appropria-tion	President's Budget	Appropri-ation	President's Budget
Army	557.8	772.8	249.9	337.2	167.9
Navy	152.2	181.2	119.7	137.2	82.4
Air Force	318.2	343.2	291.9	288.4	217.1
Defense-wide	306.1	352.6	321.7	353.7	316.6
Total	1334.3	1649.8	962.4	1116.5	789.7

Source: CRS correspondence with the Office of the Under Secretary of Defense for Research and Engineering, April 30, 2024.

Note: This table does not include funding for classified programs.

Potential Advantages of HPM Weapons

In addition to deep magazines, low costs per shot, fast engagement times, and graduated responses, potential advantages of HPM weapons include the following:

- *Temporary or system-specific effects.* HPM weapons can generate waves at different frequencies and power levels to temporarily or permanently disrupt targeted electronic systems while leaving others unaffected.
- *Broad effects.* HPM weapons can destroy a wide array of unshielded electronic systems, including both military and commercial systems. In addition, they are capable of disabling any unshielded electronic system within their electromagnetic cone (i.e., they can disable numerous systems, including swarms of UAS, at once).
- *Nonlethal applications.* Certain HPM weapons, such as "heat rays," could provide a nonlethal anti-personnel capability in circumstances in which lethal force might otherwise be used.
- *Limitation of collateral damage.* HPM weapons would generate little to no collateral damage of physical structures.[105] This feature could make them attractive weapons in urban areas or in situations "short of war."

[105] Anti-personnel HPM weapons could not, however, discriminate between military personnel and civilians and could therefore impact civilians within the weapon's electromagnetic cone. Similarly, HPM weapons used against military electronic equipment could disable unshielded civilian equipment.

Chapter 5

Directed Energy Weapons: Science and Tech Spotlight*

United States Government Accountability Office

Why This Matters

There is a surge in interest in directed energy weapons from several nations—including the U.S.—primarily for counter drone missions. These weapons use electromagnetic energy to cause effects ranging from deterrence to destruction. They offer capabilities that conventional weapons may not, but challenges have so far prevented widespread operational use.

The Technology

What is it? Directed energy weapons (DEW) use concentrated electromagnetic energy to combat enemy forces and assets. These weapons include high energy lasers and other high-power electromagnetics—such as millimeter wave and high-power microwave weapons. Unlike weapons that fire bullets or missiles, DEWs can respond to a threat in different ways. For example, they can temporarily degrade electronics on a drone or physically destroy it. See our 2022 Spotlight for more information on counter-drone technology.

How does it work? Each type of DEW uses a different region of the electromagnetic spectrum (see figure 1). This spectrum describes all of the kinds of light—including those the human eye cannot see—and classifies them according to wavelength. Different types of electromagnetic energy have different properties. For example, the wavelength affects what the directed energy can penetrate—such as metal or human skin.

All DEWs emit energy at the speed of light, and are often discussed in terms of their power output—the amount of electromagnetic energy transferred over time. While DEWs use electromagnetic energy similar to everyday items, such as household microwaves, their power output is vastly higher, as described below.

High energy lasers produce a very narrow beam of light, usually in the infrared to visible region, and are typically used on one target at a time. The beam can be pulsed or continuous,

* This is an edited, reformatted and augmented version of the United States Governmen Accountability Office Science, Technology Assessment, and Analytics Publication No. GAO-23-106717, dated May 2023.

In: Emerging Military Technologies
Editor: Jerry Shields
ISBN: 979-8-89530-613-0

generating a power output of at least 1 kilowatt. This output is 200,000 times greater than a typical laser pointer and is capable of melting steel.

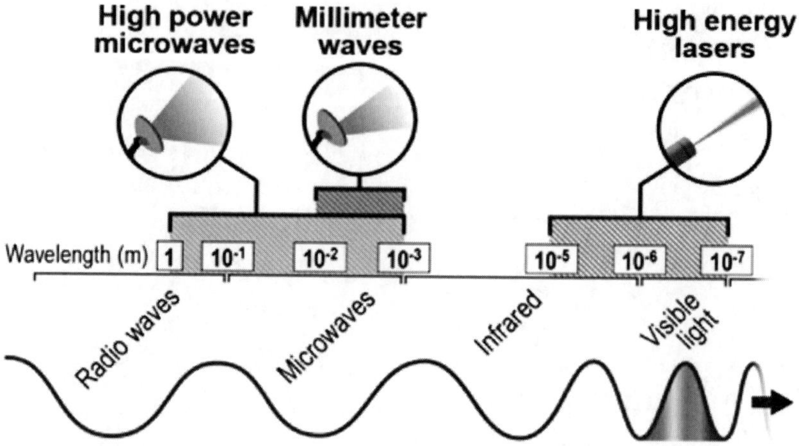

Source: GAO. | GAO-23-106717.

Figure 1. Location of directed energy weapons on the electromagnetic spectrum.

Millimeter wave weapons have wavelengths between 1 and 10 millimeters and generate more than 1 kilowatt of power. Millimeter wave weapons have a larger beam size than high energy lasers and therefore can affect multiple targets at once.

High power microwave weapons produce microwaves, which have longer wavelengths than high energy lasers and millimeter wave weapons. These weapons can produce more than 100 megawatts of power, which is nearly 150,000 times more powerful than the average household microwave. Like millimeter wave weapons, they can also affect multiple targets because of their larger beam size.

Each DEW can produce a range of effects from nonlethal to lethal, depending on factors such as the time on target, the distance to the target, and even the part of the target on which the DEW is focused. DEWs can use this range of effects to graduate responses to a threat. A graduated response could start with temporarily preventing use of an asset or its access to an area and increase to destruction of the asset if necessary (see figure 2).

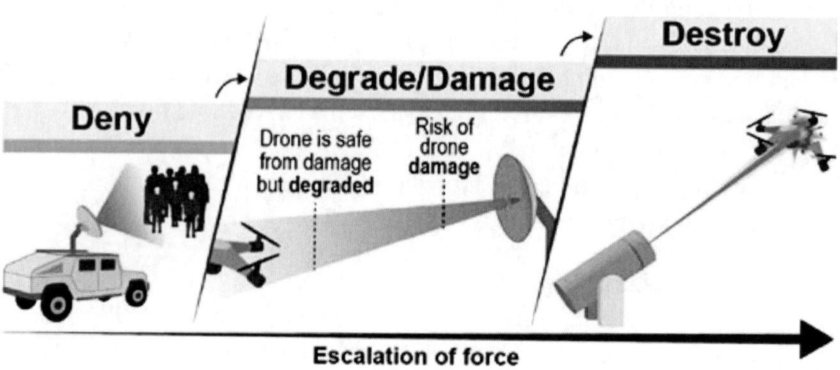

Source: GAO. | GAO-23-106717.

Figure 2. Examples of graduated responses using directed energy weapons.

DEWs can *deny* entrance to an area or prevent enemy forces or assets from functioning within an area. DEWs used for denial do not cause long- term damage to targets, and when enemy forces or assets leave the area, they typically regain function or the effect is mitigated. For example, the Department of Defense's (DOD) Active Denial System uses millimeter waves that interact with the water and fat molecules in a person's skin to create a heating sensation. During testing, the discomfort persuaded individuals to move away from the area.

DEWs can also degrade the efficacy of an enemy's assets. For example, high energy lasers can temporarily overwhelm a person or a sensor's ability to see or sense by emitting a glare—called dazzling. Dazzling can act as a non-verbal warning before resorting to increased force.

If a greater amount of force is required, DEWs can also damage or destroy enemy assets. To do this, a high energy laser can emit electromagnetic energy with a wavelength the target material absorbs most effectively, melting the material. The laser could focus on a sensor and damage a drone, or focus on a fuel tank or battery and destroy it.

How mature is it? DEWs range in maturity from research projects to prototypes tested in the field. DOD named DEWs as a technology critical to enabling the 2018 National Defense Strategy and reported spending about $1 billion annually for the last 3 years on research and development. The U.S. military has tested a variety of DEW prototypes since 2014, primarily for counter-drone missions. For example, the Air Force's prototype Tactical High Power Microwave Operational Responder (THOR) recently completed 2 years of testing. DOD is researching ways to increase the power output of DEWs to engage more powerful targets—like missiles. However, as GAO recently reported, the U.S. military faces challenges bridging the gap between DEW development and acquisition, potentially limiting widespread operational use.

Why now? DEW research and development has been ongoing for decades in many countries—including the U.S.—and is currently experiencing a surge worldwide. This surge stems in part from advances in technology and a desire to maintain competitiveness on the battlefield. Technological innovations, such as the development of smaller lasers that are safer to operate, enable modern DEWs to be much more portable and practical. For example, a four-wheel all-terrain vehicle can now hold a high energy laser powerful enough to damage drones. The U.S. and 30 other nations are developing DEWs, most for counter-drone missions, according to a 2021 Air Force report.

Opportunities

- Complement to conventional weapons. DEWs use energy fired at the speed of light, making them faster and potentially less costly per shot than missiles. Some DEWs have virtually unlimited ammunition and can fire as long as they have power.
- Ease of graduated response. DOD can tailor DEWs to meet mission needs from nonlethal to lethal responses. For example, the longer a laser is focused on target, the more damage or destruction will occur.
- Advancing other uses. Research and development for DEWs could also benefit civilian uses. For example, the development of higher energy lasers could help projects that use directed energy to transport or "beam" power to remote and disadvantaged locations.

Challenges

- Technological limitations. DEWs are generally less effective the farther they are from the target, and atmospheric conditions and cooling requirements can limit their effectiveness. For example, fog and storms can reduce laser beam range and quality.
- Battlefield use. Decisions about how and when to use DEWs or conventional weapons may be challenging. For example, wider beam DEWs, such as high power microwave or millimeter wave weapons, affect all assets in an area, whether friend or foe.
- Ethical and health concerns. Although there are potentially relevant international laws and guidelines, their applicability to
- DEWs is not always well defined. Uncertainty around long-term health effects of DEWs on people either intentionally or unintentionally exposed to directed energy has led to concerns regarding the ethics of using DEWs.

Policy Context and Questions

- As the technology matures, what actions could policymakers take to help bridge the gap between DEW development and acquisition?
- What actions could policymakers take to ensure there is appropriate guidance for using DEWs as the technology matures?
- What are the trade-offs of using nonlethal DEW technologies before potential health effects are fully understood?

Selected References

Air Force Research Laboratory, Directed Energy Futures 2060: 2021.

Congressional Research Service, Department of Defense Directed Energy Weapons: Background and Issues for Congress, R46925 (Washington, D.C.: 2022).

Chapter 6

Directed Energy Weapons: DOD Should Focus on Transition Planning*

United States Government Accountability Office

Abbreviations

AFRL	Air Force Research Laboratory
DE	directed energy
DOD	Department of Defense
HEL	high energy laser
HPM	high power microwave
kW	kilowatt
PEO	Program Executive Office
STP	soldier touch point
TRL	Technology Readiness Level
UAS	uncrewed aerial systems

Why GAO Did This Study

DOD spends about $1 billion annually on directed energy—concentrated electromagnetic energy—weapons, including high energy lasers and high power microwaves. DOD has pursued these potentially transformative technologies for decades because they could provide considerable advantages. They can deliver destructive or disruptive effects to targets at the speed of light and have potentially significant advantages over kinetic weapons, such as missiles, including lower per-use cost.

A Senate report includes a provision for GAO to review DOD's directed energy work including the technologies, industrial base and related infrastructure, and transition efforts. This report (1) describes the status of DOD and military department directed energy weapon efforts, and (2) assesses challenges with transitioning directed energy weapon efforts from prototyping. GAO selected seven directed energy efforts to obtain insights on a variety of types of efforts,

* This is an edited, reformatted and augmented version of the United States Government Accountability Office Report to Congressional Committees, Publication No. GAO-23-105868, dated April 2023.

In: Emerging Military Technologies
Editor: Jerry Shields
ISBN: 979-8-89530-613-0

intended uses environment, and military departments. GAO reviewed DOD documentation and interviewed DOD officials and industry representatives.

What GAO Recommends

GAO is making four recommendations to DOD, including that the Navy and the Air Force develop transition agreements between prototype developers and planned transition partners. DOD concurred with three recommendations and partially concurred with one recommendation. GAO continues to believe that the recommendation should be fully implemented.

What GAO Found

The Department of Defense (DOD) is currently developing directed energy weapons with the goal of defeating a range of threats, including drones and missiles. However, GAO found that, even as DOD makes progress developing these capabilities, its efforts to transition prototypes to acquisition programs face challenges.

DOD and the military departments have efforts underway to develop directed energy weapons. For example, DOD and military departments developed multiple laser weapon system demonstrators and prototypes, which have been used in live fire demonstrations to successfully shoot down drones. DOD and the military departments are also developing higher-powered laser weapons to counter bigger threats. Additionally, the departments developed a range of high power microwave capabilities for purposes such as engaging missile or drone swarm attacks against a military base.

Notional Depiction of High Power Microwave and High Energy Laser Defending an Installation

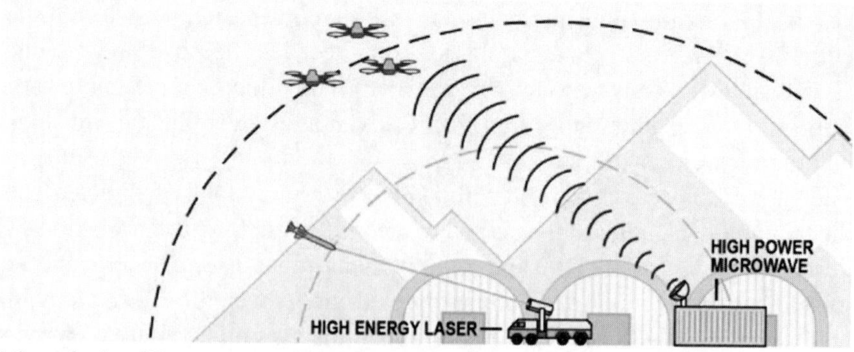

Source: GAO analysis of Department of Defence information. | GAO-23-105868.

However, DOD has long noted a gap—sometimes called “the valley of death”— between its development and its acquisition communities that impede technology transition. For example, the acquisition community may require a higher level of technology maturity than the development community is able to produce.

For prototypes that a military department expects to eventually transition to a new or existing acquisition program, it needs to identify a transition partner that can support the further development of the new technology. To support transition, the Army developed a detailed plan describing schedules and stakeholder roles to build supporting activities around the use of directed energy weapons and early capabilities documents. However, while the Navy fielded several directed energy weapon prototypes and identified a potential transition partner, it does not have documented transition agreements for the directed energy programs that GAO reviewed. The Air Force has not consistently prioritized establishing transition partners, which makes planning for future transition even more challenging. Without these transition planning steps, the Navy and Air Force risk developing directed energy weapons that may be misaligned with operational needs.

April 17, 2023 Congressional Committees

Directed energy (DE) is a potentially transformative technology for the military. Unlike kinetic weapons, such as bullets and missiles that generally rely on physical impact to attack a target, DE weapons use concentrated electromagnetic energy to deliver destructive or disruptive effects to targets at the speed of light. In addition, DE weapons have potentially significant advantages over kinetic weapons, including lower per-use cost. In 2018, the Department of Defense (DOD) identified DE as one of 11 technologies critical to enabling the 2018 National Defense Strategy. Over the past 3 years, DOD has reported spending an average of $1 billion annually on DE weapons development efforts. According to DOD, the goal for current DE development efforts is to improve the capability to defeat a range of threats, including drones and cruise missiles, through an integrated and layered defense with DE weapons complementing kinetic weapons.[1]

Senate Report 117-39 accompanying the National Defense Authorization Act for Fiscal Year 2022 includes a provision for us to review DOD's DE work, including the technologies, industrial base and related infrastructure, and transition efforts.[2] This report (1) describes the status of DOD and military department DE weapon efforts; and (2) assesses the challenges with transitioning DE weapon efforts from prototyping and actions the military departments have taken to mitigate the challenges.

To describe the status of DE weapons efforts, we reviewed the DOD's 2021 Directed Energy Roadmap, the fiscal year 2022 and 2023 President's Budgets, and briefings provided by DOD and the military departments. To assess the challenges with transitioning DE weapon efforts from prototyping, we selected a non-generalizable sample of seven DE weapon efforts from the Army, Navy, and Air Force based on criteria such as the type of technology and the intended use environment to review a variety of efforts. For each of these seven efforts, we reviewed relevant documentation and compared development activities associated with these programs to GAO's prior work on technology transition, GAO's leading practices for product development, and DOD's guidance on prototyping.

---

[1] To better defend against threats like cruise missiles, an integrated and layered defense includes sensors to detect and track threats, threat interceptors like surface-to-air missiles, guns, or directed energy, and a command and control system.

[2] S. Rep. No. 117-39, at 154.

For both objectives, we conducted site visits and interviewed officials within DOD and the Army, Navy, and Air Force responsible for DE weapon efforts to understand the status of the systems and any plans to transition these systems from prototyping to acquisition programs. See appendix I for a more detailed description of our objectives, scope, and methodology. In addition, see a list of Related GAO Products at the end of this report.

We conducted this performance audit from February 2022 to April 2023 in accordance with generally accepted government auditing standards.

Those standards require that we plan and perform the audit to obtain sufficient, appropriate evidence to provide a reasonable basis for our findings and conclusions based on our audit objectives. We believe that the evidence obtained provides a reasonable basis for our finding and conclusions based on our audit objectives.

Background

DE Weapons Technology

For decades, DOD has been developing DE weapons technologies that use electromagnetic energy to deny, degrade, damage, destroy, or deceive enemy weapons, equipment, facilities, and personnel. For many years, DOD's focus on DE weapons development was through chemical laser weapons development. Most notably, DOD developed the Airborne Laser Program, a chemical laser that achieved destructive power output. But the weapon required the use of hazardous chemicals to generate power, had significant issues with size, weight, and power due to heavy physical components, and required a large team of 15 to 18 operators.

More recently, DOD's laser development efforts have focused on solid- state lasers, which use a medium such as a fiber-optic cable to carry the generated electromagnetic energy, and are lighter and can be operated by a single person.

The two primary DE technologies, high energy lasers (HEL) and high power microwaves (HPM), have some similarities. For example, HEL and HPM both project electromagnetic energy to affect their targets, require a power source, must be connected to a command, control and communication center, and are mounted on a platform of some kind.[3]

However, the two technologies also have some fundamental differences.

HEL tightly focuses a beam, or beams, of energy on a single target to damage sensors or cause destructive heating. HEL systems include components such as

- adaptive optics to compensate for atmospheric interference to improve HEL range,
- sophisticated sensors for identifying and tracking targets,
- advanced thermal management systems to cool the weapon as it is in use, and
- software modeling to assist with identifying aimpoints on targets.[4]

---

[3] DOD reports command, control, and communications systems are fundamental to all military operations, delivering the critical information necessary to plan, coordinate, and control forces and operations across the full range of DOD missions.

[4] An aimpoint is a precise point, based on an area of vulnerability, associated with a given target and assigned for a specific weapon to achieve the intended objective and level of destruction.

Table 1. Characteristics of High Energy Laser (HEL) and High Power Microwave (HPM) weapons

Features	High energy laser	High power microwave
Function	Delivers energy to affect physical elements of a target like wings or optical sensors. HEL beams can cut through materials such as steel and aluminum in a matter of seconds. HEL systems engage targets one by one.	Delivers energy to affect electronics by overwhelming critical components intended to carry electrical currents such as circuit boards, power systems, or sensors. HPM systems engage targets over an area within its wider beam and can penetrate solid objects.
Physical characteristics[a]	Uses electromagnetic power measured in kilowatts directed, combined, and focused into a beam. Lasers are usually infrared and visible light.	Measured in gigahertz with wavelengths 10,000 times longer than lasers.
Testing	The effect of lasers changes according to environmental and other factors, making testing in an operational environment over a broad range of conditions particularly important.	DOD officials report that the effects of HPM are "nearly linear" so that testing can be done in a lab where tests at small scale and lower power levels can accurately predict effects at higher power levels. Full system testing is important to understand component limitations outside of a laboratory environment.
Lethality	Lethality depends on the amount of energy transferred which, in turn, is a function of the amount of energy delivered to the target and time on target. The amount of energy that can be delivered to the target depends on power output, purity, and concentration of energy on the target (beam quality), target range, and physical or atmospheric obstructions. The time on target depends on tracking the specific aimpoint on the target and continuing to keep the laser focused on it.[b]	Lethality depends on peak power output, the rate of microwave pulses, and the frequency used. HPM waves can travel through non-conductive solid material such as walls. HPM requires matching to specific types of targets and uses specific radio frequencies that will affect wires or circuit boards.

Source: GAO analysis of Department of Defense (DOD) and military department documents, and information provided. | GAO-23-105868.

[a] Electromagnetic energy includes radio waves, microwaves, infrared, visible light, ultraviolet, x-rays, and gamma rays. Laser is an acronym for "light amplification by stimulated emission of radiation." Diffraction of any electromagnetic radiation beam is based on the wavelength and aperture size. For the same aperture size, lasers diffract 10,000 times less than microwaves. This allows the beam to reach farther ranges while maintaining a small spot size of concentrated energy on the target. Lasers are preferred in specific scenarios because of minimal diffraction.

[b] The ability to keep the laser on target typically refers to tracking and "jitter control." Tracking is the weapon's ability to follow its target. Jitter control refers to the HEL's ability to focus the beam on the specified target area.

By contrast, HPM radiate waves of energy that can disrupt or damage electrical components of multiple targets within the entire targeted area simultaneously. HPM systems include components such as

- capacitators to enable the weapon system to store and release energy,
- an antenna to direct the microwaves toward the target area, and
- a vacuum technology to reduce the amount of trapped gasses within the device to support the generation of high power microwaves.

Figure 1 provides an illustration of the basic design of HEL and HPM weapons.

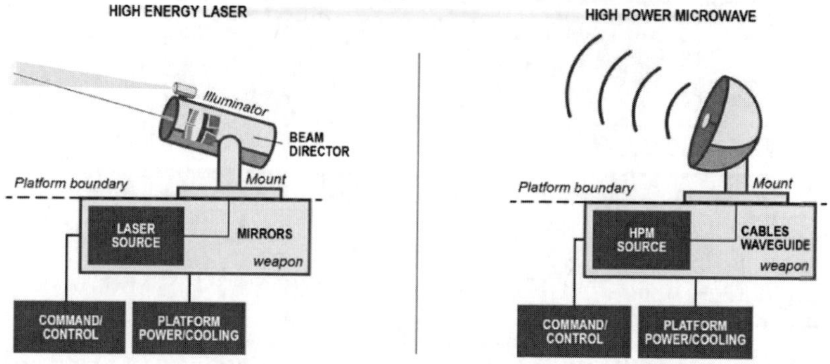

Source: GAO analysis of Department of Defence information. | GAO-23-105868.

Figure 1. Illustration of High Energy Laser (HEL) and High Power Microwave (HPM) Systems.

Potential Benefits of DE Weapons

Some advantages of DE weapons include the following:

- *Low cost per shot.* While the upfront cost of some DE weapons can be significant, DE weapon systems' costs per engagement can be much lower. For example, DOD officials said that the cost to fire a DE weapon is approximately the cost of fuel needed to generate the system's power for firing and cooling, or about $1-$10 per engagement. By comparison, defensive missiles can cost millions of dollars per shot.[5]
- *Magazine depth.* Because DE weapons do not rely on kinetic ammunition and are electrically recharged, they do not require mechanical reloading. Instead, they rely on power and cooling supported by a platform's overall power supply. HEL can repeatedly fire as long as the systems can generate sufficient power and manage the heat created by firing. HPMs do not generate as much heat as HELs and are able to continue firing with sufficient power. Platforms use fuel to generate the necessary electrical power and cooling.
- *HPM time on target.* HPM may have an immediate effect on a target with systems that DOD officials say fire the equivalent of hundreds or thousands of shots per second.

Possible Limitations of DE Weapons

Some possible limitations of DE weapons include the following:

- *Distance to target.* HPM and HEL are generally less effective the farther they are from the target. HPM power on target decreases over range. For HEL, distance may cause beams to be distorted, or energy may become absorbed by particles and water vapor in the air.

[5] For example, Army budget materials report that the Patriot Advanced Capability 3 Missile Segment Enhancement, which can intercept ballistic missiles, cruise missiles, and uncrewed aerial systems, costs approximately $4.1 million per unit for fiscal year 2023.

- *Atmospheric effects.* The effectiveness of DE weapons can be affected by environmental factors. For example, HEL has reduced effectiveness under certain atmospheric conditions, such as severe wind turbulence. Air particles and water vapor can also absorb or distort HEL beams. To mitigate this issue, a DOD official said DOD takes steps to select wavelengths that are less affected by water vapor. Adaptive optics help HEL weapons remain focused and "see" through atmospheric turbulence. Still, there are environmental conditions where DE weapons would not be effective and could not be used by the warfighter. DOD officials said HPMs that operate at high frequency and long range are also affected by the environment in similar ways to communication devices already in use, such as radios.
- *Delicate components.* Because HELs involve sending light through optical lenses and off mirrors, any contaminants or debris on those optics will interact with the beam and potentially damage the lens. For example, debris on the lens can damage the lens when the laser is fired, because the particle on the lens will capture some of the heat. Because of this, components are maintained or fixed in a specialized facility called a clean room to avoid contamination.[6]

Technology Development Process

For decades, DOD has prioritized investing in early-stage research and development of technologies. As part of the DOD effort to initiate, rather than react to, strategic technological surprises, DOD and the military departments invest in research and development to evaluate the feasibility or usefulness of a technology or concept.[7] DOD's science and technology community—including DOD research laboratories and test facilities, industry, and academia—conducts initial research, development, and testing of new technologies to evaluate the potential to improve military operations and strive for technological superiority over potential adversaries.

Establishing funding for technology development may include a number of challenges and may take years. We have reported that it may be challenging for new technologies to compete against established acquisition programs in obtaining funding.[8] Lengthy budgeting timelines also pose a challenge to developing new technologies aimed at responding to emerging threats. For example, under DOD's programming, budgeting, and execution process, a project idea may take at least a year and a half to get initial funding after the idea is first conceived.

As part of the technology development process, DOD groups research and development into seven budget activity categories for its budget estimates and the President's Budget. The categories follow a mostly sequential path for developing technologies, ranging from basic research (funded through budget activity 6.1) to operational system development (budget activity 6.7). Requested funding for prototyping is mostly found in budget activities for

---

[6] A "clean room" is an actively cleaned and isolated space engineered to reduce contamination from airborne particles. Clean room classifications vary by levels of airborne particles within the area.

[7] GAO, *Defense Advanced Research Projects Agency: Key Factors Drive Transition of Technologies, but Better Training and Data Dissemination Can Increase Success,* GAO-16-5 (Washington, D.C.: Nov. 18, 2015).

[8] For more on DOD's budgeting process for technology development, see GAO, Weapon Systems: Prototyping Has Benefitted Acquisition Programs, but More Can Be Done to Support Innovative Initiatives, GAO-17-309 (Washington, D.C.: June 27, 2017).

advanced technology development (6.3) and advanced component development (6.4). Requested funding in budget activity 6.3 is not directly tied to acquisition programs, whereas budget activity 6.4 is typically associated with an acquisition program.

Specifically, 6.4 funding is for advanced component development, system specific evaluations of integrated technologies, representative models, or prototype systems in a realistic operating environment. Activities under this level of funding focus on proving component and subsystem maturity prior to integration into major systems.

Prototyping

A key component of technology development is prototyping. Prototypes allow for rapid learning and accelerated demonstration of the value of new concepts, technologies, components, systems, and applications earlier in a technology development process than typically would have been possible. The DOD Prototyping Guidebook defines prototyping as a model built to evaluate and inform feasibility or usefulness.[9] Prototypes can be described relative to when they occur in a program's life cycle, including

- Conceptual—supports proof of concept analysis and may be demonstrated in a lab environment.
- Development—explores the technical feasibility of capability and may explore its operational value and is demonstrated in a relevant environment.
- Operational—offers technical and operational value and upon completion can be employed in an operational environment.

Prototyping is a tool to support the development of innovative technologies—that may be more risky than known designs or existing technology—prior to investing in an acquisition program. We have previously reported on two types of innovation: incremental and disruptive.[10] Incremental innovation seeks to gradually improve existing products and capabilities. Disruptive innovation attempts to shift the balance of military power by providing new capabilities, potentially unforeseen by military departments. The capabilities can be a result of new technologies, new ways to integrate existing technologies, or changes to how systems are employed. Examples of potentially disruptive technologies include DE, artificial intelligence, and hypersonics.[11]

The DOD Prototyping Guidebook states that successful prototyping begins with effective planning. Planning should include a clear articulation of the problem or need to be addressed, a description of the future decision to be made, the data to be generated by the project, and an explanation of how the data will be used to inform a future decision.

---

[9] Department of Defense, *Department of Defense Prototyping Guidebook,* Version 2.0 (Washington, D.C.: November 2019).

[10] GAO-17-309.

[11] GAO, *Hypersonic Weapons: DOD Should Clarify Roles and Responsibilities to Ensure Coordination across Development Efforts*, GAO-21-378 (Washington, D.C.: Mar. 22, 2021); and *Artificial Intelligence: Status of Developing and Acquiring Capabilities for Weapon Systems,* GAO 22-104765 (Washington, D.C.: Feb. 17, 2022). GAO has additional ongoing work on artificial intelligence.

Technology Transition

Once technologies are determined to be sufficiently mature, the acquisition community typically manages product development, in which technologies are further advanced and system development begins.[12] Although not precisely defined, technology transition to an acquisition program generally occurs at the point when advanced technology development ends and new product development begins.[13]

Prototyping efforts may result in different outcomes including transitioning successfully demonstrated technologies to operational use, leveraging the technology for other uses, or terminating unsuccessful efforts and associated technology research. Transitioning the prototype to operational use may include rapid fielding of the effort, transitioning the technology for use by an existing acquisition program, or beginning a new acquisition program. Figure 2 illustrates these potential transition pathways.

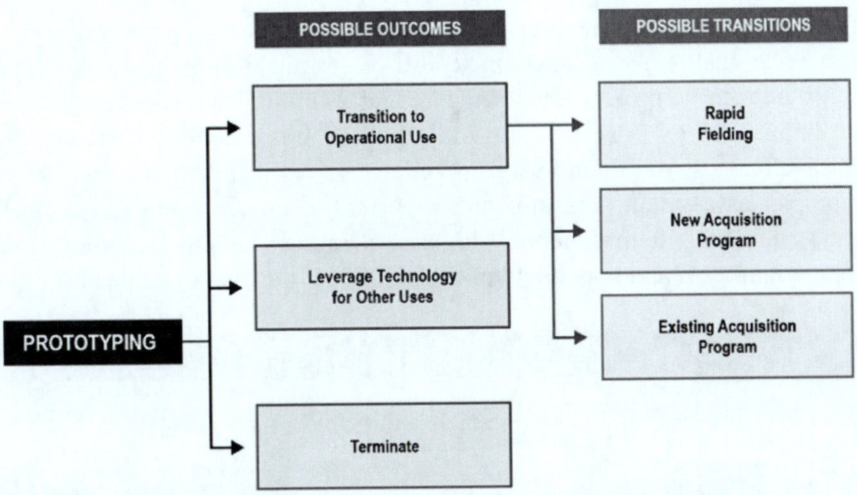

Source: GAO analysis of Department of Defence information. | GAO-23-105868.

Figure 2. Department of Defense Potential Prototype Outcomes.

For transitioning from prototyping to operational use, the DOD Prototyping Guidebook states that transition planning should be initiated as early as possible through collaborative efforts by the innovator, project management, and the warfighter. This collaboration is essential for clearly understanding the need and criteria to be met by the prototype for a successful demonstration in an operational environment. Our work also found that active collaboration with potential transition partners contributed to transition success.[14]

---

[12] A way of describing technological maturity is through Technology Readiness Levels (TRL), based on assessed maturity and numbered 1-9. TRLs range from a point where basic principles are observed (TRL 1) to full systems demonstrated in an operational environment under operational mission conditions (TRL 9).

[13] Advanced technology development can encompass subsystem or component evaluation in a laboratory or relevant environment, and new product development may involve system or prototype development and demonstration for use in an operational environment.

[14] GAO-16-5.

Our work found that DOD recognizes the difficulties involved in transitioning technologies to operational use.[15] DOD has long noted the existence of a chasm between its technology community and its acquisition community that impedes consistent technology transition.

Department insiders often refer to this as the "valley of death," which exists because the acquisition community may require a higher level of technology maturity than the science and technology community is able to fund and develop. Technologies may not leave the lab for several reasons including

- The technology's potential has not been adequately demonstrated or recognized.
- The military departments may not be willing to budget for funding final stages of development.

Despite the challenges for transitioning technologies, prior DOD and GAO work found that this gap can be bridged through cooperative efforts from both communities. Technology development officials can make decisions that balance needs, resources, and technical feasibility in a way that is responsive to the end-user. Acquisition programs and intended end-users can provide early project endorsement, and communicate measurable performance metrics for the technology to achieve. Overall, this requires early and frequent collaboration among the developer, acquirer, and user.[16] We have found that formal transition agreements and other project assessment measures are important to ensure projects stay on track and stakeholders sustain their commitments.[17] Figure 3 illustrates DOD's technology management and development process and transition vulnerability over the "valley of death."

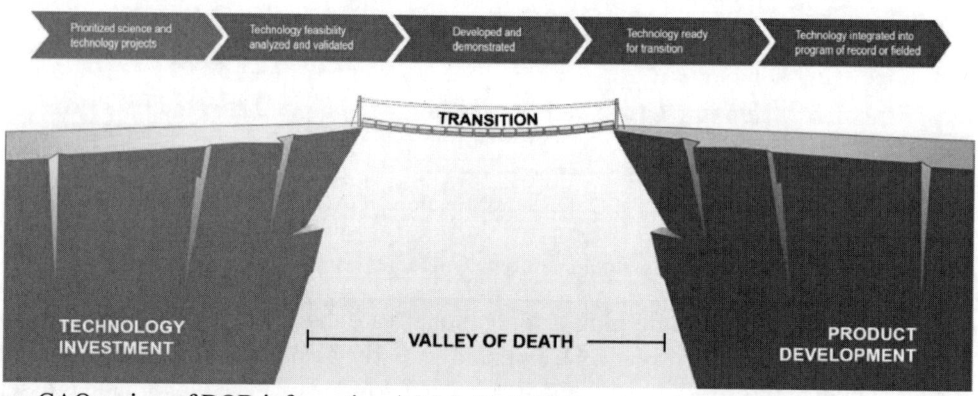

Source: GAO review of DOD information. | GAO-23-105868.

Figure 3. Department of Defense (DOD) Technology Management Process and Depiction of Transition Vulnerability over the "Valley of Death."

---

[15] GAO-13-286, Defense Technology Development: Technology Transition Programs Support Military Users, but Opportunities Exist to Improve Measurement of Outcomes, (Washington, D.C.: Mar. 7, 2013).

[16] GAO-16-5.

[17] GAO-13-286.

DOD and Military Departments Have Undertaken Steps to Coordinate on and Develop Multiple Types of DE Weapons

DOD and the military departments have undertaken several efforts to support the development of DE weapons. This includes taking steps to coordinate, test, and develop processes to support DE technology development, developing and demonstrating DE technologies in operational environments, and initiating an HEL scaling initiative to increase power output of future technologies.

DOD and Military Departments Have Undertaken Steps to Coordinate, Test, and Develop Processes to Support DE Technology Development

DOD and military departments have undertaken several initiatives related to supporting the development of DE weapons. These initiatives include extensive department-wide coordination and planning efforts, testing procedures and technologies, and developing a review process for prototypes to be used in the field.

Coordination and planning. DOD engaged in several coordination and planning efforts, such as creating the Joint DE Transition Office, which oversees multiple department-wide DE technical area working groups.[18] These groups include multiple representatives from the Army, Navy, and Air Force that discuss crosscutting DE technology investments, such as for beam control and tracking.

Additionally, DOD developed the 2021 DOD Directed Energy Roadmap as a way to communicate the department's overall DE goals and help facilitate coordination among the military departments. The Roadmap describes coordinated efforts, goals, initiatives, types of technologies, and department-wide DE-related challenges.

DOD officials and industry representatives described instances of coordination among the military departments with the sharing of effective technologies across DE weapon systems. For example, DE weapon systems within the Army and Air Force share a targeting and tracking system because of demonstrated effectiveness. Additionally, each of the military departments has participated in sharing where prototypes or components developed at one department were used to develop prototypes at another. For example, Army officials said the Air Force developed an HPM that the Army considered pursuing. While the Army did not ultimately pursue that technology, Army officials said that they leveraged that technology in their own prototype development. This coordination means that improvements to the system can be shared across DE weapons.

A number of entities across DOD are involved in identifying requirements for, developing, and testing DE efforts, as listed in table 2.

---

[18] The Joint DE Transition Office was originally the High Energy Lasers Joint Technology Office and was formed in 2001. Additionally, the William M. (Mac) Thornberry National Defense Authorization Act for Fiscal Year 2021 directed DOD to establish a Directed Energy Working Group to (1) analyze and evaluate the current and planned DE programs of each military department; (2) make recommendations to the Secretary of Defense; (3) identify methods of quickly fielding DE capabilities and programs; and (4) develop a compendium on the effectiveness of DE weapon systems and integrate the compendium into an overall manual. Pub. L. No. 116-283, § 215.

Table 2. Selected DOD organizations involved in directed energy efforts

Organization	Organizational component
Army	Space and Missile Defense Command
	Army Test and Evaluation Command
	Rapid Capabilities and Critical Technologies Office
	Armament Research, Development, and Engineering Center
	Joint Counter-small Unmanned Aerial Systems Office
Navy	Naval Sea Systems Command, including: Naval Surface Warfare Center Dahlgren Division, and Program Executive Office – Integrated Warfare Systems
	Office of Naval Research
	Surface Warfare Directorate
Air Force	Air Force Research Laboratory
	Air Force Life Cycle Management Center, including the Digital Directorate
	Air Force Special Operations Command
Other/DOD-wide	Missile Defense Agency
	Directed Energy Joint Transition Office
	Office of the Undersecretary of Defense for Research & Engineering
	Test Resource Management Center
	Director, Operational Test and Evaluation
	Combatant Commands

Source: GAO analysis of Department of Defense (DOD) information. | GAO-23-105868.

Testing procedures and technologies. DOD and the military departments planned for and invested in DE weapon testing capabilities. To support the development of infrastructure necessary to test DE efforts, the Test Resource Management Center developed a DE testing roadmap, updated annually with input from more than 75 participants across DOD. The testing roadmap helps to prioritize investments in testing infrastructure, advanced instrumentation, and test targets representative of modern threats. DOD testing officials said that it is important that the appropriate testing infrastructure is in place to determine the effect that DE weapons have on targets. For example, to test HEL, DOD needs testing instrumentation to measure the beam output as well as the magnitude of the effect on the target. To test HPM, DOD is developing instrumentation to measure the susceptibility of various electronic technologies to HPM.

DE weapon review process. To ensure DE weapons that are fielded are adequate, feasible, and acceptable, each prototype undergoes a review process called the DE Weapon Review and Approval Process before fielding.[19] This process requires that a Combatant Commander initiates the use of DE weapons and develops concepts of operations for its use. This review process allows stakeholders to determine, among other things, whether DE weapons will likely perform as intended. Stakeholders include representatives from military departments, the Office of the Secretary of Defense, and Joint Staff. DOD officials said a process like this is typical for new technologies and that, as DE weapons become more widely used and understood, the requirement to conduct these extensive reviews may no longer be necessary.

[19] Chairman of the Joint Chiefs of Staff Manual (CJCSM), Directed Energy Weapon Initial Operational Employment Review and Approval Process, 3230.01A (Oct. 7, 2015).

DOD Has Operationally Demonstrated Multiple DE Weapons Capabilities

DOD has demonstrated multiple DE weapons capabilities, including a range of technologically mature systems and components. Specifically, each of the military departments has developed DE weapons that have been demonstrated in relevant or operational environments, a key developmental event in demonstrating technology maturation.

Additionally, DOD matured component technologies that support DE weapons, and each of the military departments developed a range of technologies.

To respond to certain threats, DOD and the military departments have developed multiple types of DE weapons. Over the last decade, this has included demonstrating and prototyping more than 20 DE weapon systems. Most of the HELs developed during this time are in the 10-to-60 kilowatt (kW) class power range. DOD officials told us that this class of laser has been used in live fire demonstrations to successfully disable, or shoot down, uncrewed aerial systems (UAS), also referred to as drones.[20] DOD DE development officials say drones are uniquely challenging because they can be inexpensive enough for adversaries to deploy in large quantities in combat zones; and traditional defensive tools, such as kinetic weapons, are expensive to use against them. These officials also note that countering drones in civilian areas with traditional kinetic kill weapons could pose risks of casualties, because the use of these can involve large explosions and debris. In contrast, DE weapons could be useful because they can disable the drone with less risk of collateral damage. Other DE weapons under development have been designed to have improved capability with power ranges up to 300kW. Specifically, HEL in the 300kW class range are designed to respond to targets such as rockets, artillery, and mortars. In addition, the military departments developed a range of HPM capabilities for the purposes of engaging missile or drone swarm attacks against a base. Some details of HPM efforts are classified.

Each of the military departments has made specific contributions to developing DE technologies. This process of developing DE technologies has meant that warfighters have access to this technology to address needs and allow for learning in an operational environment.

Army. The Army developed a broad range of capabilities, including:

- a 20kW counter-UAS HEL,
- a 50kW HEL to use on a maneuver platform,
- a laboratory tested 300kW HEL, and
- an HPM system.

In addition, the Army has made and plans to continue to make investments into the manufacturing industrial base for DE weapons. Examples of these investments include improving cost and speed of production of critical DE components, such as optics and mirrors. The Army identified several improvements already, including reducing the speed of producing

---

[20] According to the DOD Fiscal Year 2009-2034 Unmanned Systems Integrated Roadmap, DOD classifies UAS into five groups based on weight, operating altitude, and airspeed. Group 1 UAS are the lightest at 0-20 pounds at takeoff, operate at altitudes around 1,000 feet above ground level, and fly at speeds around or below 100 knots. Group 5 UAS can weigh thousands of pounds, operate above 18,000 feet, and have any airspeed. For the purposes of this review, we are using the gender-neutral term uncrewed as a replacement for the term unmanned except when referring to the proper name of a DOD document or program.

a component from 6 months to 3 months and increasing production of another component from four units to 20 units per week.

Navy. The Navy developed several DE efforts that can be used on ships, including:

- a low kW dazzler to disrupt drones,
- a 150kW laser to counter small boats, and
- several HPMs.

Additionally, to support integration of DE weapons on ships, the Navy is developing energy magazines—an energy storage device—to enable DE weapons with higher power requirements than can be supported by the systems currently installed on some ships on their own.[21] When activated, DE weapon systems and sensors can suddenly increase energy demand from a platform, which could exceed what the onboard generators can produce. These generators are operated at a continuous power level for efficiency and reliability and may not be able to boost output to support a large demand over a short period of time. The Navy's energy magazine is being designed to provide a shipboard-ready power system to support DE weapon, sensor, and electronic warfare power needs during a conflict by providing power during an engagement. The energy magazine is being designed to charge when systems are not in use so that generators can continue to operate efficiently and not be strained by erratic levels of high demand. In addition, the Navy's energy magazine will be designed to be scaled to meet power requirements and supplement ship generators. This would give the Navy the ability to make existing vessels capable of using future DE weapons and sensors.

Air Force. The Air Force developed a range of DE capabilities that could potentially be used in either air or land, include the following:

- a 50kW HEL system that can be mounted to tactical aircraft,
- a 10kW HEL system integrated with a light ground vehicle,
- a counter-UAS HPM system, and
- several other HPM efforts.

In addition to developing DE weapons, the Air Force has invested in technologies to address specific DE-related challenges. For example, flight at high speeds generates vibrations that requires DE weapon designs to be ruggedized to a flight environment. In addition, the typical turbulence caused by airflow moving around the aircraft or a DE weapons system at high speeds—including supersonic speeds—would make it difficult to use HEL weapons, because the beam could get distorted or disrupted by this turbulence. The Air Force invested in developing technology to reduce turbulence affecting these DE systems. This investment and development may support the Air Force's effort to mount HEL onto aircraft moving at high speeds and maneuvering. The Air Force also sponsored an experimentation campaign from 2020 through 2022 that used modeling and simulations and war-gaming exercises.

DOD and the military departments are also investing in component technologies that improve DE weapon capability. These include the following:

---

[21] An energy magazine refers to a system used to store energy for later use by a weapons system. Such a system is used to support sustained engagement and, ideally, would allow for infinite use as long as fuel is available.

- Identifying more effective aimpoints for potential DE weapon targets. Each military department conducts vulnerability assessments for various threats, which are coordinated by the Joint DE Technology Office. These assessments are used to develop target descriptions and support a DOD initiative to build a database of those targets to help inform the best way for a DE weapon to defeat them.
- Using adaptive optics to improve beam quality—the measure of how well a beam can be focused—so that systems remain effective over distance.

Additionally, the military departments are also integrating DE weapons into command and control systems. Improved integration means DE weapons will have real-time information on incoming threats, which enables DE weapons like HEL to quickly identify and track precise aimpoints.

DOD Is Pursuing Higher Power Weapons for Improved Capabilities

DOD, in coordination with each of the military departments, is pursuing higher power laser weapons for improved capability to engage a broader range of threats. In 2019, DOD introduced the High Energy Laser Scaling Initiative to leverage current technologies, increase the power output of HEL, and expand the potential range of HEL mission sets. The Office of the Under Secretary for Research and Engineering leads the initiative, which includes three HEL development efforts. DOD and each of the military departments are working toward building lasers with output in the 300-, 500-, and 1,000kW power ranges. Such systems could eventually enable HEL to engage powerful targets such as cruise missiles.

According to DOD, the initiative supports the improvement of directed energy weapon capability by focusing on increasing output power, refining energy delivery to targets, and developing efficient power and laser generation systems.

In addition, the initiative aims to develop the industrial base so that the military departments can make additional purchases of HEL at 300kW and above. As part of this initiative, DOD is supporting three different technology development approaches. DOD officials said they expect the laser scaling initiative to result in an HEL prototype in the 300kW class delivered in fiscal year 2023. After delivery and integration of the HEL prototypes and testbeds, DOD officials said that the weapon systems may participate in counter-cruise missile demonstrations and exercises.

While a DOD official confirmed DOD does not have a similar coordinated effort to develop higher power HPMs, some military departments continue to invest in more powerful HPM systems for greater capability, including higher power and increased range.

Two Military Departments Have Not Consistently Taken Steps to Support DE Technology Transition

The Army, Navy, and Air Force reported in budget requests that they expect to transition selected DE prototypes to acquisition programs when the technology is ready. The Army has taken actions that provide a foundation to transition DE technologies that meet the Army's warfighter and acquisition community needs. However, the Navy and Air Force have not consistently applied prototype transition practices, such as identifying transition partners early, drafting transition agreements, and documenting their user feedback process.

DE Weapons Technology Faces Barriers to Transition

For prototypes that are expected to transition to an acquisition program, the unique nature of DE weapons technology may present novel variations to the traditional barriers to transition.[22] DOD officials pointed out that, DE technology is fundamentally different than the kinetic weapons the military departments are accustomed to using. Therefore, not having long-term experience with them limits the understanding of how to integrate DE weapons into missions which may, in turn, make obtaining "buy-in" from potential users more challenging.

Near-term challenges with transitioning DE technologies beyond the development stage include space, weight, and power; testing; and meeting the defense industrial base needs.

Space, weight, and power. DE weapons require integration into existing platforms such as vehicles, ships, and aircraft. This integration may be as simple as plugging into a radar, or as challenging as removing and repurposing sections of a Navy ship. Further, DE weapons require power that may be integrated into the weapon or may need to be provided directly from the platform. HEL also generate heat that must be managed. A cooling mechanism to absorb or transfer this heat may take up significant space. These system requirements may pose challenges on platforms with limited space and power, such as ground vehicles or aircraft. Specifically, DOD officials said some HEL weapons have about a 25 to 40 percent efficiency rate, which means that about 25 to 40 percent of power used by the weapon is used in the laser. Some of the rest of the energy is released as heat that must be absorbed or transferred elsewhere, often through some cooling system. Although DOD has experience managing space, weight, and power considerations for new weapons, integrating the specific needs of DE weapons onto platforms requires advanced planning.

Testing limitations. Testing DE weapons to verify and validate performance observed in laboratories can be limited for several reasons, including access to operational environments. For example, testing HEL against airborne targets could inadvertently damage or destroy sensitive components on satellites. Because of this, developers must consult a laser clearing house to determine when it is safe to shoot an HEL weapon above the horizon in a testing, training, or maintenance environment.

Unlike testing kinetic weapons, this restriction tightly controls HEL testing and may prohibit an HEL weapon from firing in a test range. This may make it challenging for potential transition to acquisition or fielding because the full capability or limitations of a DE weapon system may not be fully understood ahead of transition.

Defense industrial base. Representatives of companies within the defense industrial base have expressed hesitation to invest in developing DE technologies without a government commitment. Unlike other emerging technologies where DOD may leverage commercial products to build military products, there are no current commercial applications for DE

---

[22] As discussed earlier, prototypes can result in a variety of outcomes, including transitioning for further development or into acquisition programs, or termination. For the purposes of this discussion, we are focusing on only those DE efforts that have been identified by the services for transition into an acquisition program, and therefore potential transition partners include weapon system programs or DOD components.

weapons or key technologies. As a result, contractor representatives emphasized the need for a clear "demand signal" from DOD about future investments in DE technologies to support internal investments. Officials pointed out that current costs to develop DE weapons are high because of the relatively low number of initial development efforts. Therefore if DE weapons were produced in larger quantities, costs may be spread out over more units, lowering costs per unit.

Longer term, there are challenges with battle management; tactics, techniques, and procedures; maintainability; and user safety concerns.

Battle management. Battle management refers to how commanders at all echelons of conflict make decisions based on available information. Commanders make decisions about weapon selection with the objective of best achieving operational goals. Oftentimes, there is an emphasis on quick decision-making after processing large quantities of information.

Commanders' decisions to use DE weapons will likely require considering the weapons' benefits, such as lower cost, and limitations, such as time on target. Further, compared to kinetic weapons, DE weapons may not have the same range capability. Thus, the decision about whether to use them, and when, may be challenging. Figure 4 shows an example of the way DE weapons may be used as part of a layered defense strategy.

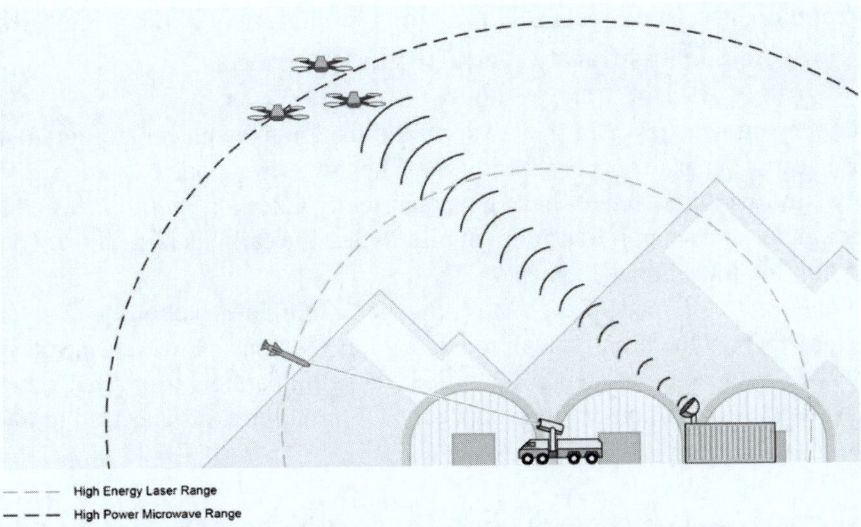

Source: GAO analysis of Department of Defense information. | GAO-23-105868.

Figure 4. Notional Depiction of Long Range High Power Microwave and Short Range High Energy Laser Used to Defend a Military Installation.

Tactics, techniques, and procedures.[23] As a novel technology, DE weapons require the development of new tactics, techniques, and procedures; processes by which the warfighter knows how best to use a particular technology in an operational environment. For traditional kinetic weapon systems like missiles, their use is well understood, and changes to missile systems improve capability but do not require a fundamental change in use. For example, officials said the tactics envisioned for using a DE weapon by the developer may not be how a

[23] Tactics are the employment and ordered arrangement of forces in relation to each other. Techniques are non-prescriptive ways or methods used to perform missions, functions, or tasks. Procedures are standard, detailed steps that prescribe how to perform specific tasks.

warfighter would employ the system. Because of this, tactics would need to be reworked for the DE system, with any necessary design changes. The tactics, techniques, and procedures to support use of DE weapons for a broad range of uses are still being developed.

Maintainability. The internal mechanisms for DE weapons are sensitive, and typically require a specialized clean room for repairs. For example, DOD officials said that one DE weapon fielded to an operational environment encountered challenges with battery charge and cooling, and had to be returned to the manufacturer in the United States for repairs. Ultimately, this challenge reduced system availability, which is key to a successful weapon. Developing maintenance and sustainment processes for weapon systems is a key role for an acquisition program office. The ease of maintaining a weapon may factor into an acquisition program office's decision to pursue that technology compared with another.

User safety concerns. Several DOD and military department officials described a possible apprehension among users regarding potential biological effects of DE weapons, such as a fear that personnel might be blinded by a laser. DOD reported undertaking rigorous testing and other measures to ensure safety of personnel, and DOD officials said education among users may help encourage adoption of this new technology.

Navy and Air Force Leadership Have Not Consistently Taken Key Steps to Support Expected Transition to Acquisition Programs

Two military departments reported they expect to transition to acquisition programs when the technology is ready, but have not consistently taken key steps to support transition. Specifically, leaders of two military departments have not consistently taken key steps identified in DOD's prototyping guidance to support transitioning prototypes: the early identification of a transition partner and drafting transition agreements.

With the support of leadership, Army engages multiple stakeholders and documents transition plans early in the prototyping process for DE weapons. However, the Navy and Air Force leadership have not consistently identified transition partners or drafted agreements to support transition to acquisition programs once the DE prototype was expected to transition.

Army Leadership Takes Steps to Support Expected Transition of DE Weapons to Acquisition Programs

For prototypes the Army has reported it expects to transition to acquisition programs, Army leadership supported the implementation of the two key steps to support transition: identifying a transition partner and drafting a transition agreement. In 2019, the Army Board of Directors initiated the Army's current approach to developing DE through the Multi-Mission High Energy Laser Technology Maturation Initiative. At that time, the Board directed the Rapid Capabilities and Critical Technologies Office to develop combat-capable DE weapon prototypes as part of modernization efforts for Army Air and Missile Defense.[24] Once

[24] DE weapon prototypes are part of four modernization efforts: (1) Army Air and Missile Defense integrated battle management, (2) Counter Unmanned Aerial Systems, (3) Maneuver – Short Range Air Defense, and (4) Indirect Fire Protection.

successful prototyping has been completed, the Army expects to proceed to a program of record. Program Executive Office (PEO) Missiles and Space will be the office responsible for doing so.

We have found early planning, including establishing military demand and active collaboration with transition partners, as key indicators of successful transition from prototype development to use by the warfighter.[25] Further, the DOD Prototyping Guidebook states that some prototyping efforts will transition into their own acquisition program. If this pathway is expected from the outset of the prototyping project planning, it would be prudent to begin planning and coordinating the transition as early as possible. Specifically, the 2019 DOD Prototyping Guidebook lists drafting a transition agreement between the program manager and the transition partner within the first year of the project as a best practice.

DOD's 2022 version of the guidebook notes that a transition partner should be identified and an agreement drafted as soon as possible.[26]

We have previously reported that documented technology transition agreements with technology and business readiness metrics can help transition partners track the progress of new technologies.[27] Additionally, transition planning agreements are not meant to weigh down either the innovators or the potential transition partners. Instead, these agreements are not static and, according to the DOD Prototyping Guidebook, should be regularly revisited by the developer and the transition partner as prototype projects progress.

The Army drafted transition agreements to establish the conditions that enable transition of DE efforts to acquisition programs beginning in fiscal year 2025, when the prototype is ready. To support this effort, Army officials described embedding transition teams in the prototyping effort to monitor technology development progress and relay information to the transition partner, so that the acquisition program office understands the capabilities it is receiving and can provide feedback. As part of the transition agreement, the Army engaged in several planning activities, including budget planning and determining how funding obligations will transition development to the PEO, all in an effort to be ready once the transition decision is made.

The Army developed a detailed plan, called an "octagon," describing schedules and stakeholder roles to build supporting activities around the use of DE weapons. This includes semiannual meetings to coordinate not only development of the prototypes themselves, but to plan for the next steps in using a DE weapon. This includes developing what DOD refers to as DOTMLPF-P, or "doctrine, organization, training, materiel, leadership, personnel, facilities, and policy for DE weapons" to support future use.

Additionally, the Army developed early planning documents used to support acquisition programs of record. For example, for one prototype, Army officials described developing an "abbreviated Capabilities Development Document," a document that specifies requirements for

---

[25] GAO-16-5.

[26] Department of Defense, *Department of Defense Prototyping Guidebook,* Version 2.0 (Washington, D.C.: November 2019). This Guidebook has a more recent version, released in October 2022, which includes the best practices identified in Version 2. In 2019, Version 2.0 of this guide stated that: "to increase the likelihood that the project will transition as planned, a transition agreement between the program manager and transition partner should be drafted within the first year of the project and all affected parties should sign the agreement." We refer to this earlier version because the 2022 update, Version 3.0, did not provide specific and measurable time frames for drafting transition agreements.

[27] GAO, *Best Practices: Stronger Practices Needed to Improve DOD Technology Transition Processes,* GAO-06-883 (Washington, D.C.: Sept. 14, 2006).

the system that will deliver the capability to meet the performance criteria specified by operational need.

Army officials said that no final decision has been made about whether any current Army prototypes will transition to programs of record. These officials said that they will make that decision based on Army needs, and the ability of the technology to meet those needs. Should the Army decide to pursue a program of record, having these actions in place will help support transition when ready.

Navy Leadership Has Not Taken Key Actions to Support Expected Transition of DE Capability

The Navy reported that it expects to transition several DE technologies to acquisition programs, but has not drafted transition agreements to support the transition of systems we reviewed. The Navy developed DE prototypes to demonstrate DE weapon technical feasibility on ships or in response to urgent operational needs. However, it has not taken a key action identified in the DOD Prototyping Guidebook to draft transition agreements for selected efforts, even though—as indicated in the Navy budget documents—the Navy expects to transition these prototypes.

The Navy reported that it identified potential transition partners for DE weapons. Overseeing transition, the Assistant Secretary of the Navy for Research, Development and Acquisition is responsible for the development and acquisition of Navy platforms and weapon systems. In general, Navy officials said the Navy's current process for developing and transitioning DE technologies is through a partnership between the Navy science and technology community and the PEO, Integrated Warfare Systems, which supports fielding, testing, and maintaining DE weapon prototypes. This includes working with Navy sailors and vessels to schedule DE weapon installment onto ships, and recording data on DE weapon usage for the development community. In addition to the technology developers and the PEO, the Surface Warfare Directorate supports DE efforts by funding the installation of the weapons onto ships and developing requirements for DE weapons from warfighters and through strategic documents.

The Navy's strategic documents describe future plans to use DE weapons capable of defeating anti-ship cruise missile threats, but the Navy has not taken a key step to enable the transition of the DE efforts into acquisition programs. Specifically, even with plans for increasing DE power output to meet counter anti-ship cruise missile requirements and demonstrating and fielding additional DE weapon prototypes in the next several fiscal years, the Navy has not documented transition agreements, including funding and requirements to facilitate future transition efforts for prototypes that are expected to transition to acquisition programs. Navy officials said they have identified transition partners and have ongoing discussions with the Surface Warfare Directorate and PEO Integrated Warfare System throughout the development of the effort, but they do not have documented agreements.

Although the Navy reports that selected prototypes are expected to transition to an acquisition program in fiscal year 2024, the Navy has not drafted a transition agreement among the potential transition partners.

Navy officials said they are waiting on additional testing to ensure the capability could meet the Navy's needs to defeat anti-ship cruise missile threats before generating agreements between developer and the acquisition community. Navy officials also said that they are undergoing a broader mission analysis through an initiative called the Navplan Implementation Framework to better determine how weapons, including DE weapons, fit into future battle management.

Drafting transition agreements early could better support the Navy's efforts to eventually transition DE technologies when ready. Navy officials reported that current and future Navy ships would not be able to provide the necessary power and cooling to a DE weapon with significant power output. Specifically, the Navy cited several integration challenges that require early planning to address including ventilation, fire suppression, and cooling. Although the Navy has efforts underway to address the power required for future weapon systems, the Navy also reported that the unique power requirements of DE weapons make energy storage a challenge, and that current mission energy demands outpace ship capability, even without factoring in DE needs. Documenting a transition agreement can provide developers with information such as the criteria that need to be met to transition the technology. Without early transition planning and drafting transition agreements, the Navy risks developing technology that is misaligned with operational needs.

Air Force Leadership Has Not Consistently Taken Key Steps to Support Expected Transition of DE Capability

The Air Force reported that it expects to transition DE counter-UAS technologies to acquisition programs, but has not consistently taken key steps to support this transition. Specifically, the Air Force does not consistently take early steps, per DOD's Prototyping Guide, to identify transition partners and draft transition agreements for prototypes it expects to transition to an acquisition program when ready.

Air Force officials reported that technology transition partners typically include the technology developers, the acquisition community, including PEOs, and users in the operational community, such as a major command. The Office of the Assistant Secretary of the Air Force for Acquisition, Technology, and Logistics oversees Air Force research, development, acquisition, and program sustainment activities. In general, the Air Force's current process for developing and transitioning DE weapons is through partnership between the Air Force science and technology community, including the Air Force Research Laboratory (AFRL) and the Air Force major commands, which support fielding, testing, and maintaining DE weapon prototypes. Air Force major commands have sponsored experimentation plans and technology demonstrators executed by AFRL. Additionally, AFRL has funding for early technology and prototype development it leverages for selected technologies.

In one effort we reviewed, the Air Force identified a transition partner and documented requirements agreed upon by developers and the transition partner. This effort was an advanced technology demonstration, or pre-prototyping, airborne laser effort, where officials initiated a technology transition plan between the technology developers and an Air Force major command. In this instance, the technology transition plan included types of technology to be developed, goals, and criteria for continued development efforts. The plan was signed by both developers and a transition partner, and included flexibility to adjust as needed. The plan acknowledged that, given the factors outside of parties' control, the commitment for future funding was not binding. Officials said having this early coordination and agreement helped establish parameters for developing the technology, but acknowledged that some additional information would be necessary for such an effort to move into prototyping. Specifically, officials said that, without knowing whether the Air Force plans to use airborne lasers in a crewed or uncrewed aircraft, developers do not know whether to design a laser mounted to the outside of an aircraft, such as in a pod affixed to the body of the aircraft, or integrated with the aircraft itself.

For other efforts the Air Force expects to transition to acquisition programs, though the Air Force has not taken key steps to support transition. Specifically, one counter-UAS HEL prototype we reviewed was under development and testing for more than 3 years—with the expectation of transitioning into an acquisition program when ready— before a transition partner was identified. Air Force officials said that it was unclear who the transition partner should be, even though a transition partner, PEO Digital, was identified in 2021. However, the prototype did not fully align to program office needs, and the office did not have funding planned—nor transition agreements in place—to support initiating a DE weapon acquisition. Officials from PEO Digital, the identified transition partner, said that DE is one of a large number of technologies they are considering to meet their counter-UAS mission.

The future of DE weapons in the Air Force is unclear. Although the Air Force developed a number of technologies that have been leveraged across DOD, Air Force leadership has not incorporated DE efforts into funding planning over the next few years, and there are no current agreements to transition any DE efforts. Over the long term, the Air Force has expressed interest in using DE weapons but has not consistently taken key steps to support transition. Air Force officials said work is still underway to understand the technological maturity DE weapons need to achieve, as well as the way to effectively apply DE weapons during a conflict. Further, officials cite ongoing technical challenges when considering how DE weapons fit into Air Force mission needs. Without identifying transition partners early and drafting transition agreements, the Air Force risks developing technology that is misaligned with operational needs.

Navy and Air Force Do Not Consistently Document How They Collect, Track, and Incorporate Feedback

Each of the military departments report soliciting user feedback throughout the design and development of demonstrators and prototypes. The Army also has detailed tracking mechanisms to gauge the types and amount of feedback received, and how it is incorporated. However, the Air Force and Navy do not have a formal process for collecting, tracking, and incorporating feedback during the design and prototyping phases of DE development.

Officials from each military department described collecting feedback from warfighters in a variety of ways, including informal discussions, allowing warfighters to test prototypes, modeling and simulation exercises, and ongoing discussions with operators using prototypes in the field. Each military department reported receiving helpful feedback, such as on equipment user interfaces. Specifically, each military department reported receiving feedback that helped it leverage a video game controller-like interface to make operating DE weapons more intuitive to warfighters. Warfighters also gave practical feedback, including that operators may be eating while using the DE weapon, so having a touch screen is problematic.

GAO's leading practices for product development, including prototyping, emphasize the importance of collecting customer feedback to inform improvements or to identify challenges that need to be addressed.[28] DOD best practices recognize end-user involvement as a best practice for evaluating prototypes. Figure 5 shows that feedback can be incorporated at multiple points throughout the development of prototypes.

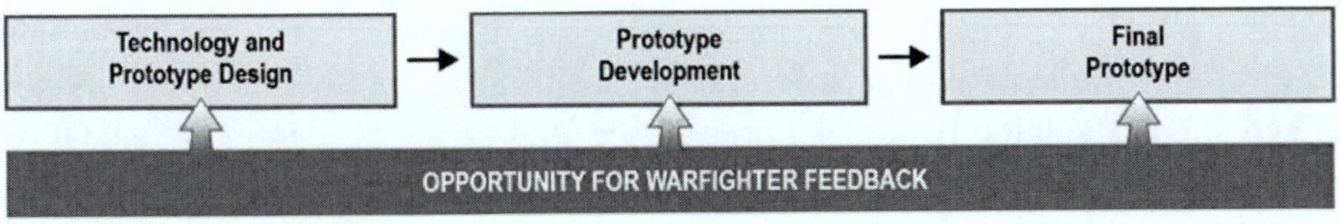

Source: GAO analysis of Department of Defense information. | GAO-23-105868.

Figure 5. Opportunities for Warfighter Feedback throughout Prototype Development.

Army Tracks How User Feedback Is Collected and Incorporated

The Army documents how user feedback for DE development efforts is collected and incorporated throughout the development process through recorded "soldier touch points" (STP) that represent the number of hours a system has been used by soldiers. This includes incorporating feedback throughout the design and development of a prototype, which helps to ensure feedback is gathered and incorporated early in the process in a variety of areas. For one system we reviewed, Army officials said they documented thousands of STP hours and tracked the number of hours spent on specific aspects and focus areas of development. For example, for one system, Army recorded 179 STP hours focused on developing training tools, 16 STP hours on software development focused on command and control, and 40 STP hours focused on design reviews. Overall, for this system alone, Army reported more than 6,200 STP hours.

The Army also tracks the number of comments received and whether they have been incorporated into design. Army officials reported that this process resulted in improvements to its DE technologies under development. In some cases, users identified design defects that expert designers overlooked. For example, in one instance, development officials reported receiving feedback from a soldier about the design for a laser mounted to a vehicle. Officials said that developers originally positioned the laser to fire through what would be the likely pathway of the vehicle's exhaust. This would have created an atmospheric effect that might have distorted the laser's beam, rendering it significantly less effective. Development officials said this design problem had been overlooked by the developers, and they were able to integrate the feedback into the design, positioning the laser through another path.

Officials said making this change at design would help them avoid costs associated with redesign had the issue been identified later.

[28] GAO, *Leading Practices: Agency Acquisition Policies Could Better Implement Key Product Development Principles,* GAO-22-104513 (Washington, D.C.: Mar. 10, 2022). Our prior work has demonstrated that, while structural differences between the private sector and government can affect outcomes, key principles from the private sector can be thoughtfully applied to government acquisition to improve outcomes, even with the different cultures and incentives.

Navy and Air Force Do Not Consistently Document How They Collect, Track, and Incorporate User Feedback

Navy officials described receiving operator feedback on prototypes but did not formally document the way they collect, track, and incorporate user feedback. The Navy described ongoing efforts to collect regular informal feedback from prototype operators that was incorporated into design modifications. For example, Navy officials described one DE weapon includes eight prototypes that have been installed and used in operational environments over the course of 2 years, starting in 2020. During this time, officials report that operators provided regular feedback that has been incorporated as the prototypes were finalized and installed. Operators we spoke with said that they received a great deal of support from the DE weapon developers, and that they typically do not experience such close support and contact from developers of new technology.

However, Navy officials reported that, although they collect and incorporate feedback into prototype designs, they do not specifically document the feedback from operators because they said it wasn't necessary. The informal nature of collecting and using feedback means that there may not be a standard approach that ensures the same level of effort will be used on future DE technology development.

Air Force officials described inconsistent processes for collecting feedback, and an instance where feedback was not incorporated into technology design. Air Force officials described one formal effort to collect feedback, a semiannual modeling and simulation exercise that focused on specific scenarios. Goals of this exercise include improved understanding of HEL capabilities and exposing warfighter to DE capabilities. However, some efforts to collect information are not as formal. For one of the Air Force DE weapons we reviewed, there was no transition partner identified until well into prototype development. Because there was no transition partner to identify intended end-users that could give feedback on their needs, officials described informal efforts to get feedback from available warfighters. These efforts included looking to "see who was around" to have them test out the demonstrators or prototype. This ad hoc approach risks gathering feedback from warfighters that may not be reflective of the needs of ultimate end-users. Further, Air Force officials described a program where feedback collected early in technology design and development was not incorporated into the design of the DE prototype. Specifically, program officials described collecting user feedback from pilots during the design phase of an airborne HEL technology demonstrator. Officials said that, during this simulation of airborne lasers, pilots reported positive feedback on features of the design and capability. For example, the magazine depth meant they did not have to stop to reload the weapons. However, DOD officials also reported that pilots expressed concerns about the need to fly low and steady while using the laser, which made them more vulnerable. Air Force officials said that, although they understood the concern, they did not factor that feedback into design because the transition partner, the Air Force Air Combat Command, had already provided design requirements for the demonstrator. Officials explained that the effort was a technology demonstration and that the program was planning a fiscal year 2024 flight test.[29] However, because the Air Force does not consistently document how it collects, tracks, and

[29] The Air Force subsequently decided not to proceed with the flight demonstration and ended development due to cost increases and other technology challenges.

incorporates feedback on design efforts, the military department risks developing DE designs that do not meet the needs of the ultimate end-user.

Conclusion

DOD has long pursued DE weapons but could do more to facilitate the transition of these efforts. DE could provide considerable advantages to the warfighter to address challenging threats, and DOD has invested significant time and resources into developing prototypes. However, the Air Force and Navy have not taken certain steps that could support transition. Specifically, the Air Force does not consistently identify transition partners, and neither the Air Force nor the Navy documents transition agreements to support the transition of DE technologies. Having documented transition agreements with partners helps align technology development with warfighter needs, which ultimately helps ensure DE weapons are developed to meet warfighter needs. Additionally, the Navy and the Air Force do not consistently document how they collect, track, and incorporate user feedback into prototype efforts. Consistently documenting how they collect, track, and incorporate feedback into prototype design and development will better ensure significant issues, such as those that affect how a DE weapon can be used on a mission, are addressed early to align with warfighter needs, and support future DE weapon transition efforts.

Recommendations for Executive Action

We are making a total of four recommendations, including two to the Navy and two to the Air Force. Specifically:

The Secretary of the Navy should ensure the Assistant Secretary of the Navy for Research, Development and Acquisition, for efforts the Navy expects to transition to an acquisition program, develop transition agreements between prototype developers and identified transition partners within the first year of a DE prototype project, or as soon as possible. (Recommendation 1)

The Secretary of the Air Force should ensure the Assistant Secretary of the Air Force for Acquisition, Technology, and Logistics, for efforts the Air Force expects to transition to an acquisition program, identify transition partners early and develop transition agreements between prototype developers and identified transition partners within the first year of a DE prototype project, or as soon as possible. (Recommendation 2)

The Secretary of the Navy should ensure the Assistant Secretary of the Navy for Research, Development and Acquisition documents how the Navy collects, tracks, and incorporates DE prototype user feedback during development and testing. (Recommendation 3)

The Secretary of the Air Force should ensure the Assistant Secretary of the Air Force for Acquisition, Technology, and Logistics documents how the Air Force collects, tracks, and incorporates DE prototype user feedback during development and testing. (Recommendation 4)

Agency Comments and Our Evaluation

We provided a draft of this report to DOD in February 2023 for review and comment. DOD provided written comments in March 2023 that we reproduced in appendix II and summarized below. DOD concurred with three recommendations and partially concurred with one recommendation.

DOD concurred with our second, third, and fourth recommendations. DOD partially concurred with our first recommendation to the Navy to develop transition agreements between prototype developers and identified transition partners. While the comments indicate that DOD partially concurred, the comments did not clearly state what aspects of our recommendation they did not concur with. Rather, they noted that the Navy has some degree of coordination built into the way they have organized their directed energy efforts.

We acknowledge that the way the Navy has organized development of DE weapons has facilitated collaboration among developers and transition partners within PEO Integrated Warfare Systems. However, to support eventual transition to an acquisition program when technology is ready, the Navy can better articulate a pathway by documenting plans with key decision makers such as the Surface Warfare Directorate, including the ways that the transition partners plan to identify requirements and transition funding responsibilities.

We are sending copies of this report to the appropriate congressional committees, the Secretary of Defense, the Under Secretary of Defense for Research and Development, the Secretary of the Army, the Secretary of the Air Force, and the Secretary of the Navy. In addition, the report is available at no charge on the GAO website at https://www.gao.gov.

If you or your staff have any questions about this report, please contact me at (202) 512-4841 or ludwigsonj@gao.gov. Contact points for our Offices of Congressional Relations and Public Affairs may be found on the last page of this report. GAO staff who made key contributions to this report are listed in appendix III.

Jon Ludwigson
Director, Contracting and National Security Acquisitions

List of Committees

The Honorable Jack Reed
Chairman

The Honorable Roger Wicker
Ranking Member
Committee on Armed Services
United States Senate

The Honorable Jon Tester
Chair

The Honorable Susan Collins
Ranking Member Subcommittee on Defense Committee on Appropriations
United States Senate

The Honorable Mike Rogers
Chairman

The Honorable Adam Smith
Ranking Member
Committee on Armed Services
House of Representatives

The Honorable Ken Calvert
Chair

The Honorable Betty McCollum
Ranking Member
Subcommittee on Defense Committee on Appropriations
House of Representatives

Appendix I: Objectives, Scope, and Methodology

This report (1) describes the status of Department of Defense (DOD) and military department Directed Energy (DE) weapon efforts; and (2) assesses the challenges with transitioning DE efforts from prototyping and actions the military departments have taken to mitigate the challenges.

For this engagement, we reviewed DOD and military departments' DE development efforts, as described in DOD's 2021 Directed Energy Roadmap. To understand the status of DOD and the Army, Navy, and Air Force's (military departments) DE weapons development efforts, we reviewed DE planning documents, such as the Roadmap and budget documents for fiscal years 2021 and 2022. We used the Roadmap to identify current and planned DE efforts, the status of development, and responsible military departments. We reviewed program briefs for DE efforts within each military department as well as DOD policy on the operational deployment of DE weapons. We also interviewed officials responsible for creating and overseeing the policy.

We interviewed relevant officials from the military departments responsible for the development and fielding of DE weapons. We also interviewed officials responsible for fielding and requirements in the Marine Corps; however, because the Marine Corps does not have independent development efforts, we did not include them in this review. In addition, we interviewed officials from DOD components, such as the Office of the Under Secretary of Defense for Research and Engineering; the Office of the Under Secretary of Defense for Acquisition and Sustainment; Director, Operational Test and Evaluation's Center for Counter Measures; and the Test Resource Management Center. Other DOD agencies, such as the Missile Defense Agency and the Defense Advanced Research Agency, have some DE-related science and technology development efforts. However, we did not include those efforts in our review because DOD reported they are nascent, or the efforts and funding are being moved to other DOD entities.

We conducted site visits to military department installations and contractors responsible for developing DE weapons. Specifically, we met with science and technology, program, and other relevant officials from each military department. In addition, we toured laboratories and viewed prototypes in Virginia for the Navy, in New Mexico for the Air Force, and Alabama for the Army. We also toured White Sands Missile Range to discuss DE testing efforts in an operational testing environment. Finally, we observed a DE weapon prototype installed on a U.S. Navy ship and spoke with operators to better understand issues related to training on, operating, and maintaining the device. We met multiple contractor representatives and toured facilities doing DE development, including both prime and subcontractors, and large and small contractors.

To evaluate challenges associated with transitioning DE efforts, we selected a non-generalizable sample of seven total DE efforts. The selection included efforts from each military department, and both HEL and HPM efforts. Additional selection criteria included the type of technology, such as power levels and the intended use environment. For these selected efforts, we reviewed, if available, transition plans and other documentation describing schedules, funding, and development activities. We then interviewed military department officials responsible for the efforts to gain an understanding of transition plans, and any challenges encountered. We also interviewed the acquisition program office and resource sponsor officials to gain an understanding of transition challenges and the efforts underway to address those challenges. We identified several challenges associated with transitioning DE efforts.

To assess the extent of the challenges, we evaluated the military departments' transition planning for the selected DE efforts based on the DOD Prototyping Guidance from 2019 and 2022 and our prior work on technology transition.[30] We also reviewed documentation and interviewed military department officials to understand the process for soliciting and incorporating warfighter feedback into DE weapon development.

We conducted this performance audit from February 2022 to April 2023 in accordance with generally accepted government auditing standards.

Those standards require that we plan and perform the audit to obtain sufficient, appropriate evidence to provide reasonable basis for our findings and conclusions based on our audit objectives. We believe that the evidence obtained provides a reasonable basis for our findings and conclusions based on our audit objectives.

[30] GAO-16-5.

Appendix II: Comments from the Department of Defense

OFFICE OF THE ASSISTANT SECRETARY OF DEFENSE
3600 DEFENSE PENTAGON
WASHINGTON, DC 20301-3600

ACQUISITION

MAR 2 9 2023

Mr. Jon Ludwigson
Director, Contracting and National Security Acquisitions
U.S. Government Accountability Office
441 G Street, NW
Washington, DC 20548

Dear Mr. Ludwigson,

This is the Department of Defense (DoD) response to the GAO Draft Report GAO-23-105868, "DIRECTED ENERGY WEAPONS: DOD Should Focus on Transition Planning," dated February 22, 2023 (GAO Code 105868). Detailed comments on the report recommendations are enclosed.

Sincerely,

Tanya M. Skeen

Tanya M. Skeen
Performing the Duties of Assistant Secretary
of Defense for Acquisition

Enclosure(s):
As Stated

GAO DRAFT REPORT DATED FEBRUARY 22, 2023
GAO-23-105868 (GAO CODE 105868)

"DIRECTED ENERGY WEAPONS: DOD Should Focus on Transition Planning"

DEPARTMENT OF DEFENSE COMMENTS
TO THE GAO RECOMMENDATIONS

The Secretary of the Navy should ensure the Assistant Secretary of the Navy for Research, Development and Acquisition, for efforts the Navy, expects to transition to an acquisition program, develop transition agreements between prototype developers and identified transition partners within the first year of a DE prototype project, or as soon as possible. (Recommendation 1)

DoD RESPONSE: Partially Concur: The Department agrees that the DoD Prototyping Guidance states that transition planning should start early, and as appropriate, include the transition partner. We also recognize that the Navy has organized directed energy prototyping efforts differently than the other Services. For the Navy, PEO IWS 2 is both the prototype developer and the transition partner. PEO IWS 2. is the Technology Development (BA-4 R&D) to Program of Record (BA-5 R&D, OPN, O&MN) transition/development office for directed energy and works closely with the Office of Naval Research (S&T R&D).

The Secretary of the Air Force should ensure the Assistant Secretary of the Air Force for Acquisition, Technology, and Logistics, for efforts the Air Force expects to transition to an acquisition program, identify transition partners early and develop transition agreements between prototype developers and identified transition partners within the first year of a DE prototype project, or as soon as possible. (Recommendation 2)

DoD RESPONSE: Concur

The Secretary of the Navy should ensure the Assistant Secretary of the Navy for Research, Development and Acquisition, documents how they collect, track, and incorporate DE prototype user feedback during development and testing. (Recommendation 3)

DoD RESPONSE: Concur

The Secretary of the Air Force should ensure the Assistant Secretary of the Air Force for Acquisition, Technology, and Logistics documents how they collect, track, and incorporate DE prototype user feedback during development and testing. (Recommendation 4)

DoD RESPONSE: Concur

Appendix III: GAO Contacts and Staff Acknowledgments

Contacts

Jon Ludwigson at (202) 512-4841 or ludwigsonj@gao.gov.

Staff Acknowledgments

In addition to the contact name above, the following staff members made key contributions to this report: Raj Chitikila (Assistant Director), Mary Diop (Analyst-in-Charge), Rose Brister, Joe Neumeier, Christine Pecora, Andrew Stavisky, Jay Tallon, Alyssa Weir, and Adam Wolfe.

Chapter 7

Hypersonic Weapons: Background and Issues for Congress*

Kelley M. Sayler

Summary

The United States has actively pursued the development of hypersonic weapons—maneuvering weapons that fly at speeds of at least Mach 5—as a part of its conventional prompt global strike program since the early 2000s. In recent years, the United States has focused such efforts on developing hypersonic glide vehicles, which are launched from a rocket before gliding to a target, and hypersonic cruise missiles, which are powered by high-speed, air-breathing engines during flight. As former Vice Chairman of the Joint Chiefs of Staff and former Commander of U.S. Strategic Command General John Hyten has stated, these weapons could enable "responsive, long-range, strike options against distant, defended, and/or time-critical threats [such as road-mobile missiles] when other forces are unavailable, denied access, or not preferred." Critics, on the other hand, contend that hypersonic weapons lack defined mission requirements, contribute little to U.S. military capability, and are unnecessary for deterrence.

Funding for hypersonic weapons has been relatively restrained in the past; however, both the Pentagon and Congress have shown a growing interest in pursuing the development and near-term deployment of hypersonic systems. This is due, in part, to the advances in these technologies in Russia and China, both of which have a number of hypersonic weapons programs and have likely fielded operational hypersonic glide vehicles—potentially armed with nuclear warheads. Most U.S. hypersonic weapons, in contrast to those in Russia and China, are not being designed for use with a nuclear warhead. As a result, U.S. hypersonic weapons will likely require greater accuracy and will be more technically challenging to develop than nuclear-armed Chinese and Russian systems.

The Pentagon's FY2025 budget request for hypersonic research was $6.9 billion—up from $4.7 billion in the FY2023 request. The Pentagon declined to provide a breakout of funding for hypersonic-related research in FY2024, but requested $11 billion for long-range fires—a category that includes hypersonic weapons. The Missile Defense Agency additionally requested $182.3 million for hypersonic defense in FY2025, down from its $190.6 million request in FY2024 and $225.5 million request in FY2023. At present, the Department of Defense (DOD) has not established any programs of record for hypersonic weapons, suggesting that it may not have approved either mission requirements for the systems or long-term funding plans. Indeed, as former Principal Director for Hypersonics (Office of the Under Secretary of Defense for Research and Engineering) Mike White has stated,

* This is an edited, reformatted and augmented version of Congressional Research Service Publication No. R45811, updated August 7, 2024.

In: Emerging Military Technologies
Editor: Jerry Shields
ISBN: 979-8-89530-613-0

DOD has not yet made a decision to acquire hypersonic weapons and is instead developing prototypes to assist in the evaluation of potential weapon system concepts and mission sets. As Congress reviews the Pentagon's plans for U.S. hypersonic weapons programs, it might consider questions about the rationale for hypersonic weapons, their expected costs, and their implications for strategic stability and arms control.
Potential questions include the following:

- What mission(s) will hypersonic weapons be used for? Are hypersonic weapons the most cost-effective means of executing these potential missions? How will they be incorporated into joint operational doctrine and concepts?
- Given the lack of defined mission requirements for hypersonic weapons, how should Congress evaluate funding requests for hypersonic weapons programs or the balance of funding requests for hypersonic weapons programs, enabling technologies, and supporting test infrastructure? Is an acceleration of research on hypersonic weapons, enabling technologies, or hypersonic missile defense options both necessary and technologically feasible?
- How, if at all, will the fielding of hypersonic weapons affect strategic stability?
- Is there a need for risk-mitigation measures, such as expanding New START, negotiating new multilateral arms control agreements, or undertaking transparency and confidence-building activities?

Introduction

The United States has actively pursued the development of hypersonic weapons as a part of its conventional prompt global strike program since the early 2000s.[1] In recent years, it has focused such efforts on hypersonic glide vehicles and hypersonic cruise missiles with shorter and intermediate ranges for use in regional conflicts. Although funding for these programs has been relatively restrained in the past, both the Pentagon and Congress have shown a growing interest in pursuing the development and near-term deployment of hypersonic systems. This is due, in part, to advances in these technologies in Russia and China, leading to a heightened focus in the United States on the strategic threat posed by hypersonic flight. Open-source reporting indicates that both China and Russia have conducted numerous successful tests of hypersonic glide vehicles and likely fielded an operational capability.

Experts disagree on the potential impact of competitor hypersonic weapons on both strategic stability and the U.S. military's competitive advantage. Nevertheless, former Under Secretary of Defense for Research and Engineering (USD[R&E]) Michael Griffin has testified to Congress that the United States does not "have systems which can hold [China and Russia] at risk in a corresponding manner, and we don't have defenses against [their] systems."[2] Although the John S. McCain National Defense Authorization Act for Fiscal Year 2019 (FY2019 NDAA, P.L. 115- 232) accelerated the development of hypersonic weapons, which USD(R&E) identifies as a priority research and development area, the United States is unlikely to field an operational system before FY2025. However, most U.S. hypersonic weapons

---

[1] For details, see CRS Report R41464, Conventional Prompt Global Strike and Long-Range Ballistic Missiles: Background and Issues, by Amy F. Woolf.

[2] U.S. Congress, Senate Committee on Armed Services, "Testimony of Michael Griffin," Hearing on New Technologies to Meet Emerging Threats, April 18, 2018, at https://www.armed-services.senate.gov/imo/media/doc/18- 40_04-18-18.pdf.

programs, in contrast to those in Russia and China, are not being designed for potential use with a nuclear warhead.[3] As a result, U.S. hypersonic weapons will likely require greater accuracy and will be more technically challenging to develop than nuclear-armed Chinese and Russian systems.

In addition to accelerating development of hypersonic weapons, Section 247 of the FY2019 NDAA required that the Secretary of Defense, in coordination with the Director of the Defense Intelligence Agency, produce a classified assessment of U.S. and adversary hypersonic weapons programs, to include the following elements:

1) An evaluation of spending by the United States and adversaries on such technology.
2) An evaluation of the quantity and quality of research on such technology.
3) An evaluation of the test infrastructure and workforce supporting such technology.
4) An assessment of the technological progress of the United States and adversaries on such technology.
5) Descriptions of timelines for operational deployment of such technology.
6) An assessment of the intent or willingness of adversaries to use such technology.[4]

This report was delivered to Congress in July 2019. Similarly, Section 1689 of the FY2019 NDAA requires the Director of the Missile Defense Agency (MDA) to produce a report on "how hypersonic missile defense can be accelerated to meet emerging hypersonic threats."[5] The findings of these reports could hold implications for congressional authorizations, appropriations, and oversight.

The following report reviews the hypersonic weapons programs in the United States, Russia, and China, providing information on the programs and infrastructure in each nation, based on unclassified sources. It also provides a brief summary of the state of global hypersonic weapons research development. It concludes with a discussion of the issues that Congress might address as it considers the Department of Defense's (DOD's) funding requests for U.S. hypersonic technology programs.

Background

Several countries are developing hypersonic weapons, which fly at speeds of at least Mach 5 (five times the speed of sound).[6] There are two primary categories of hypersonic weapons:

---

[3] Until recently, the United States was not believed to be considering the development of nuclear-armed hypersonic weapons; however, a since-revoked Air Force solicitation sought ideas for a "thermal protection system that can support [a] hypersonic glide to ICBM ranges." Senior defense officials responded to news reports of the revocation, stating that DOD "remains committed to non-nuclear role for hypersonics." See Steve Trimble, "USAF Errantly Reveals Research on ICBM-Range Hypersonic Glide Vehicle," Aviation Week, August 18, 2020, at https://aviationweek.com/defense-space/missile-defense-weapons/ usaf-errantly-reveals-research-icbm-range- hypersonic-glide.

[4] P.L. 115-232, Section 2, Division A, Title II, §247.

[5] P.L. 115-232, Section 2, Division A, Title XVI, §1689.

[6] At a minimum, the United States, Russia, China, Australia, India, France, Germany, and Japan are developing hypersonic weapons technology. See Richard H. Speier et al., *Hypersonic Missile Proliferation: Hindering the Spread of a New Class of Weapons*, RAND Corporation, 2017, at https://www.rand.org/pubs/research_reports/ RR2137.html; and Mike Yeo, "Japan unveils its hypersonic weapons plans," *Defense News,* March 14, 2020.

- *Hypersonic glide vehicles* are launched from a rocket before gliding to a target.[7]
- *Hypersonic cruise missiles* are powered by high-speed, air-breathing engines, or "scramjets," after acquiring their target.

Unlike ballistic missiles, hypersonic weapons do not follow a ballistic trajectory and can maneuver en route to their destination. As former Vice Chairman of the Joint Chiefs of Staff and former Commander of U.S. Strategic Command General John Hyten has stated, hypersonic weapons could enable "responsive, long-range, strike options against distant, defended, and/or time-critical threats [such as road-mobile missiles] when other forces are unavailable, denied access, or not preferred."[8] Conventional hypersonic weapons use only kinetic energy—energy derived from motion—to destroy unhardened targets or, potentially, underground facilities.[9]

Hypersonic weapons could challenge detection and defense due to their speed, maneuverability, and low altitude of flight.[10] For example, terrestrial-based radar cannot detect hypersonic weapons until late in the weapon's flight.[11] Figure 1 depicts the differences in terrestrial-based radar detection timelines for ballistic missiles versus hypersonic glide vehicles.

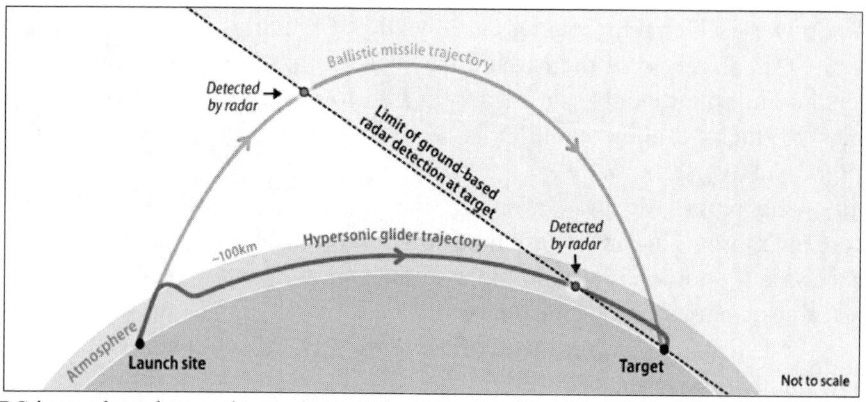

Source: CRS image based on an image in "Gliding Missiles That Fly Faster Than Mach 5 Are Coming," The Economist, April 6, 2019, https://www.economist.com/ science-and-technology/2019/04/06/ gliding-missiles-that-fly-faster-than-mach-5-are-coming.

Figure 1. Terrestrial-Based Detection of Ballistic Missiles vs. Hypersonic Glide Vehicles.

This delayed detection compresses the timeline for decisionmakers assessing their response options and for a defensive system to intercept the attacking weapon—potentially permitting only a single intercept attempt.[12]

---

[7] When hypersonic glide vehicles are mated with their rocket booster, the resulting weapon system is often referred to as a hypersonic boost-glide weapon.

[8] U.S. Congress, Senate Committee on Armed Services, "Testimony of John E. Hyten," Hearing on United States Strategic Command and United States Northern Command, February 26, 2019, at https://www.armed-services.senate.gov/imo/media/doc/Hyten_02-26-19.pdf.

[9] Richard H. Speier et al., Hypersonic Missile Proliferation: Hindering the Spread of a New Class of Weapons, p. 13.

[10] See Department of Defense, *2019 Missile Defense Review*, at https://www.defense.gov/ Portals/1/Interactive/2018/11-2019-Missile-Defense-Review/The%202019%20MDR _Executive%20Summary.pdf.

[11] Richard H. Speier et al., *Hypersonic Missile Proliferation: Hindering the Spread of a New Class of Weapons.*

[12] Bradley Perrett et al., "U.S. Navy Sees Chinese HGV as Part of Wider Threat," *Aviation Week*, January 27, 2014.

Furthermore, U.S. defense officials have stated that both terrestrial- and current space-based sensor architectures are insufficient to detect and track hypersonic weapons, with former USD(R&E) Griffin noting that "hypersonic targets are 10 to 20 times dimmer than what the U.S. normally tracks by satellites in geostationary orbit."[13] Some analysts have suggested that space- based sensor layers—integrated with tracking and fire-control systems to direct high-performance interceptors or directed energy weapons[14]—could theoretically present viable options for defending against hypersonic weapons in the future.[15] Indeed, the *2019 Missile Defense Review* notes that "such sensors take advantage of the large area viewable from space for improved tracking and potentially targeting of advanced threats, including [hypersonic glide vehicles] and hypersonic cruise missiles."[16]

Other analysts have questioned the affordability, technological feasibility, and/or utility of wide- area hypersonic weapons defense.[17] As physicist and nuclear expert James Acton explains, "point- defense systems, and particularly [Terminal High-Altitude Area Defense (THAAD)], could very plausibly be adapted to deal with hypersonic missiles. The disadvantage of those systems is that they can only defend small areas. To defend the whole of the continental United States, you would need an unaffordable number of THAAD batteries."[18] In addition, some analysts have argued that the United States' current command and control architecture would be incapable of "processing data quickly enough to respond to and neutralize an incoming hypersonic threat."[19]

---

[13] David Vergun, "DOD Scaling Up Effort to Develop Hypersonics," *DoD News*, December 13, 2018, at https://dod.defense.gov/News/Article/Article/1712954/dod-scaling-up-effort-to-develop-hypersonics/. See also U.S. Congress, Senate Committee on Armed Services, "Testimony of Michael Griffin," Hearing on New Technologies to Meet Emerging Threats, April 18, 2018, at https://www.armed-services.senate.gov/imo/media/doc/18-40_04-18-18.pdf, and U.S. Congress, Senate Committee on Armed Services, "Testimony of John E. Hyten," Hearing on United States Strategic Command and United States Northern Command, February 26, 2019, at https://www.armed- services.senate.gov/imo/media/doc/ Hyten_02-26-19.pdf.

[14] Section 1664 of the FY2022 NDAA (P.L. 117-81) granted the "Director of the Missile Defense Agency the authority to budget for, direct, and manage directed energy programs applicable for ballistic and hypersonic missile defense missions, in coordination with other directed energy efforts of the Department of Defense."

[15] U.S. Congress, Senate Committee on Armed Services, "Testimony of Michael Griffin," Hearing on New Technologies to Meet Emerging Threats, April 18, 2018, at https://www.armed-services.senate.gov/imo/media/doc/18-40_04-18-18.pdf; and U.S. Congress, Senate Committee on Armed Services, "Testimony of John E. Hyten," Hearing on United States Strategic Command and United States Northern Command, February 26, 2019, at https://www.armed- services.senate.gov/imo/media/doc/Hyten_02-26-19.pdf.

[16] Department of Defense, *2019 Missile Defense Review*, p. XVI, at https://www.defense.gov/Portals/1/Interactive/2018/11-2019-Missile-Defense-Review/The%202019%20MDR_Executive%20Summary.pdf.

[17] See James M. Acton, "Hypersonic Weapons Explainer," Carnegie Endowment for International Peace, April 2, 2018, at https://carnegieendowment.org/2018/04/ 02/hyper sonic-weapons-explainer-pub-75957; and Margot van Loon, "Hypersonic Weapons: A Primer."

[18] Acton, "Hypersonic Weapons Explainer."

[19] Margot van Loon, "Hypersonic Weapons: A Primer" in *Defense Technology Program Brief: Hypersonic Weapons*, American Foreign Policy Council, May 17, 2019. Some analysts have suggested that future command and control systems may require autonomous functionality to manage the speed and unpredictability of hypersonic weapons. See John L. Dolan, Richard K. Gallagher, and David L. Mann, "Hypersonic Weapons Are Literally Unstoppable (as in America Can't Stop Them)," *Real Clear Defense*, April 23, 2019, at https://www.realcleardefense.com/articles/2019/04/23/hypersonic_weapons a_threat_to_national_security_114358.html.

United States

DOD is currently developing hypersonic weapons under the Navy's Conventional Prompt Strike (CPS) program, which is intended to provide the U.S. military with the ability to strike hardened or time-sensitive targets with conventional warheads, as well as through several Air Force, Army, and Defense Advanced Research Projects Agency (DARPA) programs.[20] Those who support these development efforts argue that hypersonic weapons could enhance deterrence, as well as provide the U.S. military with an ability to defeat capabilities such as advanced air and missile defense systems that form the foundation of U.S. competitors' anti-access/area denial strategies.[21] In recognition of this, the 2018 National Defense Strategy identifies hypersonic weapons as one of the key technologies "[ensuring the United States] will be able to fight and win the wars of the future."[22] Similarly, the House Armed Services Committee's bipartisan Future of Defense Task Force Report notes that hypersonic weapons could present challenges to the United States in the years to come.[23]

Programs

Unlike programs in China and Russia, U.S. hypersonic weapons are to be conventionally armed. As a result, U.S. hypersonic weapons will likely require greater accuracy and will be more technically challenging to develop than nuclear-armed Chinese and Russian systems. Indeed, according to one expert, "a nuclear-armed glider would be effective if it were 10 or even 100 times less accurate [than a conventionally armed glider]" due to nuclear blast effects.[24]

According to open-source reporting, the United States is conducting research, development, test, and evaluation (RDT&E) on a number of offensive hypersonic weapons and hypersonic technology programs, including the following (see *Table 1*):

- U.S. Navy—Conventional Prompt Strike (CPS);
- U.S. Navy—Offensive Anti-Surface Warfare Increment 2 (OASuW Inc 2), also known as Hypersonic Air-Launched OASuW (HALO);
- U.S. Army—Long-Range Hypersonic Weapon (LRHW); and
- U.S. Air Force—Hypersonic Attack Cruise Missile (HACM).

---

[20] For a full history of U.S. hypersonic weapons programs, see CRS Report R41464, *Conventional Prompt Global Strike and Long-Range Ballistic Missiles: Background and Issues*, by Amy F. Woolf.

[21] Roger Zakheim and Tom Karako, "China's Hypersonic Missile Advances and U.S. Defense Responses," Remarks at the Hudson Institute, March 19, 2019. See also Department of Defense Fiscal Year (FY) 2020 Budget Estimates, Army Justification Book of Research, Development, Test and Evaluation, Volume II, Budget Activity 4, p. 580.

[22] Department of Defense, "Summary of the 2018 National Defense Strategy of The United States of America," p. 3, at https://dod.defense.gov/Portals/1/Documents/pubs/2018-National-Defense-Strategy-Summary.pdf.

[23] House Armed Services Committee, *Future of Defense Task Force Report 2020*, September 2020, at https://armedservices.house.gov/_cache/files/2/6/26129500-d208-47ba-a9f7-25a8f8 2828b0/424EB2008281A3C79BA8C7EA71890AE9.future-of-defense-task-force-report.pdf.

[24] James M. Acton, "China's Advanced Weapons," Testimony to the U.S. China Economic and Security Review Commission, February 23, 2017, at https://carnegieendowment.org/ 2017/02/23/china-s-advanced-weapons-pub-68095.

These programs are intended to produce operational prototypes, as there are currently no programs of record for hypersonic weapons.[25]

U.S. Navy

In a June 2018 memorandum, DOD announced that the Navy would lead the development of a Common Hypersonic Glide Body for use across the services.[26] The glide body is being adapted from a Mach 6 Army prototype warhead, the Alternate Re-Entry System. The Navy's CPS is expected to pair the glide body with a booster system to create a common All Up Round (AUR) for use by both the Navy and Army. The first test of the AUR, conducted in June 2022, resulted in failure.[27] Two subsequent flight tests, planned for March and September 2023, did not occur due to failed preflight checks.[28] DOD completed a test of the AUR in June 2024.[29]

According to the Navy's FY2025 budget documents, the service intends to deploy CPS on Zumwalt-class destroyers by the end of FY2025.[30] Although Navy officials have previously noted plans to achieve "limited operating capability" on Ohio-class submarines as early as 2025[31] and on Virginia-class submarines by FY2028, as well as to eventually field hypersonic weapons on Burke-class destroyers, such plans are not reflected in current budget documents.[32] The Navy is requesting $903.9 million for CPS RDT&E in FY2025—an increase from the

---

[25] Steve Trimble, "New Long-Term Pentagon Plan Boosts Hypersonics, but Only Prototypes," *Aviation Week*, March 15, 2019, at https://aviationweek.com/defense/new-long-term-pentagon-plan-boosts-hypersonics-only-prototypes.

[26] The services coordinate efforts on a Common Hypersonic Glide Body Board of Directors with rotating chairmanship. Sydney J. Freedberg Jr., "Army Ramps up Funding for Laser Shield, Hypersonic Sword," *Breaking Defense*, February 28, 2020, at https://breakingdefense. com/2020/02/army-ramps-up-funding-for-laser-shield-hypersonic-sword/.

[27] Jon Herskovitz and Anthony Capaccio, "US Hypersonic Missile Fails in Test in Fresh Setback for Program," *Bloomberg*, June 29, 2022, at https://www.bloomberg.com/news/articles/ 2022-06-30/us-hypersonic-missile-fails-in- test-in-fresh-setback-for-program.

[28] Director, Operational Test and Evaluation, *FY2023 Annual Report*, January 2024, p. 168, at https://www.dote.osd. mil/Portals/97/pub/reports/FY2023/other/2023annual-report.pdf?ver= d7gusiIrcbYmxM0oDkPSFg%3d%3d.

[29] See Department of Defense, "DOD Completes Flight Test of Hypersonic Missile," June 28, 2024, at https://www.defense.gov/News/Releases/Release/Article/3821376/dod-completes-flight-test-of-hypersonic-missile/.

[30] Department of Defense Fiscal Year (FY) 2025 Budget Estimates, Navy Justification Book of Research, Development, Test and Evaluation, Volume II, Budget Activity 4, p. 1395, at https://www.secnav. navy.mil/fmc/fmb/Documents/25pres/RDTEN_BA4_Book.pdf. Reports indicate that Zumwalt- class destroyers could carry up to 12 missiles each. See Sam LaGrone, "Navy Awards HII Planning Contract for Zumwalt Hypersonic Upgrades," *USNI News*, January 9, 2023, at https://news.usni.org/2023/01/09/navy-awards-hii-planning-contract-for-zumwalt-hypersonic-upgrades?utm_campaign=dfn-ebb&utm_medium =email&utm_source=sailthru&SToverlay=2002c2d9-c344-4bbb-8610-e5794efcfa7d.

[31] See Department of Defense Fiscal Year (FY) 2022 Budget Estimates, Navy Justification Book of Research, Development, Test and Evaluation, Volume II, Budget Activity 4, p. 1466, at https://www.secnav.navy.mil/ fmc/fmb/Documents/22pres/RDTEN_BA4_Book.pdf; Department of the Navy, "Highlights of the Department of the Navy FY 2021 Budget," February 10, 2020, at https://www.secnav.navy.mil/fmc/fmb/Documents/ 21pres/High lights_book.pdf; and Megan Eckstein, "Navy Says Hypersonic Weapons Coming to Subs in 5 Years," *USNI News*, November 17, 2020, at https://news.usni.org/2020/11/17/navy-says-hypersonic-weapons-coming-to-subs-in-5-years.

[32] David B. Larter, "All US Navy Destroyers Will Get Hypersonic Missiles, Says Trump's National Security Adviser," *Defense News*, October 21, 2020, at https://www. defensenews.com/naval/2020/10/21/all-us-navy-destroyers-will-get- hypersonic-missiles-trumps-national-security-advisor-says/. Budget documents do indicate that the Navy is continuing "non-recurring engineering (NRE) efforts associated with CPS payload hosting on Block V Virginia platforms." See Department of Defense Fiscal Year (FY) 2025 Budget Estimates, Navy Justification Book of Research, Development, Test and Evaluation, Volume II, Budget Activity 4, p. 1378, at https://www.secnav.navy.mil/fmc/fmb/ Documents/25pres/RDTEN_BA4_Book.pdf.

FY2024 request of $901.1 million.[33] The Navy did not request funding for CPS procurement in FY2025.[34]

The Navy is also developing the Offensive Anti-Surface Warfare Increment 2 (OASuW Inc 2), also known as Hypersonic Air-Launched OASuW (HALO)—a new start in FY2023.[35] Although few details about the program have been released publicly, HALO is likely to be compatible with the Navy's F/A-18 fighter jet.[36] The Navy is requesting $178.6 million for HALO RDT&E in FY2025.[37]

U.S. Army

The Army's Long-Range Hypersonic Weapon (LRHW) program, also known as Dark Eagle, is expected to pair the common glide vehicle with the Navy's booster system.[38] The system is intended to have a range of over 1,725 miles and "provide the Army with a prototype strategic attack weapon system to defeat A2/AD capabilities, suppress adversary Long Range Fires, and engage other high payoff/time sensitive targets."[39] The Army is requesting $538 million for LRHW RDT&E in FY2025[40] and $744.2 million for the procurement of LRHW ground support equipment and AURs and canister.[41] The Army has fielded prototype LRHW equipment and

---

[33] Department of Defense Fiscal Year (FY) 2025 Budget Estimates, Navy Justification Book of Research, Development, Test and Evaluation, Volume II, Budget Activity 4, p. 1369, at https://www.secnav.navy.mil/fmc/fmb/Documents/25pres/RDTEN_BA4_Book.pdf.

[34] Department of Defense Fiscal Year (FY) 2025 Budget Estimates, Navy Justification Book of Weapons Procurement, p. 1, at https://www.secnav.navy.mil/fmc/fmb/Documents/ 25pres/WPN_Book.pdf. The Navy's FY2024 request of$304 million was to support the procurement of eight CPS weapons.

[35] Department of Defense Fiscal Year (FY) 2023 Budget Estimates, Navy Justification Book of Research, Development, Test and Evaluation, Volume II, Budget Activity 4, p. 1373, at https://www.secnav.navy.mil/fmc/fmb/Documents/23pres/RDTEN_BA4_Book.pdf. Some reports indicate that, despite its name, HALO may not travel at hypersonic speeds, and thus may not qualify as a hypersonic weapon. See, for example, Jon Harper, "Navy's Future HALO 'Hypersonic' Missile Might not Actually Be Hypersonic," *Defense Scoop*, April 3, 2023, at https://defensescoop.com/2023/04/03/navys-future-halo-hypersonic-missile-might-not-actually-be-hypersonic/. It is included here for reference.

[36] Joseph Trevithick, "Hypersonic Anti-Ship Cruise Missile Has To Be Ready by 2028 Navy Says," *The Drive*, April 23, 2022, at https://www.thedrive.com/the-war-zone/hypersonic-anti-ship-cruise-missile-has-to-be-ready-by-2028- navy-says.

[37] Department of Defense Fiscal Year (FY) 2025 Budget Estimates, Navy Justification Book of Research, Development, Test and Evaluation, Volume II, Budget Activity 4, p. 1285, at https://www.secnav.navy.mil/fmc/fmb/Documents/25pres/RDTEN_BA4_Book.pdf.

[38] For additional information about LRHW, see CRS In Focus IF11991, *The U.S. Army's Long-Range Hypersonic Weapon (LRHW)*, by Andrew Feickert.

[39] Sydney J. Freedberg Jr., "Army Discloses Hypersonic LRHW Range Of 1,725 Miles; Watch Out China," *Breaking Defense*, May 12, 2021, at https://breakingdefense.com/2021/05/ army-discloses-hypersonic-lrhw-range-of-1725-miles-watch-out-china/; https://breakingdefense.com/2019/03/army-sets-2023-hypersonic-flight-test-strategic-cannon- advances/; and Department of Defense Fiscal Year (FY) 2022 Budget Estimates, Army Justification Book of Research, Development, Test and Evaluation, Volume II, Budget Activity 4, p. 639, at https://www.asafm.army.mil/Portals/72/ Documents/Budget Material/2022/Base%20Budget/rdte/RDTE_BA_4_FY_2022_PB.pdf.

[40] Department of Defense Fiscal Year (FY) 2025 Budget Estimates, Army Justification Book of Research, Development, Test and Evaluation, Volume II, Budget Activity 5D, p. 82, at https://www.asafm.army.mil/Portals/72/Documents/BudgetMaterial/2025/Base%20Budget/Research,%20Development,%20Test%20and%20Evaluation/RDTE%20-%20Vol%202% 20-%20Budget%20Activity%205D.pdf; and Department of Defense Fiscal Year (FY) 2025 Budget Estimates, Army Justification Book of Research, Development, Test and Evaluation, Volume II, Budget Activity 4B, p. 264, at https://www.asafm.army.mil/Portals/ 72/Documents/BudgetMaterial/2025/Base%20Budget/Research,%20Development,%20Test%20and%20Evaluation/RDTE%20-%20Vol%202%20-%20Budget%20Activity%204B.pdf.

[41] Department of Defense Fiscal Year (FY) 2025 Budget Estimates, Army Justification Book of Missile Procurement, p. 104, at https://www.asafm.army.mil/Portals/72/Documents/ BudgetMaterial/2025/Base%20Budget/Procurement/Missile-Procurement-Army.pdf.

"intends to field two additional batteries of LRHW" by FY2027.[42] Army officials have previously expressed plans to transition LRHW to a program of record in the fourth quarter of FY2024—a timeline that officials have termed "very, very aggressive" and that would require the program to take on "a lot of risk."[43]

U.S. Air Force

The AGM-183 Air-Launched Rapid Response Weapon (ARRW, pronounced "arrow") was to leverage DARPA's Tactical Boost Glide (TBG) technology to develop an air-launched hypersonic glide vehicle prototype capable of travelling at average speeds of between Mach 6.5 and Mach 8 at a range of approximately 1,000 miles.[44] ARRW successfully completed a "captive carry" test flight in June 2019. It then experienced three successive failures before completing three successful flight tests in 2022.[45] Although the first test of the full operational ARRW prototype in December 2022 was successful, ARRW's flight testing record since then appears to have been mixed, with at least one 2023 test flight failure.[46] The Air Force declined to comment on the outcome of a second 2023 test, noting only that it "gained valuable new insights into [ARRW's] capabilities."[47] Following the March 2023 failure, Secretary of the Air Force Frank Kendall stated that the Air Force is "more committed to HACM at this point in time than [it is] to ARRW."[48] The Air Force conducted its final test of ARRW in March 2024 but

---

[42] Director, Operational Test and Evaluation, *FY2023 Annual Report*, January 2024, p. 123, at https://www.dote.osd.mil/Portals/97/pub/reports/FY2023/other/2023annual-report.pdf?ver= d7gusiIrcbYmxM0oDkPSFg%3d%3d.

[43] Jon Harper, *DefenseScoop*, October 10, 2022, at https://defensescoop.com/2022/10/10/army-assuming-a-lot-of-risk- as-it-moves-to-field-hypersonic-weapons-by-end-of-fiscal-year/. See also Department of Defense Fiscal Year (FY) 2023 Budget Estimates, Army Justification Book of Research, Development, Test and Evaluation, Volume II, Budget Activity 4, pp. 705-709, at https://www.asafm.army.mil/Portals/72/Documents/Budget Material/2023/Base%20Budget/ rdte/vol_2-Budget_Activity_4.pdf.

[44] ARRW is expected to be launched initially from the B-52H strategic bomber. Thomas Newdick, "Air Force Says New Hypersonic Missile Will Hit Targets 1,000 Miles Away in Under 12 Minutes," *The Drive*, October 13, 2020, at https://www.thedrive.com/the-war-zone/37045/air-force-says-new-hypersonic-missile-will-hit-targets-1000-miles-away-in-under-12-minutes.

[45] Oriana Pawlyk, "Air Force's Hypersonic ARRW Missile Fails First Flight Test," *Military.com*, April 6, 2021, at https://www.military.com/daily-news/2021/04/06/air-forces-hypersonic-arrw-missile-fails-first-flight-test.html#:~:text=In%20June%202019%2C%20the%20 service,early%202020s%2C%20the%20release%20states; John A. Tirpak, "Hypersonic ARRW Flies Successfully for Second Time, Completing Booster Tests," *Air Force Magazine*, July 13, 2022, at https://www.airforcemag.com/hypersonic-arrw-flies-success fully-for-second-time-completing-booster-tests/.

[46] Ilka Cole, "Air Force Conducts First ARRW Operational Prototype Missile Test," U.S. Air Force, December 12, 2022, at https://www.af.mil/News/Article-Display/Article/3243194/ air-force-conducts-first-arrw-operational-prototype- missile-test/; and Stephen Losey, "US Air Force Fires Hypersonic ARRW in First Test Since March Failure," *C4ISRNet*, August 21, 2023, at https://www.c4isrnet.com/battlefield-tech/space/2023/08/21/us-air-force-fires-arrw- hypersonic-in-first-test-since-march-failure/.

[47] Stephen Losey, "US Air Force Fires Hypersonic ARRW in First Test Since March Failure," *C4ISRNet*, August 21, 2023, at https://www.c4isrnet.com/battlefield-tech/space/2023/08/ 21/us-air-force-fires-arrw-hypersonic-in-first-test- since-march-failure/.

[48] John A. Tirpak, "Kendall: Air Force 'More Committed' to HACM After Latest Unsuccessful ARRW Test," *Air and Space Forces Magazine*, March 28, 2023, at https://www. airandspaceforces.com/kendall-air-force-hacm-unsuccessful- arrw-test/.

also declined to discuss the results of that test.[49] The Air Force did not request funds for ARRW in FY2025 and budget documents characterized the program as "completed."[50]

In February 2020, the Air Force announced that it had cancelled its second hypersonic weapon program, the Hypersonic Conventional Strike Weapon (HCSW), which had been expected to use the common glide vehicle and booster system, due to budget pressures that forced it to choose between ARRW and HCSW.[51] Then-Air Force acquisition chief Will Roper explained that ARRW was selected because it was more advanced and gave the Air Force additional options. "[ARRW] is smaller; we can carry twice as many on the B-52, and it's possible it could be on the F-15," he explained.[52] A senior Air Force official has since noted that a B-52 could potentially carry four ARRWs.[53]

Finally, in FY2022, the Air Force launched the Hypersonic Attack Cruise Missile (HACM) program to develop a hypersonic cruise missile that integrates Air Force and DARPA technologies.[54] Some reports indicate that HACM is intended to be launched from both bombers and fighter aircraft,[55] with a senior Air Force official noting that a B-52 could potentially carry 20 HACMs or more.[56] According to the Air Force, "the ability to execute HACM development is contingent upon fully funded and successful predecessor capability development efforts."[57] The Air Force requested $517 million for HACM in FY2025, up from the $382 million request in FY2024.[58]

The Air Force is also developing the Expendable Hypersonic Air-Breathing Multi-Mission Demonstrator Program, alternatively known as Project Mayhem. According to then-Principal Director for Hypersonics Mike White, "Project Mayhem is to look at the next step in what the opportunity space allows relative to hypersonic cruise missile systems" and is intended to be

---

49 See, for example, Michael Marrow, "Air Force conducts final test of ARRW hypersonic missile, won't discuss 'specific' results," *Breaking Defense*, March 20, 2024, at https://breakingdefense.com/2024/03/air-force-conducts-final-test-of-arrw-hypersonic-missile-wont-discuss-specific-results/.

50 Department of Defense Fiscal Year (FY) 2025 Budget Estimates, Air Force Justification Book of Research, Development, Test and Evaluation, Volume II, p. 183, at https://www.saffm. hq.af.mil/LinkClick. aspx?fileticket=jQCmIF-YLMg%3d&portalid=84.

51 Valerie Insinna, "US Air Force Kills One of Its Hypersonic Weapons Programs," *Defense News*, February 10, 2020, at https://www.defensenews.com/smr/federal-budget/2020/02 /10/the-air-force-just-canceled-one-of-its-hypersonic- weapons-programs/.

52 John A. Tirpak, "Roper: The ARRW Hypersonic Missile Better Option for USAF," *Air Force Magazine*, March 2, 2020, at https://www.airforcemag.com/arrw-beat-hcsw-because-its-smaller-better-for-usaf/. Tirpak additionally notes that "the F-15 could accelerate the ARRW to Mach 3 before launch, potentially reducing the size of the booster needed to get the weapon to hypersonic speed."

53 John A. Tirpak, "Air Force Will Try Again to Launch ARRW Hypersonic Missile in July," *Air Force Magazine*, June 3, 2021, at https://www.airforcemag.com/air-force-july-launch-arrw-hypersonic-missile/.

54 According to Air Force budget documents, "the program leverages Southern Cross Integrated Flight Research Experiment (SCIFiRE) investment, a bi-lateral U.S./Australian air-breathing hypersonic cruise missile prototyping effort which is a prelude to HACM." Department of Defense Fiscal Year (FY) 2025 Budget Estimates, Air Force Justification Book of Research, Development, Test and Evaluation, Volume II, p. 191, at https://www.saffm.hq.af.mil/LinkClick.aspx?fileticket=jQCmIF-YLMg%3d&portalid=84.

55 FY2023 Air Force budget documents note that "the HACM program will prioritize integration on the F-15E platform to enable quick entry into flight test."

56 John A. Tirpak, "Air Force Will Try Again to Launch ARRW Hypersonic Missile in July," *Air Force Magazine*, June 3, 2021, at https://www.airforcemag.com/air-force-july-launch-arrw-hypersonic-missile/.

57 Department of Defense Fiscal Year (FY) 2022 Budget Estimates, Air Force Justification Book of Research, Development, Test and Evaluation, Volume II, p. 148, at https://www. saffm.hq.af.mil/Portals/84/documents/FY22/RDTE_/FY22%20DAF%20J-Book%20-%203600%20-%20AF%20RDT%20and%20E%20Vol%20II.pdf?ver=KpJJbVq68o32d Svkjuv_Iw%3d%3d.

58 Department of Defense Fiscal Year (FY) 2025 Budget Estimates, Air Force Justification Book of Research, Development, Test and Evaluation, Volume II, p. 191, at https://www.saffm. hq.af.mil/LinkClick.aspx?fileticket=jQCmIF-YLMg%3d&portalid=84.

capable of flying "significantly longer ranges than what we're doing today."[59] Some reports indicate that Project Mayhem may be developing an uncrewed hypersonic bomber capable of flying at Mach 10 and performing both strike and intelligence, surveillance, and reconnaissance missions.[60]

DARPA

DARPA, in partnership with the Air Force, conducted tests of TBG, a wedge-shaped hypersonic glide vehicle capable of Mach 7+ flight that "[aimed] to develop and demonstrate technologies to enable future air-launched, tactical-range hypersonic boost glide systems."[61] TBG "also [considered] traceability, compatibility, and integration with the Navy Vertical Launch System" and is planned to transition to both the Air Force and the Navy. DARPA did not request funds for TBG in FY2025, describing the program as "completed."[62]

DARPA's Operational Fires reportedly sought to leverage TBG technologies to develop a ground- launched system that will enable "advanced tactical weapons to penetrate modern enemy air defenses and rapidly and precisely engage critical time sensitive targets." OpFires completed its first flight test in July 2022.[63] The OpFires program concluded in FY2022.[64]

DARPA has similarly concluded work on the Hypersonic Air-breathing Weapon Concept (HAWC), which, with Air Force support, sought "to develop and demonstrate critical technologies to enable an effective and affordable air-launched hypersonic cruise missile."[65] DARPA successfully tested HAWC in March and July 2022 and in January 2023, launching the missile from a B-52 bomber.[66] Former Principal Director for Hypersonics Mike White has stated that hypersonic cruise missiles like HAWC would be smaller than hypersonic glide

---

[59] Mike White, Remarks at the Center for Strategic and International Studies, "Hypersonic Strike and Defense: A Conversation with Mike White," June 10, 2021, at https://www.csis.org/ analysis/hypersonic-strike-and-defense- conversation-mike-white.

[60] See, for example, Darren Orf, "Project Mayhem, the Air Force's Secret Hypersonic Bomber, Has Begun Cooking," *Popular Mechanics*, January 20, 2023, at https://www.popular mechanics.com/military/aviation/a42589676/air-forces-secret-hypersonic-bomber-project-mayhem-updates/; and Howard Altman and Joseph Trevithick, "Future Of Mayhem Hypersonic Strike-Recon Aircraft Program Murky," *The War Zone*, February 16, 2024, at https://www.twz.com/news-features/future-of-mayhem-hypersonic-strike-recon-aircraft-program-murky.

[61] "Tactical Boost Glide (TBG) Program Information," DARPA, https://www.darpa.mil/ program/tactical-boost-glide; and Guy Norris, "U.S. Air Force Plans Road Map to Operational Hypersonics," *Aviation Week*, July 27, 2017, at https://aviationweek.com/ defense/us-air-force-plans-road-map-operational-hypersonics.

[62] Department of Defense Fiscal Year (FY) 2025 Budget Estimates, Defense Advanced Research Projects Agency, Defense-Wide Justification Book 1 of 5, p. 167, at https://comptroller. defense.gov/Portals/45/Documents/ defbudget/FY2025/budget_justification/pdfs/03_RDT_and_E/R DTE_Vol1_DARPA_MasterJustificationBook_PB_2025.pdf.

[63] DARPA, "Operational Fires Program Successfully Completes First Flight Test," July 13, 2022, at https://www.darpa.mil/news-events/2022-07-13a.

[64] Department of Defense Fiscal Year (FY) 2023 Budget Estimates, Defense Advanced Research Projects Agency, Defense-Wide Justification Book 1 of 5, p. 172, at https://comptroller. defense.gov/Portals/45/Documents/ defbudget/fy2023/budget_justification/pdfs/03_RDT_and_E/RDTE_Vol1_DARPA_MasterJustificationBook_P B_2023.pdf.

[65] "Hypersonic Air-breathing Weapon Concept (HAWC) Program Information," DARPA, at https://www.darpa.mil/ program/hypersonic-air-breathing-weapon-concept.

[66] Oren Liebermann, "US Tested Hypersonic Missile in mid-March but Kept It Quiet to Avoid Escalating Tensions with Russia," *CNN*, April 5, 2022, at https://us.cnn.com/2022/ 04/04/politics/us-hypersonic-missile-test/index.html; Courtney Albon, "Raytheon Hypersonic Scramjet Missile Has Another Successful Flight Test," *Defense News*, July 19, 2022, at https://www.defensenews.com/battlefield-tech/2022/07/19/darpas-hypersonic-scramjet-missile-logs-another-flight-test-success/?utm_source=sailthru&utm_medium= email&utm_campaign=dfn-ebb&SToverlay=2002c2d9-c344-4bbb-8610-e5794efcfa7d; and DARPA, "Final Flight of HAWC Program Screams Through the Sky," January 30, 2023, at https://www.darpa.mil/news-events/2023-01-30.

vehicles and could therefore launch from a wider range of platforms. Former Principal Director White has additionally noted that HAWC and other hypersonic cruise missiles could integrate seekers more easily than hypersonic glide vehicles.[67] HAWC's successor program, More Opportunities with HAWC (MOHAWC), similarly sought to develop technologies for use in future air-launched hypersonic cruise missiles.[68] DARPA did not request funds for MOHAWC in FY2025, describing the program as "completed."[69]

Table 1. Summary of selected U.S. hypersonic weapons RDT&E funding

Title	FY2024 Request ($ in millions)	FY2024 Enacted ($ in millions)	PB2025 ($ in millions)	Schedule
Conventional Prompt Strike (CPS)	901	901	904	Platform deployment in FY2025
Hypersonic Air-Launched OASuW (HALO)	96	96	179	Achieve early operational capability by FY2029
Long-Range Hypersonic Weapon (LRHW)	943	943	538	Field two operational batteries by FY2027
Hypersonic Attack Cruise Missile (HACM)	382	382	517	Complete test and development in FY2027; continue follow-on development through FY2029

Source: Program information taken from U.S. Navy, Army, Air Force, and DARPA FY2024 and FY2025 Justification Books, available at https://comptroller.defense. gov/Budget-Materials/.
Note: MOHAWC, a new start in FY2023, is the successor program to HAWC, which concluded in 2023.

Table 2. Summary of U.S. hypersonic weapons procurement funding

Title	FY2024 Request ($ in millions)	FY2024 Enacted ($ in millions)	PB2025 ($ in millions)	Status
CPS	341	341	0	FY2024 request was to procure eight AURs
LRHW	157	157	744	Request would procure LRHW ground support equipment and eight AURs plus canister

Source: Program information taken from U.S. Navy and Army FY2024 and FY2025 Justification Books, available at https://comptroller.defense.gov/Budget-Materials/.

---

67 "Department of Defense Press Briefing on Hypersonics," March 2, 2020, at https://www.defense.gov/Newsroom/Transcripts/Transcript/Article/2101062/department-of-defense-press-briefing-on-hypersonics/.

68 Ibid.

69 Department of Defense Fiscal Year (FY) 2025 Budget Estimates, Defense Advanced Research Projects Agency, Defense-Wide Justification Book 1 of 5, p. 168, at https://comptroller. defense.gov/Portals/45/Documents/defbudget/FY2025/budget_justification/pdfs/03_RDT_and_E/RDTE_Vol1_DARPA_MasterJustification Book_PB_2025.pdf.

Hypersonic Missile Defenses[70]

DOD is also investing in counter-hypersonic weapons capabilities. In September 2018, MDA—which in 2017 established a Hypersonic Defense Program pursuant to Section 1687 of the FY2017 NDAA (H.Rept. 114-840)—commissioned 21 white papers to explore hypersonic missile defense options, including interceptor missiles, hypervelocity projectiles, laser guns, and electronic attack systems.[71] In January 2020, MDA issued a draft request for prototype proposals for a Hypersonic Defense Regional Glide Phase Weapons System interceptor intended to be fielded in the mid-2030s; however, the program was later cancelled in favor of an alternative solution, the Glide Phase Intercept (GPI).[72] According to MDA FY2024 budget documents, the agency sought to field a regional, sea-based GPI capability in FY2034.[73] Section 1666 of the FY2024 NDAA (P.L. 118-31) directed MDA to accelerate this timeline to achieve initial operational capability by December 31, 2029, and full operational capability by December 31, 2032; however, MDA's FY2025 budget documents state that GPI is to be delivered in FY2035.[74] On May 15, 2024, MDA announced that it had formalized a Cooperative Development Project Arrangement to co-develop GPI with Japan.[75]

In addition, MDA is developing the Hypersonic and Ballistic Tracking Space Sensor (HBTSS) in an effort to improve the agency's ability to detect and track incoming missiles.[76] MDA requested $76 million for HBTSS in FY2025 and $182.3 million for hypersonic defense.[77] Finally, DARPA is working on a program called Glide Breaker, which "will develop critical component technology to support a lightweight vehicle designed for precise engagement of

---

[70] For additional information about hypersonic missile defense, see CRS In Focus IF11623, *Hypersonic Missile Defense: Issues for Congress*, by Jennifer DiMascio and Kelley M. Sayler.

[71] H.Rept. 114-840, Section 2, Division A, Title XVI, §1687; Hudson and Trimble, "Top U.S. Hypersonic Weapon Program"; and Steve Trimble, "A Hypersonic Sputnik?" p. 21.

[72] Missile Defense Agency, "Draft Request for Prototype Proposal: Hypersonic Defense Regional Glide Phase Weapon System," January 30, 2020, p. 8; and Steve Trimble, "MDA Unveils GPI In Retooled Counter-Hypersonic Plan," *Aviation Week*, February 4, 2021, at https://aviationweek.com/defense-space/missile-defense-weapons/mda-unveils- gpi-retooled-counter-hypersonic-plan.

[73] Steve Trimble, "MDA Unveils GPI In Retooled Counter-Hypersonic Plan," *Aviation Week*, February 4, 2021, at https://aviationweek.com/defense-space/missile-defense-weapons/mda-unveils-gpi-retooled-counter-hypersonic-plan.

[74] Department of Defense Fiscal Year (FY) 2025 Budget Estimates, Missile Defense Agency, Defense-Wide Justification Book 2a of 5, p. 643, at https://comptroller.defense.gov/ Portals/45/Documents/defbudget/FY2025/budget_justification/pdfs/03_RDT_and_E/R DTE_Vol2_MDA_RDTE_PB25_Justification_Book.pdf.

[75] Per the arrangement, MDA is to "provide hypersonic missile defense capability during the glide-phase portion of hypersonic flight," while Japan is to "lead development of rocket motors and propulsion components of GPI." DOD, "U.S. Department of Defense Statement on the Signing of the Glide Phase Interceptor Cooperative Development," May 15, 2024, at https://www.defense.gov/News/Releases/Release/Article/3775546/us-department-of-defense-statement-on- the-signing-of-the-glide-phase-intercepto/.

[76] Department of Defense Fiscal Year (FY) 2023 Budget Estimates, Missile Defense Agency, Defense-Wide Justification Book 2a of 5, p. 841, at https://comptroller.defense.gov/ Portals/45/Documents/defbudget/fy2023/budget_justification/pdfs/03_RDT_and_E/RDTE_Vol2_MDA_RDTE_PB23_Justification_Book.pdf; and *Defense Budget Overview: United States Department of Defense Fiscal Year 2023 Budget Request*, Office of the Under Secretary of Defense (Comptroller)/Chief Financial Officer, April 2022, p. 2-15, at https://comptroller.defense.gov/Portals/45/Documents/defbudget/FY2023/FY2023_Budget_Request_Overview_Book.pdf; and Sandra Erwin, "Pentagon agencies team up in upcoming launch of hypersonic tracking satellites," *Space News*, December 28, 2023, at https://spacenews.com/pentagon-agencies-team-up-in-upcoming-launch-of-hypersonic-tracking-satellites/.

[77] Department of Defense Fiscal Year (FY) 2025 Budget Estimates, Missile Defense Agency, Defense-Wide Justification Book 2a of 5, pp. 643 and 852, at https://comptroller. defense.gov/Portals/45/Documents/defbudget/FY2025/budget_justification/pdfs/03_RDT_and_E/R DTE_Vol2_MDA_RDTE_PB25_Justification_Book.pdf.

hypersonic threats at very long range."[78] DARPA requested $38 million for Glide Breaker in FY2025.[79]

Infrastructure

According to a study mandated by the FY2013 National Defense Authorization Act (P.L. 112-239) and conducted by the Institute for Defense Analyses (IDA),[80] the United States had 48 critical hypersonic test facilities and mobile assets in 2014 needed for the maturation of hypersonic technologies for defense systems development through 2030.[81] These specialized facilities, which simulate the unique conditions experienced in hypersonic flight (e.g., speed, pressure, heating),[82] included 10 DOD hypersonic ground test facilities, 11 DOD open-air ranges, 11 DOD mobile assets, 9 National Aeronautics and Space Administration (NASA) facilities, 2 Department of Energy (DOE) facilities, and 5 industry or academic facilities.[83] In its 2014 evaluation of U.S. hypersonic test and evaluation infrastructure, IDA noted that "no current U.S. facility can provide full-scale, time-dependent, coupled aerodynamic and thermal-loading environments for flight durations necessary to evaluate these characteristics above Mach 8."

Since the 2014 study report was published, there have been a number of changes in U.S. hypersonic test infrastructure. For example, the University of Notre Dame has opened Mach 6 and Mach 10 quiet wind tunnels, Purdue University has opened a Mach 8 quiet wind tunnel, and at least one hypersonic testing facility has been inactivated.[84] In addition, the University of Arizona modified one of its wind tunnels to enable Mach 5 testing, while Texas A&M University—in partnership with Army Futures Command—is constructing a kilometer-long Mach 10 wind tunnel.[85] The United States also uses the Royal Australian Air Force Woomera

---

[78] Department of Defense Fiscal Year (FY) 2021 Budget Estimates, Defense Advanced Research Projects Agency, Defense-Wide Justification Book 1 of 5, p. 164.

[79] Department of Defense Fiscal Year (FY) 2025 Budget Estimates, Defense Advanced Research Projects Agency, Defense-Wide Justification Book 1 of 5, p. 160, at https://comptroller. defense.gov/Portals/45/Documents/defbudget/FY2025/budget_justification/pdfs/03_RDT_and_E/RDTE_Vol1_DARPA_MasterJustificationBook_PB_2025.pdf.

[80] P.L. 112-239, Section 2, Division A, Title X, §1071.

[81] A more recent report by the Government Accountability Office states that there are "26 DOD, DOE, NASA, and private U.S. wind tunnel facilities capable of supporting hypersonic research." Government Accountability Office, *Hypersonic Weapons: DOD Should Clarify Roles and Responsibilities to Ensure Coordination Across Development Efforts*, GAO-21-378, March 22, 2021, p. 15, at https://www.gao.gov/products/gao-21-378.

[82] These conditions additionally require the development of specialized materials such as metals and ceramics.

[83] This list is taken directly from a 2014 Institute for Defense Analysis report and, therefore, may not be current. See (U//FOUO) Paul F. Piscopo et al., (U) Study on the Ability of the U.S. Test and Evaluation Infrastructure to Effectively and Efficiently Mature Hypersonic Technologies for Defense Systems Development: Summary Analysis and Assessment, Institute for Defense Analyses, September 2014. Permission to use this material has been granted by the Office of Science and Technology Policy.

[84] Jessica Sieff, "University of Notre Dame Adds Two New Hypersonics Research Facilities," *Notre Dame News*, June 6, 2022, at https://news.nd.edu/news/university-of-notre-dame-adds-two-new-hypersonics-research-facilities/; and Evamarie Socha, "Purdue Applied Research Institute Opens $41M Hypersonics and Applied Research Facility," *Purdue University News*, June 7, 2023, at https://www.purdue.edu/newsroom/releases/2023/Q2/purdue-applied-research-institute-opens-41m-hypersonics-and-applied-research-facility.html.

[85] University of Arizona, "Mach 5 Quiet Ludwieg Tube," at https://transition.arizona.edu/facilities/qlt5?_ga=2.62515882.768526379.1582843192-983632914.1582843192; and Ashley Tressel, "Army to Open Hypersonic Testing Facility at Texas A&M," Inside Defense, October 13, 2019, https://insidedefense.com/daily-news/army-open- hypersonic-testing-facility-texas-am. Additional universities such as the University of Maryland, the California Institute of Technology, the Georgia Institute of Technology,

Test Range in Australia and the Andøya Rocket Range in Norway for flight testing.[86] (For an illustrative list of U.S. hypersonic test assets and their capabilities, see the *Appendix.*)

In February 2022, DOD's Office of Inspector General announced that it had concluded its two- year-long evaluation of current ground test and evaluation facilities to determine if the capability and capacity would be sufficient to execute DOD's planned test schedule; however, DOD did not release the evaluation to the public.[87] Similarly, the *FY2022 Director, Operational Test & Evaluation (DOT&E) Annual Report* evaluated the sufficiency of U.S. hypersonic weapons test infrastructure.[88] The DOT&E report concluded that "additional missile test range modernization efforts are needed to support an increase in the tempo of testing and the development of new capabilities to measure hypersonic missile flight performance in increasingly complex threat environments."[89] Congress appropriated $47.5 million to USD(R&E) and DOT&E in FY2022 for hypersonic test infrastructure; however, the *FY2023 DOT&E Annual Report* notes that at least one hypersonic weapon program's "flight test schedule [was still] continually challenged due to the limited availability and numbers of hypersonic flight corridors, target areas, and test support assets."[90] Congress may consider whether additional funds would be required to address DOT&E's FY2022 recommendation.

DOD reportedly plans to expand hypersonic test infrastructure in the coming years. In January 2019, the Navy announced plans to reactivate its Launch Test Complex at China Lake, CA, to improve air launch and underwater testing capabilities for the CPS program.[91] DOD has also announced the development of the Multi-Service Advanced Capability Hypersonics Test Bed (MACH-TB), which is to "increase domestic capacity for hypersonic flight testing and leverage multiple commercially-available launch vehicles for ride-along hypersonic

---

the Air Force Academy, the University of Tennessee, and Virginia Polytechnic Institute and State University also maintain experimental hypersonic facilities or conduct hypersonic research.

86 (U//FOUO) Paul F. Piscopo et al., *(U) Study on the Ability of the U.S. Test and Evaluation Infrastructure.* For example, the Government Accountability Office reports that HACM is to be tested "both in Australia from Australian Air Force F-18s, and in the United States from F-15Es." See Government Accountability Office, *Hypersonic Weapons: DOD Could Reduce Cost and Schedule Risks by Following Leading Practices*, July 2024, p. 14, at https://www.gao.gov/assets/gao-24-106792.pdf.

87 See Department of Defense Office of Inspector General, "Memorandum for Distribution: Evaluation of the Ground Test and Evaluation Infrastructure Supporting Hypersonic Capabilities (Project No. D2020-DEV0SN-0106.000)," April 13, 2020, at https://media. defense.gov/2020/Apr/14/2002280826/-1/-1/1/D2020-DEV0SN-0106.000. PDF; and Department of Defense Office of Inspector General," Evaluation of the Ground Test and Evaluation Infrastructure Supporting Hypersonic Capabilities (DODIG-2022-056)," February 3, 2022, at https://www.dodig.mil/reports.html/Article/2921419/evaluation-of-the-ground-test-and-evaluation-infrastructure-supporting-hyperson/.

88 DOT&E, *FY2022 DOT&E Annual Report*, January 2023, pp. 18-19, at https://www.dote.osd.mil/Portals/97/pub/reports/FY2022/FY22DOTEAnnualReport.pdf?ver=UBO7t2O1FkRuvrB-nJDZ-g%3d%3d.

89 Ibid., p. 18.

90 Ibid., p. 19; and *DOT&E, FY2022 DOT&E Annual Report*, January 2024, pp. 18-19, at https://www.dote.osd.mil/Portals/97/pub/reports/FY2023/other/2023annual-report.pdf? ver=d7gusiIrcbYmxM0oDkPSFg%3d%3d.

91 "Update: US Navy to Develop China Lake to Support CPS Weapon Testing," *Jane's* (subscription required), February 12, 2019, at https://janes.ihs.com/Janes/Display/ FG_1644858-JMR.

payloads."[92] DOD successfully tested components of MACH-TB in November 2023.[93] According to an assessment conducted by the Government Accountability Office, DOD has dedicated approximately $1 billion to hypersonic facility modernization from FY2015 to FY2024.[94]

Congress has also continued to express interest in hypersonic weapons infrastructure. Section 222 of the FY2021 NDAA (P.L. 116-283) required the Under Secretary of Defense for Research and Engineering, in consultation with the Director of Operational Test and Evaluation, to submit to the congressional defense committees "an assessment of the sufficiency of the testing capabilities and infrastructure used for fielding hypersonic weapons, and a description of any investments in testing capabilities and infrastructure that may be required to support in-flight and ground-based testing for such weapons."[95] Section 225 of the FY2022 NDAA (P.L. 117-81) requires the Secretary of Defense to identify the hypersonic facilities and capabilities of the Major Range and Test Facility Base and brief the congressional defense committees on a plan for improvement.

Similarly, Section 237 of the FY2023 NDAA (P.L. 117-263) directs the Secretary of Defense to both assess DOD's capacity to test and evaluate hypersonic capabilities and "[identify] test facilities outside the Department of Defense that have potential to be used to expand [DOD] capacity ... including test facilities of other departments and agencies of the Federal Government, academia, and commercial test facilities." Section 218 of the FY2024 NDAA (P.L. 118-31) directs the Secretary to update this assessment at least once every two years. It additionally directs the Secretary to conduct a study to evaluate at least two possible locations in the United States that "have potential to be used as additional corridors for long-distance hypersonic system testing" and to submit to the congressional defense committees an annual report on DOD funding and investments in hypersonic capabilities.

Finally, in March 2020, DOD announced that it had established a "hypersonic war room" to assess the U.S. industrial base for hypersonic weapons and identify "critical nodes" in the supply chain.[96] DOD has also amended its "5000 series" acquisition policy in order to enhance supply chain resiliency and reduce sustainment costs.[97]

---

[92] U.S. Department of Defense, "DoD Announces New Contract to Increase Hypersonic Flight Testing Tempo," October 6, 2022, at https://www.defense.gov/News/Releases/Release/ Article/3182305/dod-announces-new-contract-to-increase-hypersonic-flight-testing-tempo/. According to a Dynetics press release, Dynetics is to lead a MACH-TB team composed of over 20 partners, including Peraton, Kratos Defense & Security Solutions, Stratolaunch, JRC Integrated Systems, NineTwelve Institute, Corvid, SpinLaunch, Varda, Kitty Hawk Technologies, Systima Division of Karman Space and Defense, Sandia National Laboratories, Oak Ridge National Laboratory, X-Bow Systems, RLNS and other hypersonic experts. See PRNewswire, "Dynetics Awarded New Contract to Increase Hypersonic Flight Testing Tempo," October 20, 2022, at https://www.prnewswire.com/news-releases/dynetics-awarded-new-contract-to-increase-hypersonic-flight-testing-tempo-301654753.html.

[93] U.S. Navy, "Department of Defense Demonstrates Advanced Hypersonic Technologies," November 16, 2023, at https://www.navy.mil/Press-Office/News-Stories/Article/3591504/department-of-defense-demonstrates-advanced - hypersonic-technologies/.

[94] Government Accountability Office, *Hypersonic Weapons: DOD Should Clarify Roles and Responsibilities to Ensure Coordination Across Development Efforts*, GAO-21-378, March 22, 2021, p. 27, at https://www.gao.gov/products/gao- 21-378.

[95] This report was delivered to the committees on December 16, 2021.

[96] Aaron Mehta, "Pentagon Launches Hypersonic Industrial Base Study," *Defense News*, March 3, 2020, at https://www.defensenews.com/pentagon/2020/03/02/pentagon-launches-hyper sonic-industrial-base-study/.

[97] C. Todd Lopez, "Rewrite of Acquisition Regulation Helps U.S. Build Hypersonic Arsenal More Quickly," *DOD News*, October 30, 2020, at https://www.defense.gov/Explore/ News/Article/Article/2400205/rewrite-of-acquisition-regulation-helps-us-build-hypersonic-arsenal-more-quickly/.

Russia

Although Russia has conducted research on hypersonic weapons technology since the 1980s, it accelerated its efforts in response to U.S. missile defense deployments in both the United States and Europe, and in response to the U.S. withdrawal from the Anti-Ballistic Missile Treaty in 2001.[98] Detailing Russia's concerns, President Putin stated that "the US is permitting constant, uncontrolled growth of the number of anti-ballistic missiles, improving their quality, and creating new missile launching areas. If we do not do something, eventually this will result in the complete devaluation of Russia's nuclear potential. Meaning that all of our missiles could simply be intercepted."[99] Russia thus seeks hypersonic weapons, which can maneuver as they approach their targets, as an assured means of penetrating U.S. missile defenses and restoring its sense of strategic stability.[100]

Programs

Russia is pursuing two hypersonic weapons programs—the Avangard and the 3M22 Tsirkon (or Zircon)—and has reportedly fielded the Kinzhal ("Dagger"), a maneuvering air-launched ballistic missile.[101]

Avangard (Figure 2) is a hypersonic glide vehicle launched from an intercontinental ballistic missile (ICBM), giving it "effectively 'unlimited' range."[102] Reports indicate that Avangard is currently deployed on the SS-19 Stiletto ICBM, though Russia plans to eventually launch the vehicle from the Sarmat ICBM. Sarmat reportedly entered combat duty in September 2023.[103] Avangard features onboard countermeasures and will reportedly carry a nuclear warhead. It was successfully tested twice in 2016 and once in December 2018, reportedly reaching speeds of Mach 20; however, an October 2017 test resulted in failure. Russian news sources claim that Avangard entered into combat duty in December 2019.[104]

---

[98] United Nations Office of Disarmament Affairs, *Hypersonic Weapons: A Challenge and Opportunity for Strategic Arms Control*, February 2019, at https://www.un.org/ disarmament/publications/more/hypersonic-weapons-a-challenge-and-opportunity-for-strategic-arms-control/.

[99] Vladimir Putin, "Presidential Address to the Federal Assembly," March 1, 2018, at http://en.kremlin.ru/events/president/news/56957.

[100] In this instance, "strategic stability" refers to a "bilateral nuclear relationship of mutual vulnerability." See Tong Zhao, "Conventional Challenges to Strategic Stability: Chinese Perceptions of Hypersonic Technology and the Security Dilemma," Carnegie-Tsinghua Center for Global Policy, July 23, 2018, at https://carnegietsinghua.org/2018/07/23/conventional-challenges-to-strategic-stability-chinese-perceptions-of-hypersonic-technology-and-security-dilemma-pub-76894.

[101] Although the Kinzhal is a maneuvering air-launched ballistic missile rather than a hypersonic glide vehicle or hypersonic cruise missile, it is often included in reporting of Russia's hypersonic weapons program. For this reason— and because it poses defensive challenges that are similar to other hypersonic weapons—it is included here for reference.

[102] Steve Trimble, "A Hypersonic Sputnik?" *Aviation Week*, January 14-27, 2019, p. 20.

[103] Al Jazeera, "Russia Puts Advanced Sarmat Nuclear Missile System on 'Combat Duty,'" September 2, 2023, at https://www.aljazeera.com/news/2023/9/2/russia-puts-advanced-sarmat-nuclear-missile-system-on-combat-duty#:~:text=Russia%20test%2Dfired%20the%20Sarmat,in%20Russia's%20far%20east%20region. Sarmat could reportedly accommodate at least three Avangard vehicles. See Malcolm Claus, "Russia unveils new strategic delivery systems," *Jane's* (subscription required), at https://janes.ihs.com/Janes/Display/FG_899127-JIR.

[104] "First Regiment of Avangard Hypersonic Missile Systems Goes on Combat Duty in Russia," *TASS*, December 27, 2019, at https://tass.com/defense/1104297.

In addition to Avangard, Russia is developing Tsirkon, a ship-launched hypersonic cruise missile capable of traveling at speeds of between Mach 6 and Mach 8. Tsirkon is reportedly capable of striking both ground and naval targets. According to Russian news sources, Tsirkon has a maximum range of approximately 625 miles and can be fired from the vertical launch systems mounted on cruisers *Admiral Nakhimov* and *Pyotr Veliky*, Project 20380 corvettes, Project 22350 frigates, and Project 885 Yasen-class submarines, among other platforms.[105] These sources assert that Tsirkon was successfully launched from a Project 22350 frigate in January, October, and December 2020 and May 2022 and from a Project 885 Yasen-class submarine in October 2021.[106] Russia reportedly deployed Tsirkon on the Project 22350 frigate *Admiral of the Fleet of the Soviet Union Gorshkov* in January 2023 and first launched the missile into Ukraine in February 2024.[107]

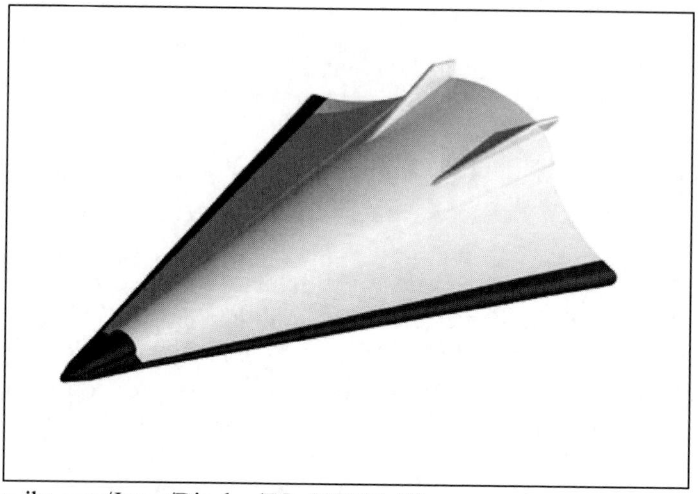

Source: https://janes.ihs.com/Janes/Display/FG_899127-JIR.

Figure 2. Artist Rendering of Avangard.

In addition, Russia has fielded Kinzhal, a maneuvering air-launched ballistic missile modified from the Iskander missile. Russia reportedly fired Kinzhal from a MiG-31 interceptor aircraft in Ukraine[108] and additionally plans to deploy the missile on the Su-34 long-range strike

[105] "Russia Makes over 10 Test Launches of Tsirkon Seaborne Hypersonic Missile," *TASS*, December 21, 2018, at http://tass.com/defense/1037426. See also *Russia Military Power: Building a Military to Support Great Power Aspirations*, Defense Intelligence Agency, 2017, p. 79, at https://www.dia.mil/portals/27/documents/news/military%20power%20 publications/russia%20military%20power%20report%202017.pdf.

[106] "TASS: Russia Conducts First Ship-Based Hypersonic Missile Test," *Reuters*, February 27, 2020, at https://www.voanews.com/europe/tass-russia-conducts-first-ship-based-hypersonic-missile-test; Samuel Cranny-Evans, "Russia conducts first submarine test launches of Tsirkon hypersonic missile," *Jane's* (subscription required), October 4, 2021; and Isabel van Brugen, "Putin to Give Navy Hypersonic Missiles as Russia Beats U.S. in Arms Race," *Newsweek*, July 18, 2022, at https://www.newsweek.com/putin-russian-navy-hypersonic-missiles-zircon-1725426.

[107] Guy Faulconbridge, "Putin Deploys New Zircon Hypersonic Cruise Missiles to Atlantic," *Reuters*, January 4, 2023, at https://www.reuters.com/world/europe/putin-sends-off-frigate-armed-with-new-hypersonic-cruise-missile-2023-01- 04/; and Brad Lendon, "Russia used an advanced hypersonic missile for the first time in recent strike, Ukraine claims," *CNN*, February 13, 2024, at https://www.cnn.com/2024/02/13/europe/ukraine-russia-zircon-hypersonic-missile-intl- hnk-ml/index.html.

[108] Roxana Tiron, "Hypersonic Weapons: Who Has Them and Why It Matters," *Washington Post*, April 6, 2022, at https://www.washingtonpost.com/business/hypersonic-weapons-who-has-them-and-why-it-matters/2022/04/05/1f6d0280-b557-11ec-8358-20aa16355fb4_story.html.

fighter[109] and the Tu-22M3 strategic bomber, although the slower-moving bomber may face challenges in "accelerating the weapon into the correct launch parameters."[110] Russian media has reported Kinzhal's top speed as Mach 10, with a range of up to 1,200 miles when launched from the MiG-31. The Kinzhal is reportedly capable of maneuverable flight, as well as of striking both ground and naval targets, and could eventually be fitted with a nuclear warhead. However, such claims regarding Kinzhal's performance characteristics have not been publicly verified by U.S. intelligence agencies, and have been met with skepticism by a number of analysts.[111]

Infrastructure

Russia reportedly conducts hypersonic wind tunnel testing at the Central Aero-Hydrodynamic Institute in Zhukovsky and the Khristianovich Institute of Theoretical and Applied Mechanics in Novosibirsk, and has tested hypersonic weapons at Dombarovskiy Air Base, the Baykonur Cosmodrome, and the Kura Range.[112]

China

According to Tong Zhao, a fellow at the Carnegie-Tsinghua Center for Global Policy, "most experts argue that the most important reason to prioritize hypersonic technology development [in China] is the necessity to counter specific security threats from increasingly sophisticated U.S. military technology," such as U.S. missile defenses.[113] In particular, China's pursuit of hypersonic weapons, like Russia's, reflects a concern that U.S. hypersonic weapons could enable the United States to conduct a preemptive, decapitating strike on China's nuclear arsenal and supporting infrastructure. U.S. missile defense deployments could then limit China's ability to conduct a retaliatory strike against the United States.[114]

As General Terrence O'Shaughnessy, then-commander of United States Northern Command (USNORTHCOM) and North American Aerospace Defense Command (NORAD), testified in a February 2020 hearing before the Senate Armed Services Committee, China is "testing a [nuclear-capable] intercontinental-range hypersonic glide vehicle" that could evade

---

109 Mark B. Schneider, "Moscow's Development of Hypersonic Missiles … and What It Means" in *Defense Technology Program Brief: Hypersonic Weapons*, American Foreign Policy Council, May 17, 2019.

110 Dave Majumdar, "Russia: New Kinzhal Aero-Ballistic Missile Has 3,000 km Range If Fired from Supersonic Bomber," *The National Interest*, July 18, 2018, at https://nationalinterest. org/blog/buzz/russia-new-kinzhal-aero-ballistic-missile-has-3000-km-range-if-fired-supersonic-bomber.

111 David Axe, "Is Kinzhal, Russia's New Hypersonic Missile, a Game Changer?," *The Daily Beast*, March 15, 2018, at https://www.thedailybeast.com/is-kinzhal-russias-new-hypersonic-missile-a-game-changer.

112 "Aerodynamics," Central Aerohydrodynamic Institute, http://tsagi.com/research/ aerodynamics/; "Russia Announces Successful Flight Test of Avangard Hypersonic Glide Vehicle," *Jane's* (subscription required), January 3, 2019, at https://janes.ihs. com/Janes/Display/FG_1451630-JMR; and "Avangard System Is Tested, Said to Be Fully Ready for Deployment," Russian Strategic Nuclear Forces, December 26, 2018, at http://russianforces.org/blog/2018/12/ avangard_system_is_tested_said.shtml.

113 Tong Zhao, "Conventional Challenges to Strategic Stability: Chinese Perceptions of Hypersonic Technology and the Security Dilemma."

114 Tong Zhao, "Conventional Challenges to Strategic Stability"; and Lora Saalman, "China's Calculus on Hypersonic Glide," Stockholm International Peace Research Institute, August 15, 2017, at https://www.sipri.org/ commentary/topical-backgrounder/2017/chinas-calculus-hypersonic-glide.

U.S. missile defense and warning systems.[115] Reports additionally indicate that China may have tested a nuclear-capable hypersonic glide vehicle[116]—launched by a Long March rocket—in August 2021.[117] In contrast to the ballistic missiles that China has previously used to launch hypersonic glide vehicles, the Long March, a fractional orbital bombardment system (FOBS), launches the hypersonic glide vehicle into orbit before the hypersonic glide vehicle deorbits to its target. This could provide China with a space-based global strike capability and further reduce the amount of target warning time prior to a strike.[118]

China has also demonstrated a growing interest in Russian advances in hypersonic weapons technology, conducting flight tests of a hypersonic glide vehicle only days after Russia tested its own system.[119] Furthermore, a January 2017 report found that over half of open-source Chinese papers on hypersonic weapons include references to Russian weapons programs.[120] This could indicate that China is increasingly considering hypersonic weapons within a regional context.

Indeed, some analysts believe that China may be planning to mate conventionally armed hypersonic glide vehicles with the DF-21 and DF-26 ballistic missiles in support of an anti-access/area denial strategy.[121]

Programs

China has conducted a number of successful tests of the DF-17, a medium-range ballistic missile specifically designed to launch hypersonic glide vehicles. U.S. intelligence analysts assess that the missile has a range of approximately 1,000 to 1,500 miles and may now be deployed.[122] China has also tested the DF-41 ICBM, which could be modified to carry a conventional or nuclear hypersonic glide vehicle, according to a report by a U.S. congressional

---

[115] General Terrence J. O'Shaughnessy, "Statement Before the Senate Armed Services Committee," February 13, 2020, at https://www.armed-services.senate.gov/hearings/20-02-13-united-states-northern-command-and-united-states- strategic-command.

[116] It is not clear if this nuclear-capable hypersonic glide vehicle is the same model as that referenced by General O'Shaughnessy.

[117] Demetri Sevastopulo and Kathrin Hille, "China Tests New Space Capability with Hypersonic Missile," October 16, 2021, at https://www.ft.com/content/ba0a3cde-719b-4040-93cb-a486e1f843fb. China's Foreign Ministry Spokesperson Zhao Lijian has stated that "this was a routine test of [a] space vehicle," rather than a test of a nuclear-capable hypersonic glide vehicle. Zhao Lijian, "Remarks at Regular Press Conference," Ministry of Foreign Affairs of the People's Republic of China, October 18, 2021, at https://www.fmprc.gov.cn/mfa_eng/xwfw_665399/s2510_665401/ t1915130.shtml.

[118] Greg Hadley, "Kendall: China Has Potential to Strike Earth from Space," *Air Force Magazine*, September 20, 2021, at https://www.airforcemag.com/global-strikes-space-china-frank-kendall/.

[119] Lora Saalman, "China's Calculus on Hypersonic Glide."

[120] Lora Saalman, "Factoring Russia into the US-China Equation on Hypersonic Glide Vehicles," SIPRI, January 2017, at https://www.sipri.org/sites/default/files/Factoring-Russia-into-US-Chinese-equation-hypersonic-glide-vehicles.pdf.

[121] Lora Saalman, "China's Calculus on Hypersonic Glide"; and Malcolm Claus and Andrew Tate, "Chinese Hypersonic Programme Reflects Regional Priorities," *Jane's* (subscription required), March 12, 2019, at https://janes.ihs.com/Janes/Display/FG_1731069-JIR.

[122] Ankit Panda, "Introducing the DF-17: China's Newly Tested Ballistic Missile Armed with a Hypersonic Glide Vehicle," *The National Interest*, December 28, 2017, at https://thediplomat.com/2017/12/introducing-the-df-17-chinas-newly-tested-ballistic-missile-armed-with-a-hypersonic-glide-vehicle/; and Bill Gertz, "China's New Hypersonic Missile," *Washington Times*, October 2, 2019, at https://www.washingtontimes.com/news/2019/oct/2/china-shows-df-7-hypersonic-missile/.

commission. The development of the DF-41 thus "significantly increases the [Chinese] rocket force's nuclear threat to the U.S. mainland," the report states.[123]

China has tested the DF-ZF hypersonic glide vehicle (previously referred to as the WU-14) at least nine times since 2014. U.S. defense officials have reportedly identified the range of the DF- ZF as approximately 1,200 miles and have stated that the vehicle may be capable of performing "extreme maneuvers" during flight.[124] China reportedly fielded the DF-ZF in 2020.[125]

According to U.S. defense officials, China also successfully tested Starry Sky-2 (or Xing Kong- 2), a nuclear-capable hypersonic vehicle prototype, in August 2018.[126] China claims the vehicle reached top speeds of Mach 6 and executed a series of in-flight maneuvers before landing.[127] Unlike the DF-ZF, Starry Sky-2 is a "waverider" that uses powered flight after launch and derives lift from its own shockwaves. Some reports indicate that the Starry Sky-2 could be operational by 2025.[128] U.S. officials have declined to comment on the program.[129]

Infrastructure

China has a robust research and development infrastructure devoted to hypersonic weapons. Then-USD(R&E) Michael Griffin stated in March 2018 that China has conducted 20 times as many hypersonic tests as the United States.[130] China tested three hypersonic vehicle models (D18-1S, D18-2S, and D18-3S)—each with different aerodynamic properties—in September 2018.[131] Analysts believe that these tests could be designed to help China develop weapons that fly at variable speeds, including hypersonic speeds. Similarly, China has used the Lingyun Mach 6+ high-speed engine, or "scramjet," test bed (Figure 3) to research thermal resistant components and hypersonic cruise missile technologies.[132]

---

123 *U.S.-China Economic and Security Review Commission 2018 Annual Report*, p. 235, at https://www.uscc.gov/sites/default/files/annual_reports/2018%20Annual%20Report%20to%20Congress.pdf.

124 "Gliding Missiles That Fly Faster Than Mach 5 Are Coming," *The Economist*, April 6, 2019, at https://www.economist.com/science-and-technology/2019/04/06/gliding-missiles-that-fly-faster-than-mach-5-are-coming; and Franz-Stefan Gady, "China Tests New Weapon Capable of Breaching US Missile Defense Systems," *The Diplomat*, April 28, 2016, at https://thediplomat.com/2016/04/china-tests-new-weapon-capable-of-breaching-u-s- missile-defense-systems/.

125 Department of Defense, *Military and Security Developments Involving the People's Republic of China 2021*, p. 60, at https://media.defense.gov/2021/Nov/03/2002885874/-1/-1/0/2021-CMPR-FINAL.PDF.

126 Office of the Secretary of Defense, *Annual Report to Congress: Military and Security Developments Involving the People's Republic of China 2019,* May 2, 2019, p. 44, at https://media.defense.gov/2019/May/02/2002127082/-1/-1/1/2019_CHINA_MILITARY _POWER_REPORT.pdf.

127 Jessie Yeung, "China Claims to Have Successfully Tested Its First Hypersonic Aircraft," *CNN*, August 7, 2018, at https://www.cnn.com/2018/08/07/china/china-hypersonic-aircraft-intl/index.html.

128 U.S.-China Economic and Security Review Commission Report 2015, p. 20.

129 Bill Gertz, "China Reveals Test of New Hypersonic Missile," *The Washington Free Beacon*, August 10, 2018, at https://freebeacon.com/national-security/chinas-reveals-test-new-hyper sonic-missile/.

130 U.S.-China Economic and Security Review Commission Report 2015, p. 20.

131 Malcolm Claus and Andrew Tate, "Chinese Hypersonic Programme Reflects Regional Priorities," Jane's (subscription required), March 12, 2019, at https://janes.ihs.com/ Janes/Display/FG_1731069-JIR.

132 Jeffrey Lin and P.W. Singer, "China's Hypersonic Military Projects Include Spaceplanes and Rail Guns," *Popular Mechanics*, June 26, 2018, at https://www.popsci.com/chinas-hypersonic-work-speeds-up.

Source: Photo accompanying Drake Long, "China Reveals Lingyun-1 Hypersonic Missile at National Science and Technology Expo," The Defense Post, May 21, 2018.

Figure 3. Lingyun-1 Hypersonic Cruise Missile Prototype.

According to Jane's Defence Weekly, "China is also investing heavily in hypersonic ground testing facilities."[133] For example, the China Aerodynamics Research and Development Center claims to have 18 wind tunnels, while the China Academy of Aerospace Aerodynamics is known to operate at least three hypersonic wind tunnels—the FD-02, FD-03, and FD-07—capable of reaching speeds of Mach 8, Mach 10, and Mach 12, respectively.[134] China also operates the JF-12 hypersonic wind tunnel, which reaches speeds of between Mach 5 and Mach 9 and the FD-21 hypersonic wind tunnel, which reaches speeds of between Mach 10 and Mach 15.[135] It reportedly completed construction of the JF-22 wind tunnel, capable of reaching speeds of Mach 30, in 2023.[136] In addition, China is known to have tested hypersonic weapons at the Jiuquan Satellite Launch Center and the Taiyuan Satellite Launch Center.

Issues for Congress

As Congress reviews the Pentagon's plans for U.S. hypersonic weapons programs during the annual authorization and appropriations process, it might consider a number of questions about the rationale for hypersonic weapons, their expected costs, budget and management, and their implications for strategic stability and arms control. This section provides an overview of some of these questions.

---

[133] Andrew Tate, "China Conducts Further Tests with Hypersonic Vehicles," *Jane's Defence Weekly* (subscription required), October 2, 2018, at https://customer.janes.com/Defence Weekly/Display/FG_1120806-JDW.

[134] Kelvin Wong, "China Claims Successful Test of Hypersonic Waverider," *Jane's* (subscription required), August 10, at 2018, https://janes.ihs.com/Janes/Display/FG_1002295-JDW; and Ellen Nakashima and Gerry Shih, "China Builds Advanced Weapons Systems Using American Chip Technology," *Washington Post*, April 9, 2021.

[135] Jeffrey Lin and P.W. Singer, "A Look at China's Most Exciting Hypersonic Aerospace Programs," *Popular Science*, April 18, 2017, at https://www.popsci.com/chinas-hypersonic-technology.

[136] Akhil Kadidal, "Chinese Hypersonic Wind Tunnel Passes Acceptance Check," *Janes Defence Weekly* (subscription required), June 8, 2023, at https://www.janes.com/defence-news/news-detail/chinese-hypersonic-wind-tunnel-passes- acceptance-check.

Global Hypersonic Weapons Programs

Although the United States, Russia, and China possess the most advanced hypersonic weapons programs, a number of other countries—including Australia, India, France, Germany, South Korea, North Korea, and Japan— are also developing hypersonic weapons technology. Since 2007, the United States has collaborated with Australia on the Hypersonic International Flight Research Experimentation (HIFiRE) program to develop hypersonic technologies. The most recent HIFiRE test, successfully conducted in July 2017, explored the flight dynamics of a Mach 8 hypersonic glide vehicle, while previous tests explored scramjet engine technologies. HIFiRE's successor, the Southern Cross Integrated Flight Research Experiment (SCIFiRE) program, is to further develop hypersonic air-breathing technologies. SCIFiRE demonstration tests are expected by the mid-2020s. In addition to the Woomera Test Range facilities—one of the largest weapons test facilities in the world—Australia reportedly operates seven hypersonic wind tunnels and is capable of testing speeds of up to Mach 30.

India has similarly collaborated with Russia on the development of BrahMos II, a Mach 7 hypersonic cruise missile. Although BrahMos II was initially intended to be fielded in 2017, news reports indicate that the program faces significant delays and is now scheduled to achieve initial operational capability between 2025 and 2028. Reportedly, India is also developing an indigenous, dual-capable hypersonic cruise missile as part of its Hypersonic Technology Demonstrator Vehicle program and successfully tested a Mach 6 scramjet in June 2019 and September 2020. India operates approximately 12 hypersonic wind tunnels and is capable of testing speeds of up to Mach 13.

France also has collaborated and contracted with Russia on the development of hypersonic technology. Although France has been investing in hypersonic technology research since the 1990s, it has only recently announced its intent to weaponize the technology. Under the V-max (Experimental Maneuvering Vehicle) program, France is modifying its air-to-surface ASN4G supersonic missile for hypersonic flight, successfully testing the modified missile in June 2023. Some analysts believe that the V-max program is intended to provide France with a strategic nuclear weapon. France operates five hypersonic wind tunnels and is capable of testing speeds of up to Mach 21.

Germany successfully tested an experimental hypersonic glide vehicle (SHEFEX II) in 2012; however, reports indicate that Germany may have pulled funding for the program. German defense contractor DLR continues to research and test hypersonic vehicles as part of the European Union's ATLLAS II project, which seeks to design a Mach 5-6 vehicle. Germany operates three hypersonic wind tunnels and is capable of testing speeds of up to Mach 11.

In addition, South Korea reportedly has been developing a ground-launched Mach 6+ hypersonic cruise missile, Hycore, since 2018. According to *Janes*, South Korea is developing the missile "in response to growing concern about North Korea military modernization" and plans to eventually develop sea- and air-launched variants.

Although North Korea tested the Hwasong-8—which it identifies as a hypersonic glide vehicle—in September 2021, reports indicate that the vehicle may have reached speeds of only Mach 3. Similarly, North Korea claims to have tested a second hypersonic weapon in January 2022; however, experts believe that that weapon may instead be a maneuvering reentry vehicle.

Finally, Japan is developing the Hypersonic Cruise Missile (HCM) and the Hyper Velocity Gliding Projectile (HVGP). It reportedly plans to field HVGPs for area suppression and neutralizing aircraft carriers. A high- supersonic HVGP is expected to enter service in 2026, with a more advanced, hypersonic version available by FY2030; HCM is expected to enter service in 2030. The Japan Aerospace Exploration Agency operates three hypersonic wind tunnels, with two additional facilities at Mitsubishi Heavy Industries and the University of Tokyo. According to DOD, Japan and the United States have agreed to conduct "a joint analysis focused on future cooperation in counter-hypersonic technology."

Other countries—including Iran, Israel, and Brazil—have conducted foundational research on hypersonic airflows and propulsion systems, but may not be pursuing a hypersonic weapons capability at this time. In addition, a number of countries are testing increasingly maneuverable systems that travel at hypersonic speeds but that do not qualify as "hypersonic weapons" as defined in this report.

Note: For information about South Korea's hypersonic weapons programs, see Jon Grevatt and Rahul Udoshi, "South Korea Develops Hycore Hypersonic Cruise Missile," *Janes* (subscription required), January 25, 2022. For information about North Korea's hypersonic weapons programs, see Choi Soo-hyang, "N. Korea's 'Hypersonic Missile' Appears to Be at Early Stage of Development: JCS," Yonhap News Agency, September 29, 2021; and Ankit Panda, "The Real Danger of North Korea's New Hypersonic Missile Is not Its Speed," NK News, January 10, 2022. For information about Japan's hypersonic weapons programs, see Gordon Arthur, "Japan reveals test launch of its hypersonic strike missile program," *Defense News*, July 10, 2024. For additional information about global hypersonic weapons programs, see Richard H. Speier et al., *Hypersonic Missile Proliferation*.

Mission Requirements

Although DOD is funding a number of hypersonic weapons programs, it has not established any programs of record, suggesting that it may not have approved requirements for hypersonic weapons or long-term funding plans.[137] Indeed, as Principal Director for Hypersonics (USD[R&E]) Mike White has stated, DOD has not yet made a decision to acquire hypersonic weapons and is instead developing prototypes to "[identify] the most viable overarching weapon system concepts to choose from and then make a decision based on success and challenges."[138]

Given the lack of mission requirements, DOD officials have expressed a number of competing perspectives about the potential costs and intended quantities of U.S. hypersonic weapons. For example, Secretary of the Air Force Frank Kendall has stated that "hypersonics are not going to be cheap anytime soon ... [and thus] we're more likely to have relatively small inventories of [hypersonic missiles] than large ones."[139] Conversely, a number of other senior

[137] Steve Trimble, "New Long-Term Pentagon Plan Boosts Hypersonics."

[138] Steve Trimble, "New Long-Term Pentagon Plan Boosts Hypersonics."

[139] John A. Tirpak, "Only Small Inventories of Hypersonic Missiles in USAF's Future, due to Cost," *Air Force Magazine*, February 15, 2022, at https://www.airforcemag.com/only-small-inventories-of-hypersonic-missiles-in-usafs- future-due-to-cost/.

defense officials have stated that DOD intends to buy large quantities of hypersonic weapons. Then-DOD Director of Defense Research & Engineering Mark Lewis has noted that DOD wants "to deliver hypersonics at scale. That means hundreds of weapons in a short period of time in the hands of the warfighter."[140] Similarly, Principal Director for Hypersonics Mike White has stated that DOD seeks to "[produce] hypersonics in mass, because you have to be able to deliver capability in meaningful numbers, even to defeat the high-end targets."[141] These perspectives appear to be grounded in differing assumptions about the affordability of hypersonic weapons. Likewise, they are likely to hold different implications for the unit cost of the weapons.

As Congress conducts oversight of U.S. hypersonic weapons programs, it may seek to obtain information about DOD's evaluation of potential mission sets for hypersonic weapons, a cost analysis of hypersonic weapons and alternative means of executing potential mission sets, and an assessment of the enabling technologies—such as space-based sensors or autonomous command and control systems—that may be required to employ or defend against hypersonic weapons. For example, Section 1671 of the FY2021 NDAA (P.L. 116-283) directs the chairman of the Joint Chiefs of Staff, in coordination with the Under Secretary of Defense for Policy, to submit to the congressional defense committees a report on strategic hypersonic weapons, including "a description of how the requirements for land and sea-based hypersonic weapons will be addressed with the Joint Requirements Oversight Council, and how such requirements will be formally provided to the military departments procuring such weapons." This report is to additionally include "the potential target sets for hypersonic weapons ... and the required mission planning to support targeting by the United States Strategic Command and other combatant commands."

Congress may also consider the conclusions of a Congressional Budget Office assessment of hypersonic weapons and their alternatives, including the following findings:

- "Both hypersonic and ballistic missiles are well-suited to operate outside potential adversaries' anti-access and area-denial (A2/AD), or 'keep-out,' zones."[142]
- "Hypersonic missiles would probably not be more survivable than ballistic missiles with maneuverable warheads in a conflict, unless the ballistic missiles encountered highly effective long-range defenses."[143]
- "Hypersonic missiles could cost one-third more to procure and field than ballistic missiles of the same range with maneuverable warheads."[144]

---

[140] Sydney J. Freedberg Jr., "Hypersonics: DoD Wants 'Hundreds of Weapons' ASAP," *Breaking Defense*, April 24, 2020, at https://breakingdefense.com/2020/04/hypersonics-dod-wants-hundreds-of-weapons-asap/.

[141] "Hypersonic Strike and Defense: A Conversation with Mike White," Center for Strategic and International Studies, June 10, 2021, at https://www.csis.org/analysis/hypersonic-strike-and-defense-conversation-mike-white. See also Jon Harper, "Just in: Pentagon to Spend Billions Mass-Producing Hypersonic Weapons," *National Defense Magazine*, March 4, 2020, at https://www.nationaldefensemagazine.org/articles/2020/3/4/pentagon-to-spend-billions-mass- producing-hypersonic-weapons.

[142] Congressional Budget Office, *U.S. Hypersonic Weapons and Alternatives*, January 2023, at https://www.cbo.gov/publication/58255.

[143] Ibid.

[144] Ibid.

Funding and Management Considerations

Principal Director for Hypersonics Mike White has noted that DOD is prioritizing offensive programs while it determines "the path forward to get a robust defensive strategy."[145] This approach is reflected in DOD's recent budget requests. For example, DOD requested $182.3 million for the hypersonic defense program element and $6.9 billion for offensive hypersonic weapons programs in FY2025.[146] Similarly, in FY2023, DOD requested $225.5 million for the hypersonic defense program element and $4.7 billion for offensive hypersonic weapons programs.[147] (Although DOD requested $190.6 million for the hypersonic defense program element in FY2024, the department has declined to provide a breakout of funding for offensive hypersonic weapons programs in FY2024.)

Although the Defense Subcommittees of the Appropriations Committees increased FY2020 appropriations for both hypersonic offense and defense above the FY2020 request, they expressed concerns, noting in their joint explanatory statement of H.R. 1158 "that the rapid growth in hypersonic research has the potential to result in stove-piped, proprietary systems that duplicate capabilities and increase costs."[148] To mitigate this concern, they appropriated $100 million for DOD to establish a Joint Hypersonics Transition Office (JHTO) to "develop and implement an integrated science and technology roadmap for hypersonics" and "establish a university consortium for hypersonic research and workforce development" in support of DOD efforts.[149]

DOD established the JHTO in April 2020 and announced on October 26, 2020, that it awarded Texas A&M University with a $20 million contract—renewable for up to $100 million—to manage a University Consortium for Applied Hypersonics (UCAH).[150] UCAH is to be overseen by a group of academic researchers from Texas A&M University, the Massachusetts Institute of Technology, the University of Minnesota, the University of Illinois

---

[145] Aaron Mehta, "Is the Pentagon Moving Quickly Enough on Hypersonic Defense?" *Defense News*, March 21, 2019, at https://www.defensenews.com/pentagon/2019/03/21/is-the-pentagon-moving-quickly-enough-on-hyper sonic- defense/.

[146] DOD has noted that the total request for all programs related to hypersonic defense was $1.9 billion in FY2025. Department of Defense Fiscal Year (FY) 2025 Budget Estimates, Missile Defense Agency, Defense-Wide Justification Book 2a of 5, p. 643, at https:// comptroller.defense.gov/Portals/45/Documents/defbudget/FY2025/ budget_justification/pdfs/03_RDT_and_E/R DTE_Vol2_MDA_RDTE_PB25_Justification_Book.pdf; and CRS correspondence with the Office of the Under Secretary of Defense (Comptroller), March 13, 2024.

[147] Department of Defense Fiscal Year (FY) 2023 Budget Estimates, Missile Defense Agency Defense-Wide Justification Book Volume 2a of 5, p. 631, at https://comptroller.defense. gov/Portals/45/Documents/ defbudget/fy2023/budget_justification/pdfs/03_RDT_and_E/RDTE_Vol2_MDA_RDTE_PB23_Justification_ Book.pdf; and Defense Budget Overview: United States Department of Defense Fiscal Year 2023 Budget Request, Office of the Under Secretary of Defense (Comptroller)/Chief Financial Officer, April 2022, p. 2-16, at https://comptroller.defense.gov/Portals/45/Documents/defbudget/FY2023/FY2023_Budget_Request_Over view_Book.pdf.

[148] "Department of Defense Appropriations Act, 2020: Joint Explanatory Statement," Defense Subcommittees of the Appropriations Committees, December 16, 2019, at https:// appropriations.house.gov/sites/democrats. appropriations.house.gov/files/HR%201158%20-%20Division%20A%20-%20Defense%20SOM%20FY20.pdf.

[149] Ibid. The Joint Hypersonic Transition Office, then called the Joint Technology Office on Hypersonics, was originally mandated by Section 218 of the FY2007 NDAA (P.L. 109-364). The office was redesignated as the Joint Hypersonics Transition Office and given additional authorities in Section 214 of the FY2018 NDAA (P.L. 115-91). Section 216 of the FY2020 NDAA (P.L. 116-92) further amended the office's authorities to include the ability to enter into agreements with institutions of higher learning. The office went unfunded until FY2020 and was not established until April 2020.

[150] David Vergun, "DOD Awards Applied Hypersonics Contract to Texas A&M University," *DOD News*, October 26, 2020, at https://www.defense.gov/Explore/News/Article/ Article/2394438/dod-awards-applied-hypersonics-contract-to-texas-am-university/.

at Urbana-Champaign, the University of Arizona, the University of Tennessee Space Institute, Morgan State University, the California Institute of Technology, Purdue University, the University of California-Los Angeles, and the Georgia Institute of Technology.[151] The consortium is to "facilitate transitioning academic research into developing systems [as well as] work with the department to reduce system development timelines while maintaining quality control standards."[152]

In addition, Section 1671 of the FY2021 NDAA (P.L. 116-283) directs the Secretary of the Army and the Secretary of the Navy to jointly submit to the congressional defense committees a report on LRHW and CPS, including total costs of the programs, "the strategy for such programs with respect to manning, training, and equipping, including cost estimates, [and] a testing strategy and schedule for such programs." It directs the Director of Cost Assessment and Program Evaluation to submit to the congressional defense committees an independent cost estimate of these programs.[153]

Given the lack of defined mission requirements for hypersonic weapons, however, it may be challenging for Congress to evaluate the balance of funding for hypersonic weapons programs, enabling technologies, supporting test infrastructure, and hypersonic missile defense.

Industrial Base and Supply Chain

U.S. government officials have expressed ongoing concern about the ability of the industrial base to support future demand for hypersonic weapons—particularly if multiple weapons programs go into production at the same time.[154] Indeed, a July 2022 DOD industry solicitation notes that "the expansion of industrial base capacity is required" [emphasis added] if DOD is to meet its goal of "[producing] the air-breathing engine constituent materials, subcomponents, components, and subsystems to support an initial integrated system production capacity of no less than 48 all-up- round (AUR) missiles (four to five units per month) and up to 72 AURs per year (six per month)."[155]

Furthermore, a DOD report issued in response to Executive Order 14017 ("America's Supply Chains") recommends investments in the hypersonic industrial base.[156] The report notes that DOD is in the process of "developing a hypersonics industrial base roadmap to inform investments over the next five years, which will guide investment decisions over this period.

---

[151] Ibid.

[152] Ibid.

[153] The Government Accountability Office notes DOD's difficulty in developing accurate cost estimates for hypersonic weapons programs. For example, between FY2019 and FY2020, estimates for CPS "almost doubled." Government Accountability Office, *Hypersonic Weapons: DOD Should Clarify Roles and Responsibilities to Ensure Coordination Across Development Efforts*, GAO-21-378, March 22, 2021, p. 21, at https://www.gao.gov/ products/gao-21-378.

[154] See, for example, Justin Katz, "Hypersonics Too Expensive, Industrial Base Too Small For Services to Go It Alone: Admiral," *Breaking Defense*, November 3, 2022, at https://breakingdefense.com/2022/11/hypersonics-too-expensive- industrial-base-too-small-for-services-to-go-it-alone-admiral/.

[155] Department of the Air Force, "Request for Information (RFI) on Supplier Based Initiative for Air-Breathing Engines for Hypersonic Systems," at https://www.businessdefense.gov/ai/dpat3/docs/AirBreathing%20 Enginesfor%20HypersonicsRFFA8650225507.pdf.

[156] Department of Defense, *Securing Defense-Critical Supply Chains: An action plan developed in response to President Biden's Executive Order 14017*, February 2022, at https://media.defense.gov/2022/Feb/24/ 2002944158/-1/-1/1/DOD-EO-14017-REPORT-SECURING-DEFENSE-CRITICAL-SUPPLY-CHAINS.PDF.

The roadmap will address sub-tier supplier development, and where appropriate, develop and retain competition that enables affordable production."[157] The report additionally recommends that DOD "identify partners and allies with capabilities to aid in the development and expansion of [the U.S.] hypersonics supply chain, especially for materials and components where domestic sources may not exist."[158] Congress may wish to conduct oversight of DOD's efforts to strengthen the industrial base and supply chain for hypersonic weapons.

Strategic Stability

Analysts disagree about the strategic implications of hypersonic weapons. Some have identified two factors that could hold significant implications for strategic stability: the weapon's short time of flight—which, in turn, compresses the timeline for response—and its unpredictable flight path—which could generate uncertainty about the weapon's intended target and therefore heighten the risk of miscalculation or unintended escalation in the event of a conflict. This risk could be further compounded in countries that co-locate nuclear and conventional capabilities or facilities.

Some analysts argue that unintended escalation could occur as a result of warhead ambiguity, or from the inability to distinguish between a conventionally armed hypersonic weapon and a nuclear-armed one. However, as a United Nations report notes, "even if a State did know that [a hypersonic glide vehicle] launched toward it was conventionally armed, it may still view such a weapon as strategic in nature, regardless of how it was perceived by the State firing the weapon, and decide that a strategic response was warranted."[159] Differences in threat perception and escalation ladders could thus result in unintended escalation. Such concerns have previously led Congress to restrict funding for CPS programs.[160]

Other analysts have argued that the strategic implications of hypersonic weapons are minimal. Pavel Podvig, a senior research fellow at the United Nations Institute for Disarmament Research, has noted that the weapons "don't … change much in terms of strategic balance and military capability."[161] This, some analysts argue, is because U.S. competitors such as China and Russia already possess the ability to strike the United States with ICBMs, which, when launched in salvos, could overwhelm U.S. missile defenses.[162] Furthermore, these analysts note that in the case of hypersonic weapons, traditional principles of deterrence hold: "it is really a stretch to try to imagine any regime in the world that would be so suicidal that it would even think threating to use—not to mention to actually use—hypersonic weapons against the United States ... would end well."[163]

---

[157] Ibid.

[158] Ibid.

[159] United Nations Office of Disarmament Affairs, *Hypersonic Weapons.*

[160] For a history of legislative activity on conventional prompt global strike, see CRS Report R41464, *Conventional Prompt Global Strike and Long-Range Ballistic Missiles: Background and Issues*, by Amy F. Woolf.

[161] Amy Mackinnon, "Russia's New Missiles Are Aimed at the U.S.," *Foreign Policy*, March 5, 2019, at https://foreignpolicy.com/2019/03/05/russias-new-missiles-are-aimed-at-you-weapons-hypersonic-putin-united-states- inf/.

[162] David Axe, "How the U.S. Is Quietly Winning the Hypersonic Arms Race," *The Daily Beast*, January 16, 2019, at https://www.thedailybeast.com/how-the-us-is-quietly-winning-the-hypersonic-arms-race. See also Mark B. Schneider, "Moscow's Development of Hypersonic Missiles," p. 14.

[163] Jyri Raitasalo, "Hypersonic Weapons Are No Game-Changer," *The National Interest*, January 5, 2019, at https://nationalinterest.org/blog/buzz/hypersonic-weapons-are-no-game-changer-40632.

Section 1671 of the FY2021 NDAA (P.L. 116-283) directs the chairman of the Joint Chiefs of Staff, in coordination with the Under Secretary of Defense for Policy, to submit to the congressional defense committees a report that examines.

How escalation risks will be addressed with regards to the use of strategic hypersonic weapons, including whether any risk escalation exercises have been conducted or are planned for the potential use of hypersonic weapons, and an analysis of the escalation risks posed by foreign hypersonic systems that are potentially nuclear and conventional dual- use capable weapons.

Arms Control

Some analysts who believe that hypersonic weapons could present a threat to strategic stability or inspire an arms race have argued that the United States should take measures to mitigate risks or limit the weapons' proliferation. Proposed measures include expanding New START, negotiating new multilateral arms control agreements, and undertaking transparency and confidence-building measures.[164]

The New START Treaty, a strategic offensive arms treaty between the United States and Russia, does not currently cover weapons that fly on a ballistic trajectory for less than 50% of their flight, as do hypersonic glide vehicles and hypersonic cruise missiles.[165] However, Article V of the treaty states that "when a Party believes that a new kind of strategic offensive arm is emerging, that Party shall have the right to raise the question of such a strategic offensive arm for consideration in the Bilateral Consultative Commission (BCC)." Accordingly, some legal experts hold that the United States could raise the issue in the BCC of negotiating to include hypersonic weapons in the New START limits.[166] However, because New START is due to expire in 2026, this may be a short-term solution.[167]

As an alternative, some analysts have proposed negotiating a new international arms control agreement that would institute a moratorium or ban on hypersonic weapon testing. These analysts argue that a test ban would be a "highly verifiable" and "highly effective" means of preventing a potential arms race and preserving strategic stability.[168] Other analysts have countered that a test ban would be infeasible, as "no clear technical distinction can be made between hypersonic missiles and other conventional capabilities that are less prompt, have shorter ranges, and also have the potential to undermine nuclear deterrence."[169] These analysts have instead proposed international transparency and confidence-building measures, such as

---

[164] See United Nations Office of Disarmament Affairs, *Hypersonic Weapon*; and Richard H. Speier et al., *Hypersonic Missile Proliferation.*

[165] In some cases, hypersonic glide vehicles may be launched from ICBMs that are already covered by New START, as is reported to be the case with Russia's Avangard hypersonic glide vehicle. See Rachel S. Cohen, "Hypersonic Weapons: Strategic Asset or Tactical Tool?"

[166] James Acton notes: "[D]uring [New START] negotiations, Russia argued that boost-glide weapons might constitute 'a new kind of strategic offensive arm,' in which case they would trigger bilateral discussions about whether and how they would be regulated by the treaty—a position [then] rejected by the United States." James M. Acton, *Silver Bullet?: Asking the Right Questions About Conventional Prompt Global Strike*, Carnegie Endowment for International Peace, 2013, p. 139, at https://carnegieendowment.org/files/cpgs.pdf.

[167] CRS Report R41219, The New START Treaty: Central Limits and Key Provisions, by Amy F. Woolf.

[168] Mark Gubrud, "Test Ban for Hypersonic Missiles?" Bulletin of the Atomic Scientists, August 6, 2015, at https://thebulletin.org/roundtable/test-ban-for-hypersonic-missiles/.

[169] Tong Zhao, "Test Ban for Hypersonic Missiles?"

exchanging weapons data; conducting joint technical studies; "providing advance notices of tests; choosing separate, distinctive launch locations for tests of hypersonic missiles; and placing restraints on sea-based tests."[170]

Appendix. U.S. Hypersonic Testing Infrastructure[171]

Table A-1. DOD hypersonic ground test facilities

Facility	Capability	Location
Air Force Arnold Engineering and Development Complex (AEDC) von Karman Gas Dynamics Facility Tunnels A/B/C	Tunnel A: 40-inch Mach 1.5-5.5; up to 290 °F Tunnel B: 50-inch Mach 6 and 8; up to 900 °F Tunnel C: 50-inch Mach 10; up to 1700 °F	Arnold AFB, TN
Air Force AEDC High-Enthalpy Aerothermal Test Arc-Heated Facilities H1, H2, H3	Simulate thermal and pressure environments at speeds of up to Mach 8	Arnold AFB, TN
Air Force AEDC Tunnel 9	59-inch Mach 7, 8,10, 14, and18; up to 2900 °F	White Oak, MD
Air Force AEDC Aerodynamic and Propulsion Test Unit	Mach 3.1-7.2; up to 1300 °F	Arnold AFB, TN
Air Force AEDC Aeroballistic Range G	Launches projectiles of up to 8 inches in diameter at speeds of up to Mach 20	Arnold AFB, TN
Holloman High Speed Test Track	59,971 ft. track; launches projectiles at speeds of up to Mach 8	Holloman AFB, NM
Air Force Research Laboratory (AFRL) Cells 18, 22	Mach 3-7	Wright-Patterson AFB, OH
AFRL Laser Hardened Materials Evaluation Laboratory (LHMEL)	High-temperature materials testing	Wright-Patterson AFB, OH
AFRL Mach 6 High Reynolds Number (Re) Facility	10-inch Mach 6	Wright-Patterson AFB, OH
Test Resource Management Center Hypersonic Aeropropulsion Clean Air Test-bed Facility	Up to Mach 8; up to 4040 °F	Arnold AFB, TN

Source: (U//FOUO) Paul F. Piscopo et al. Air Force AEDC Tunnel 9 was upgraded in 2019 to enable Mach 18 testing. See "Department of Defense Press Briefing on Hypersonics," March 2, 2020, at https://www.defense.gov/ Newsroom/Transcripts/Transcript/Article/2101062/department-of-defense-press-briefing-on-hypersonics/.

[170] Rajaram Nagappa, "Test Ban for Hypersonic Missiles?"; see also James M. Acton, *Silver Bullet?* pp. 134-138.

[171] The following information is largely derived from the 2014 report (U//FOUO) Paul F. Piscopo et al., *(U) Study on the Ability of the U.S. Test and Evaluation Infrastructure*, and therefore, may not be current. Permission to use this material has been granted by the Office of Science and Technology Policy. Additional information has been provided by Dee Howard Endowed Assistant Professor Dr. Christopher S. Combs (The University of Texas at San Antonio).

Table A-2. DOD open-air ranges

Range	Location
Ronald Reagan Ballistic Missile Defense Test Site	Kwajalein Atoll, Republic of the Marshall Islands
Pacific Missile Range Facility (PMRF)	Kauai, HI
Western Range, 30th Space Wing	Vandenberg AFB, CA
Naval Air Warfare Center Weapons (NAWC) Division	Point Mugu and China Lake, CA
White Sands Missile Range (WSMR)	New Mexico
Eastern Range, 45th Space Wing	Cape Canaveral Air Force Station/Patrick AFB/Kennedy Space Center, FL
NASA Wallops Flight Facility	Wallops Island, VA
Pacific Spaceport Complex (formerly Kodiak Launch Complex)	Kodiak Island, AK
NAWC Weapons Division R-2508 Complex	Edwards AFB, CA
Utah Test and Training Range	Utah
Nevada Test and Training Range	Nevada

Source: (U//FOUO) Paul F. Piscopo et al.

Table A-3. DOD mobile assets

Asset
Navy Mobile Instrumentation System
PMRF Mobile At-sea Sensor System
MDA Mobile Instrumentation System Pacific Collector
MDA Mobile Instrumentation System Pacific Tracker
Kwajalein Mobile Range Safety System 2
United States Navy Ship Lorenzen missile range instrumentation ship
Sea-based X-band Radar
Aircraft Mobile Instrumentation Systems
Transportable Range Augmentation and Control System
Re-locatable MPS-36 Radar
Transportable Telemetry System

Source: (U//FOUO) Paul F. Piscopo et al.

Table A-4. NASA research-related facilities

Facility	Capability	Location
Ames Research Center (ARC) Arc Jet Complex	High-temperature materials testing	Mountain View, CA
ARC Hypervelocity Free Flight Facilities	Launches projectiles at speeds of up to Mach 23	Mountain View, CA
Langley Research Center (LaRC) Aerothermodynamics Laboratory	31-inch Mach 10, 20-inch Mach 6, and 15-inch Mach 6	Hampton, VA
LaRC 8-foot High Temperature Tunnel	96-inch Mach 5 and Mach 6.5	Hampton, VA
LaRC Scramjet Test Complex	Up to Mach 8 and up to 4740 °F	Hampton, VA
LaRC HyPulse Facility	Currently inactive	Long Island, NY
Glenn Research Center (GRC) Plumbrook Hypersonic Tunnel Facility Arc Jet Facility	Mach 5, 6, and 7 and up to 3830 °F	Sandusky, OH
GRC Propulsion Systems Laboratory 4	Mach 6	Cleveland, OH
GRC 1' x 1' Supersonic Wind Tunnel	12-inch Mach 1.3-6 (10 discrete airspeeds) and up to 640 °F	Cleveland, OH

Source: (U//FOUO) Paul F. Piscopo et al.

Table A-5. Department of energy research-related facilities

Facility	Capability	Location
Sandia National Laboratories Solar Thermal Test Facility	High-temperature materials testing and aerodynamic heating simulation	Albuquerque, NM
Sandia National Laboratories Hypersonic Wind Tunnel	18-inch Mach 5, 8, and 14	Albuquerque, NM

Source: (U//FOUO) Paul F. Piscopo et al.

Table A-6. Industry/academic research-related facilities

Facility	Capability	Location
CUBRC Large Energy National Shock (LENS)-1/-II/-XX Tunnels	LENS 1: Mach 6-22 LENS II: Mach 2-12 LENS XX: Atmospheric reentry simulation	Buffalo, NY
Boeing Polysonic Wind Tunnel	48-inch up to Mach 5	St. Louis, MO
Lockheed Martin High Speed Wind Tunnel	48-inch Mach .3-5	Dallas, TX
Boeing/Air Force Office of Scientific Research (AFOSR) Quiet Tunnel at Purdue University	9.5-inch Mach 6	West Lafayette, IN
AFOSR-University of Notre Dame Quiet Tunnel	24-inch Mach 6	Notre Dame, IN
Stratolaunch Carrier Aircraft	Reusable Mach 6 test bed	Mojave, CA
University of Texas at San Antonio Hypersonic Ludwieg Tube	8-inch x 8-inch Mach 7.2	San Antonio, TX
University of Texas at Austin Blowdown Wind Tunnel	6-inch x 7-inch Mach 2 & Mach 5	Austin, TX
Southwest Research Light-Gas Gun	Quiet, flight enthalpy ballistic range up to Mach 20	San Antonio, TX
University of Texas at Arlington Aerodynamics Research Center	1.6 MW Mach 2-6 Arc Jet 13-inch Mach 4-16 Shock Tunnel	Arlington, TX
Texas A&M National Aerothermochemistry and Hypersonics Laboratory	7-inch Quiet Mach 6 36-inch Expansion Tunnel 9-inch x 14-inch variable Mach 5-8	College Station, TX
TCalifornia Institute of Technology GALCIT	12-inch Mach 5.2 T5 Reflected Shock Tunnel 6-inch Hypervelocity (up to Mach 7.1) Expansion Tube	Pasadena, CA
University of Arizona Hypersonic Ludwieg Tube	15-inch Mach 5	Tucson, AZ
Air Force Academy Ludwieg Tube	20-inch Mach 6	Colorado Springs, CO
University of Tennessee Space Institute Ludwieg Tube	18-inch x 18-inch Mach 7	Tullahoma, TN
Maryland HyperTERP Reflected Shock Tunnel	12-inch x 12-inch Mach 6	College Park, MD
Florida State Polysonic Wind Tunnel	12-inch x 12-inch Mach 0.2-5	Tallahassee, FL

Facility	Capability	Location
Princeton HyperBLaF Wind Tunnel	9-inch Mach 8	Princeton, NJ

Sources: (U//FOUO) Paul F. Piscopo et al.; Oriana Pawlyk, "Air Force Expanding Hypersonic Technology Testing"; and CRS correspondence with Dee Howard Endowed Assistant Professor Dr. Christopher S. Combs (The University of Texas at San Antonio), October 27, 2022.

Notes: Hypersonic wind tunnels are under construction at the following universities: Texas A&M University (Mach 10 quiet tunnel), Purdue University (Mach 8 quiet tunnel), and the University of Notre Dame (Mach 10 quiet tunnel). Additional universities, such as the University of Maryland, the Georgia Institute of Technology, and Virginia Polytechnic Institute and State University, also maintain experimental hypersonic facilities or conduct hypersonic research.

Chapter 8

U.S. Hypersonic Weapons and Alternatives*

Congressional Buget Office

Artist's rendering of the Air Force's Air-Launched Rapid Response Weapon with the Tactical Boost Glide vehicle at its tip. Lockheed Martin illustration.

At a Glance

The Army, Navy, and Air Force are each developing hypersonic missiles—nonnuclear offensive weapons that fly faster than five times the speed of sound and spend most of their flight in the Earth's atmosphere. Those missiles are intended to be maneuverable and capable of striking targets quickly (in roughly 15 minutes to 30 minutes) from thousands of kilometers away.

* This is an edited, reformatted and augmented version of Congressional Buget Office, nonpartisan Analysis for the US Congress Publication, dated January 2023.

In: Emerging Military Technologies
Editor: Jerry Shields
ISBN: 979-8-89530-613-0

Test launch by the Army and the Navy of the Common Hypersonic Glide Body in March 2020. U.S. Navy photo.

Artist's rendering of the Defense Advanced Research Projects Agency's (DARPA's) Hypersonic Air-Breathing Weapon Concept cruise missile. DARPA illustration.

In this report, the Congressional Budget Office analyzes the hypersonic weapons being developed by the U.S. military and compares them with less expensive existing or potential weapons that might fill similar roles, such as ballistic missiles or cruise missiles. CBO reached the following conclusions:

- *Technological challenges must still be overcome to field hypersonic missiles.* The fundamental remaining challenge involves managing the extreme heat that hypersonic missiles are exposed to by traveling at high speeds in the atmosphere for most of their flight (unlike cruise missiles, which fly in the atmosphere at lower speeds, or ballistic missiles, which mainly fly above the atmosphere). Shielding hypersonic missiles' sensitive electronics, understanding how various materials perform, and predicting aerodynamics at sustained temperatures as high as 3,000° Fahrenheit require extensive flight testing. Tests are ongoing, but failures in recent years have delayed progress.

- *Both hypersonic and ballistic missiles are well-suited to operate outside potential adversaries' antiaccess and area-denial (A2/AD), or "keep-out," zones.* The Department of Defense has developed a strategy to use accurate, long-range, high-speed missiles early in a conflict to neutralize the A2/AD zones being developed by potential adversaries, such as China and Russia. Both hypersonic missiles and ballistic missiles equipped with maneuverable warheads could provide the combination of speed, accuracy, range, and survivability (the ability to reach a target without being intercepted) that would be useful in the military scenarios CBO considered. However, many missions do not require such rapid strikes. For those missions, less costly alternatives to both hypersonic and ballistic missiles exist, including subsonic cruise missiles. Hypersonic weapons would mainly be useful to address threats that were both well-defended and extremely time-sensitive.
- *Hypersonic missiles would probably not be more survivable than ballistic missiles with maneuverable warheads in a conflict, unless the ballistic missiles encountered highly effective long-range defenses.* Hypersonic missiles can neutralize long-range (midcourse) defenses because they fly inside the atmosphere, below the altitude where midcourse ballistic missile defenses typically operate. Hypersonic weapons can also maneuver unpredictably at high speeds to counter short-range defenses near a target, making it harder to track and intercept them. Ballistic missiles are also difficult to defend against, particularly if they are equipped with countermeasures to confuse midcourse missile defenses and maneuverable warheads to defeat short-range missile defenses. Only very effective long-range defenses would be likely to threaten ballistic missiles in midcourse; to date, no potential U.S. adversaries have deployed such defenses.
- *Hypersonic missiles could cost one-third more to procure and field than ballistic missiles of the same range with maneuverable warheads.* CBO estimates that buying 300 ground- or sea-launched, intermediate-range ballistic missiles with maneuverable warheads and sustaining the missile system for 20 years would cost a total of $13.4 billion (in 2023 dollars). The same number of comparable hypersonic missiles would cost about one-third more, $17.9 billion, CBO estimates. (Neither estimate includes the cost overruns that are often associated with technically challenging programs.) The higher costs for hypersonic missiles partly reflect the complexity of building systems that can withstand the heat of hypersonic flight.

Notes

Unless this report indicates otherwise, all years referred to are federal fiscal years, which run from October 1 to September 30 and are designated by the calendar year in which they end.

Numbers in the text, tables, and figures may not add up to totals because of rounding.

Summary

After many decades of conducting basic research, the Department of Defense (DoD) recently increased its spending to develop technology for hypersonic weapons. The Air Force, Army, and Navy all plan to field hypersonic missiles within the next few years. China and Russia have stated that they are also fielding such weapons.

The term "hypersonic weapon" is often misapplied. Strictly speaking, it refers to a missile that travels at speeds at least five times the speed of sound in air (Mach 5 or above) and that spends most of its flight inside the Earth's atmosphere (rather than in space), where it can use aerodynamic design features to maneuver.

The combination of long range, high speed, and maneuverability distinguishes the hypersonic weapons that DoD is developing from the conventional strike weapons in the U.S. military's current inventory. Hypersonic missiles are considered a possible counter to the antiaccess and area-denial (A2/AD) systems that potential near-peer adversaries such as China and Russia are deploying to prevent U.S. forces from operating freely in their regions. In theory, hypersonic weapons could be launched from outside the range of those systems and could reach targets within minutes over medium to intermediate ranges (from hundreds to a few thousands of kilometers), with a high degree of accuracy and less vulnerability to defenses than existing missiles.[536]

The Congressional Budget Office analyzed the relative capabilities and possible costs of hypersonic missiles and potential alternatives in scenarios in which long-range, rapid-response weapons might be useful. CBO found that hypersonic missiles with sufficient ranges for A2/AD scenarios—at least 1,000 kilometers (km), or about 600 miles, for missiles launched from aircraft and at least 3,000 km, or about 1,900 miles, for missiles launched from the ground or sea—have the speed to be useful in the early stages of a conflict with a near-peer adversary.

Hypersonic missiles with those ranges, however, would be more expensive than similar ballistic missiles and pose much greater technical challenges (see Figure S-1). CBO estimates that hypersonic missiles would cost roughly one-third more than ballistic missiles with maneuverable warheads that had the same range and accuracy and traveled at similar speeds. (The United States does not currently field such ballistic missiles, but the technology for them is well developed.)

Hypersonic weapons could be better than ballistic missiles at penetrating long-range missile defenses that operate outside the atmosphere. So far, however, no potential U.S. adversaries have succeeded in developing such defenses. Against shorter-range defenses, it is unclear whether hypersonic missiles would have an advantage over ballistic missiles with maneuverable warheads.

Given their cost, hypersonic weapons would provide a niche capability, mainly useful to address threats that were both well-defended and extremely time-sensitive (requiring a strike in 15 minutes to 30 minutes). If time was not a concern, much cheaper cruise missiles could be used. If targets were time-sensitive but were not protected by defenses that effectively intercept

[536] CBO used kilometers in this analysis because they are the preferred unit for measuring distance in many military and engineering contexts. One kilometer equals 0.62 miles.

incoming ballistic missiles in the middle of their flight, less costly ballistic missiles with maneuverable warheads could be used.

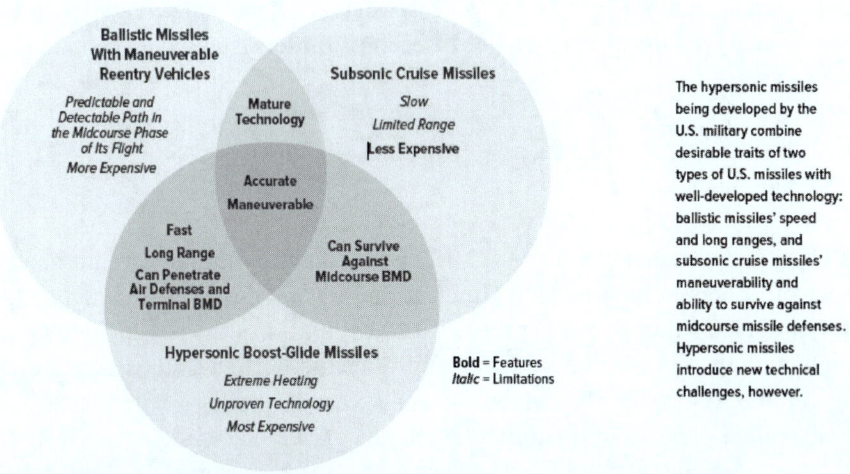

Data source: Congressional Budget Office.
BMD = ballistic missile defense.

Figure S-1. Comparison of the Features and Limitations of Hypersonic Missiles and Alternatives.

How Did CBO Compare Planned Hypersonic Weapons and Possible Alternatives?

To evaluate hypersonic missiles and potential alternatives, CBO considered possible scenarios for their use and what those scenarios would require in terms of missiles' range, time to reach their targets (responsiveness), and ability to reach those targets without being intercepted (survivability). Both China and Russia have long-range A2/AD systems that are tailored to prevent the United States from achieving air, land, and sea superiority in their areas during a conflict. A long-distance, highly responsive weapon could be useful in other circumstances, such as when fleeting intelligence was available about the location of a high-value target (a missile launcher or the head of a terrorist organization, for example). CBO concluded that in A2/AD scenarios centered on the South China Sea and the Baltic region, there would probably be a number of time-sensitive, high-value targets for which rapid strikes from longer distances could be useful. In evaluating those scenarios, CBO also considered whether potential adversaries had missile defense systems to defend against an incoming U.S. missile.

CBO identified potential weapons that could be employed in such scenarios and estimated their performance using metrics describing range, responsiveness, and survivability. The weapon systems that CBO considered include existing systems, such as subsonic cruise missiles like the Navy's Tomahawk Land Attack System, and possible future systems, such as hypersonic cruise missiles and hypersonic boost-glide missiles like those being developed by DoD. (A cruise missile is powered by an engine throughout its flight; a boost-glide missile is initially accelerated to a high altitude using a rocket and then glides at high speed through the atmosphere to its target.) CBO also considered hypothetical ballistic missiles with maneuverable warheads that would provide accuracies similar to those predicted for hypersonic boost-glide missiles.

CBO then narrowed its list of weapon systems to only those technologies that could meet the more demanding requirements for range, responsiveness, and survivability in the relevant scenarios. CBO analyzed the characteristics of each of those systems and estimated the cost to produce them.

What Did CBO Find?

To operate effectively in an adversary's A2/AD zone, the U.S. military might need to strike a small number of targets in as little as an hour at ranges that could be as long as 3,000 km to 5,000 km (about 1,900 miles to 3,400 miles), CBO estimates. Such targets include coastal air-defense systems, long-range strike systems, and over-the-horizon radars that are essential to maintaining an A2/AD zone. By degrading an adversary's long-range strike systems and defenses, the United States could limit the effects of A2/AD systems early in a conflict, allowing it to use a broader arsenal of shorter-range and less survivable weapons thereafter.

Of the potential alternatives that CBO examined for those types of missions, only ballistic missiles equipped with maneuverable reentry vehicles (MaRVs) and hypersonic boost-glide missiles would have the combination of speed and range to strike targets under the strict time constraints associated with the most challenging A2/AD scenarios.

Comparison of Hypersonic Boost-Glide Missiles and Ballistic Missiles

CBO concluded that hypersonic missiles combine many features that would present challenges to defense systems designed to intercept ballistic missiles. (Indeed, a main reason that China and Russia have been developing hypersonic weapons is a desire to defeat the United States' comparatively advanced ballistic missile defense capabilities—both short-range terminal defenses that protect forces in a theater of operations and long-range midcourse defenses that protect the United States itself.) By spending much of their flight inside the atmosphere, hypersonic boost-glide missiles would have an advantage over ballistic missiles in surviving midcourse ballistic missile defense systems, which aim to intercept ballistic missiles in the middle of flight, when those missiles are high above the atmosphere on a predictable and detectable path. That potential advantage for hypersonic boost-glide missiles might become important if a potential U.S. adversary developed midcourse ballistic missile defenses that operated outside the atmosphere and that were effective against countermeasures (such as decoys to fool interceptor missiles).

Against shorter-range missile defenses, it is unclear how much advantage hypersonic missiles would have over ballistic missiles equipped with maneuverable reentry vehicles. By flying low, hypersonic missiles could force the surface-based radar associated with short-range defenses to detect a missile later in its flight, shortening the time available for a defense system to intercept it. But because hypersonic missiles use up much of their energy while gliding toward their targets, they are likely to be traveling more slowly than a ballistic missile's MaRV as it gets close to its target. In addition, both types of missiles can perform preprogrammed maneuvers near their targets to make it harder for short-range defenses to intercept them.

Hypersonic weapons also have the potential to create uncertainty about what their ultimate target is. Their low flight profile puts them below the horizon for long-range radar and makes them difficult to track, and their ability to maneuver while gliding makes their path unpredictable. Ballistic missiles equipped with MaRVs could also create target uncertainty by maneuvering during the late stages of flight. But because hypersonic missiles can maneuver during more of their flight, their target uncertainty is greater than for a MaRV. Such uncertainty could be an advantage by making it harder for adversaries to defend against attacks, but it could also cause the target country or nearby countries to misinterpret the United States' intentions. The importance of that consideration would depend on the details of a particular scenario.

The United States does not currently field intermediate-range ballistic missiles, in part because some of those missiles were prohibited by the Intermediate-Range Nuclear Forces treaty, which the United States was a party to from 1988 to 2019. Over the years, however, the United States has mastered the technological challenges of operating such missiles. For example, in the 1980s, the U.S. military deployed Pershing II medium-range ballistic missiles with maneuverable warheads. Both the hypersonic missiles that DoD is currently developing and the ballistic missiles with MaRVs that CBO included in its analysis would probably be accurate enough to strike many fixed targets. However, both missiles would need improved targeting technology (such as a homing seeker or other type of sensor) to improve their accuracy or to enable them to strike moving targets.

Hypersonic boost-glide missiles would cost about one- third more than comparable ballistic missiles, in CBO's estimation. Specifically, CBO estimates that procuring 300 intermediate-range hypersonic boost-glide missiles like the ones being developed by the Army and the Navy and then sustaining the missile system for 20 years would cost a total of $17.9 billion in 2023 dollars (see Table S-1). By comparison, 300 MaRV-equipped ballistic missiles with the same speeds, ranges, and targeting capabilities as those hypersonic missiles would cost a total of $13.4 billion, CBO estimates. If potential adversaries developed effective midcourse defenses against ballistic missiles, the total cost for the ballistic missiles would be slightly higher because of the need to develop more advanced countermeasures to enhance the missiles' survivability.

Challenges Ahead for Hypersonic Missiles

Realizing the full potential of hypersonic weapons will depend on future success in achieving some of the improvements now being researched and developed, including missile components such as transparent communications windows that can withstand the high heat of extended flights. DoD is taking a phased approach to fielding hypersonic missiles. Early versions of those weapons will have capabilities at the lower end of the desired ranges and will not be able to locate targets autonomously or maneuver in response to attacks by missile defenses. However, those early hypersonic missiles may be able to carry out small maneuvers on preplanned trajectories to make them more difficult for defenses to track and intercept. Further enhancements would increase the risks of cost increases and schedule overruns.

What Are Some Limitations of CBO's Analysis?

The models that CBO used to estimate missile performance are based on publicly available data and do not incorporate classified technical specifications or the results of ongoing testing. To reflect uncertainty about current and likely future performance, CBO presents a range for its model results. In addition, CBO's estimates of the costs of various missile options are derived from statements about planned inventories of hypersonic weapons by public officials and program offices. Large changes in the number of missiles purchased or problems in overcoming technical challenges could alter the relative costs of the options.

Table S-1. Comparison of the missile options that CBO analyzed

Option	Launch Mode	Estimated Range (Kilometers)	Average Speed Over Estimated Range (Mach number)[a]	Number of Missiles Purchased	Estimated Procurement Cost per Missile (Millions of 2023 dollars)	Estimated Cost of Option (Billions of 2023 dollars)[b]
Intermediate-Range Hypersonic Boost-Glide Missiles (Similar to LRHW/IR-CPS)	Land or ship	3,000	10	300	41	17.9
Intermediate-Range Ballistic Missiles Equipped With MaRVs	Land or ship	3,000	9	300	26	13.4
Medium-Range Hypersonic Boost-Glide Missiles (Similar to ARRW)[c]	Aircraft	1,000	7	300 100	15 18	5.3 2.2
Short-Range Hypersonic Cruise Missiles (Similar to HACM/HALO)[d]	Aircraft	Less than 500	7	n.e.	n.e.	n.e.
Short-Range Ballistic Missiles (Similar to SM-6 Block IB)	Land or ship[e]	Less than 1,000	6	300	6	3.5

Data source: Congressional Budget Office. See www.cbo.gov/publication/58255#data.

CBO's estimates are subject to significant uncertainty about the costs of less mature technology or components, which represent long-term investments in a wide range of research areas that are associated with high risks of cost and schedule overruns. These estimates represent one possible outcome based on the specific assumptions made about technologies and capabilities (see Appendix C for details).

ARRW = Air-Launched Rapid Response Weapon (being developed by the Air Force); HACM = Hypersonic Attack Cruise Missle (being developed by the Air Force); HALO = Hypersonic Air-Launched Offensive Antisurface Warfare missile (being developed by the Navy); IR-CPS = Intermediate-Range Conventional Prompt Strike missile (being developed by the Navy); LRHW = Long-Range Hypersonic Weapon (being developed by the Army); MaRVs = maneuverable reentry vehicles;

n.e. = not estimated; SM = Standard Missile.

[a] "Mach number" refers to an object's speed relative to the speed of sound in the air at sea level; for example, Mach 5 is five times the speed of sound.

[b] Consists of the costs to procure the missiles; the costs to integrate the missiles with existing platforms and to buy associated equipment, such as launchers; and the costs to sustain the missile system for 20 years. Costs to develop the missiles are not included.

[c] CBO estimated the costs of two versions of this option: one with 300 missiles, the same as the number of hypersonic LRHW/IR-CPS missiles in Option 1, and one with 100 missiles to reflect the fact that the air-launched weapons in this option would not need to be spread among ships and submarines in different parts of the world but could be stored in a few locations and flown to where they were needed on short notice. Thus, a smaller number of the air-launched hypersonic missiles in this option could provide a worldwide availability similar to that provided by the larger number of land- or sea-launched hypersonic missiles in Option 1.

[d] The Department of Defense is early in the process of developing this missile. Little is known about its characteristics, so CBO did not have a basis for estimating the cost of the missile.

[e] The SM-6 Block IB is being developed by the Navy as a variant of its existing Block IA ballistic missile. Although the new missile would primarily be a Navy weapon, reports indicate that the Army is also planning to use it.

Part 1: Hypersonic Missiles and Their Policy Implications

A cornerstone of U.S. defense planning is the ability to establish and maintain freedom of military action anywhere around the globe if a crisis occurs. Such freedom of action is enabled by superior air and maritime forces that can neutralize defensive and offensive systems that adversaries could use to attack U.S. forces and prevent them from operating freely. Since the end of the Cold War, the United States has used its strategy of military superiority successfully against adversaries with limited military capabilities, at least during the initial combat phase of a conflict.

In recent decades, however, potential adversaries such as China and Russia have worked to diminish the United States' military advantage by developing advanced weapons with long ranges that could keep U.S. forces from operating in large areas—a strategy known as antiaccess and area denial. With the Department of Defense shifting its focus from conducting counterinsurgency operations in the Middle East and elsewhere to deterring near-peer military competitors, the challenges of operating in an A2/AD environment have come to dominate DoD's planning.

Various approaches can be used to defeat A2/AD systems. One is to develop aircraft that have stealth characteristics that can help them elude air defenses (as the Air Force did during the Cold War and continues to do today by fielding F-35 stealth fighter aircraft and developing the B-21 stealth bomber). Another approach that DoD is taking now is to develop long-range, high-speed missiles that carry conventional (nonnuclear) weapons. Long range enables the missiles' launchers to stay out of reach of offensive A2/AD systems; high speed makes the missiles harder to intercept and lets them reach their targets more quickly from long distances. The Army, Navy, and Air Force are all developing hypersonic missiles, which can fly at speeds faster than Mach 5 (five times the speed of sound) in the atmosphere during much of their flight.

Besides concern about potential adversaries' A2/AD systems, another stated rationale for developing hypersonic missiles is the sense that the United States has fallen behind those adversaries in hypersonic missile technology. The United States has performed foundational research into the technological challenges of flying at hypersonic speeds since at least the 1950s. Over the decades since then, both U.S. allies and potential adversaries have also conducted research into hypersonic missiles. In recent years, advances in aerodynamics, materials science, and computational science have enabled them to make significant progress. Both Russia and China have announced in the past few years that they have developed and fielded hypersonic missiles. Russia is believed to have used a missile that it classified as hypersonic in its war with Ukraine.

The United States has not yet fielded such weapons, for both scientific and policy reasons. Hypersonic missiles are expensive, and there have been questions about the value of the capabilities they might provide. Achieving the desired performance for those weapons would require significant investment in additional research.

Nevertheless, because of their potential, the U.S. military has established accelerated programs for hypersonic missiles, sharply increasing its support and funding, testing prototypes, and creating units in anticipation of fielding the weapons in the next few years.

Basics of Hypersonic Missiles

Missiles are considered to be hypersonic not only on the basis of their speed but also on the basis of their flight profile and means of control. The harshness of the environment around a hypersonic missile as it flies makes designing such missiles challenging. (For explanations of missile speeds and other terminology, see Box 1-1.)

Box 1-1. Missile Concepts and Terminology

Speeds

- *subsonic*: less than the speed of sound in the air at sea level— that is, less than Mach 1 (767 miles per hour)
- *supersonic*: from one to five times the speed of sound in the air at sea level—that is, Mach 1 to Mach 5 (767 miles per hour to 3,836 miles per hour)
- *hypersonic*: more than five times the speed of sound in the air at sea level—that is, more than Mach 5 (3,836 miles per hour)

Ranges

- *short range*: less than 1,000 kilometers (km, or 621 miles)
- *medium range*: between 1,000 km and 3,000 km (621 miles to 1,864 miles)
- *intermediate range*: between 3,000 km and 5,500 km (1,864 miles to 3,418 miles)
- *intercontinental range*: more than 5,500 km (3,418 miles)

Miscellaneous Characteristics

- *responsiveness:* a measure of a missile's ability to reach its target (cover its range) in a timely fashion
- *survivability:* a measure of a missile's ability to not be intercepted en route to its target
- *conventional:* nonnuclear

Types of Missiles and Components

- *ballistic missile:* a missile that consists of a warhead and a guidance system mounted on a rocket motor (or booster). At launch, the booster propels the missile up and out of the atmosphere at high speed. Once the missile is above the atmosphere and the motor has shut down, the booster is jettisoned. The warhead then travels an unpowered, arcing path to its target, much like the path of a ball thrown up in the air (see the figure).
- *hypersonic missile*: a missile that travels within the atmosphere (rather than above it) at speeds greater than Mach 5 and that can maneuver in the atmosphere for much of its flight. Hypersonic missiles can take various forms, such as boost-glide missiles or cruise missiles.
- *boost-glide missile:* a missile that consists of a rocket motor, which accelerates the missile to a high speed, and a glide body containing a warhead. After the glide body detaches from the spent rocket, it uses its kinetic and potential energy, as

well as lift generated by its movement through the air, to coast at high speed through the atmosphere and maneuver to hit its target.[537]

- *cruise missile:* a missile that typically flies at low altitudes (ranging from less than 100 meters for a subsonic cruise missile to about 30 km to 40 km for a hypersonic cruise missile) and that is powered throughout its flight. A hypersonic cruise missile is initially accelerated to speeds approaching hypersonic by a rocket booster, then accelerates and maintains speed throughout its flight using a jet engine called a supersonic combustion ramjet (or scramjet) that operates at speeds above Mach 4.
- *maneuverable reentry vehicle*: a ballistic missile warhead capable of correcting its trajectory when it reenters the atmosphere, increasing its accuracy.

Types of Launch

- *air-launched:* launched from aircraft in flight
- *ground-launched:* launched from fixed or mobile (vehicle- based) launchers on land
- *sea-launched:* launched from surface ships or submarines

Flight Phases and Paths

- *boost phase:* the period from a missile's launch until its rocket booster shuts down. During the boost phase, the missile is powered by the rocket motor. The length of this phase depends on the desired speed of the missile when the booster shuts down. For intermediate-range missiles, the boost phase can last as long as two to three minutes.
- *midcourse phase:* for a ballistic missile, the period from the end of the boost phase until the warhead reenters the atmosphere. During the midcourse phase, the warhead is coasting above the atmosphere on an arcing path. This phase is the longest part of a ballistic missile's flight. For intermediate-range ballistic missiles, it lasts roughly 10 minutes to 20 minutes.
- *terminal phase:* for a ballistic missile, the period from reentry into the atmosphere until the warhead reaches its target. This phase is very short, generally less than one minute.
- *glide phase:* for a boost-glide missile, the period during which the glide body is coasting in the atmosphere. For the intermediate-range boost-glide missiles now being developed, the glide phase may last for 10 minutes to 20 minutes.
- *minimum-energy trajectory:* the flight path that maximizes the range of a ballistic missile.
- *depressed trajectory*: a lower, but still ballistic, path that shortens the flight time and reduces the range of a ballistic missile.

---

[537] Kinetic energy is a measurement of the energy that an object has because of its motion. Kinetic energy reflects the object's mass and speed. Potential energy is a measurement of the energy stored by an object—in this case, the energy that the object derives from being in a gravitational field. That energy reflects the masses of the Earth and the object as well as the object's distance from the Earth.

Notional Flight Paths for Different Ballistic and Boost-Glide Missile Trajectories with the Same Range

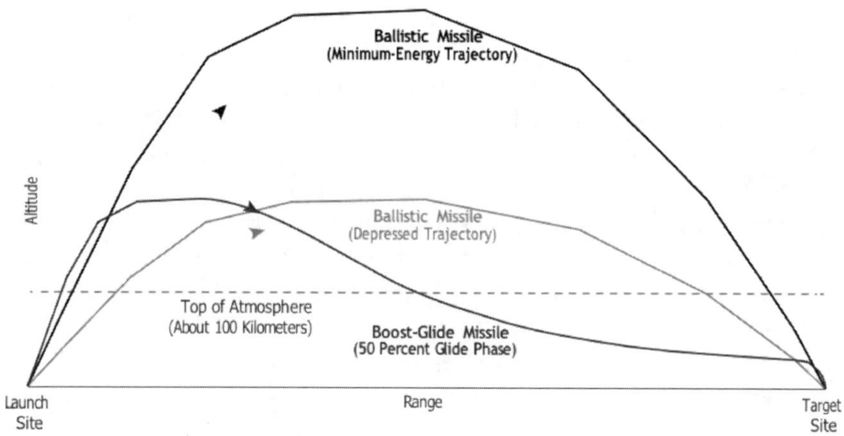

Source: Congressional Budget Office.

A missile's ballistic trajectory can be shaped in many ways. This figure shows a simple minimum-energy trajectory (MET) that is very close to a parabola, influenced mainly by gravity after the missile's rocket booster burns out. To depress the trajectory (lower the altitude), a ballistic missile could be launched at a shallower angle than the MET missile. To achieve the same range as the MET missile, the depressed-trajectory missile would require a larger rocket booster for a higher initial speed.

A boost-glide missile would start on a ballistic trajectory, but thrusters would be used in the first part of its flight to bring it back toward Earth sooner than in the ballistic flight. Control surfaces on the glide body would be used in the second part of the missile's trajectory to initiate gliding. The missile's maximum potential range would depend on its initial speed, the lift-to-drag ratio of the glide body, and the share of its trajectory that was spent inside the atmosphere.

What Is a Hypersonic Missile?

"Hypersonic" refers to speeds above Mach 5, or five times the speed of sound in air. (Although the speed of sound varies by altitude and depends on the density and temperature of the air, in most discussions of hypersonic weapons, that speed is measured at sea level under standard conditions, a convention that CBO uses as well.) Mach 5 is generally considered to describe an aerodynamic regime that is different from that of lower speeds—specifically, where heating becomes so significant that it dominates the physics of air flow around a moving object, such as an aircraft or a missile. For example, the transition between smooth flow and turbulent flow along the body of such an object becomes much more difficult to predict above Mach 5. That condition has implications for the heating of the object's surface and for the performance of aerodynamic control surfaces that allow the object to maneuver.

Thus, technically speaking, "hypersonic" refers only to travel at speeds above Mach 5 within the Earth's atmosphere, where interaction between the aircraft and molecules of air takes place. Spacecraft routinely travel at speeds well above Mach 5, and all but the shortest-range ballistic missiles also spend most of their flight at speeds above Mach 5 outside the atmosphere.

But those flights are not considered hypersonic because they do not have the associated challenges of extended interactions with the air. Spacecraft or ballistic missiles that reenter the atmosphere near the end of their flight also face extreme heating at that point, but briefly. What distinguishes hypersonic missiles is that they have to contend with thermal effects throughout most of their flight.

Another defining characteristic of hypersonic missiles is that they depend on aerodynamic control surfaces (such as wings or tail fins) to glide and maneuver, like an aircraft, rather than using thrusters, like a spacecraft. Air is necessary for control surfaces to function, so a hypersonic missile must be within the atmosphere to maneuver.

The U.S. military is developing two types of hypersonic missiles: boost-glide missiles and cruise missiles. Both types need air to operate for reasons besides maneuvering. A hypersonic boost-glide missile consists of a rocket motor that accelerates the missile to a high altitude and speed and a glide body that detaches from the spent rocket. In addition to the kinetic and potential energy from its initial acceleration, the glide body uses lift generated by its movement through the air to extend its range and maneuver to hit its target (see Box 1-1).

A hypersonic cruise missile is also initially accelerated to a high speed by a rocket booster. After that, it accelerates and maintains speed throughout its flight by using a type of jet engine known as a supersonic combustion ramjet (or scramjet). A scramjet uses oxygen from the air to burn its fuel, rather than carrying an oxidizer, as a rocket does, so it is known as an air-breather. Scramjets require supersonic air flow and only begin to operate at speeds above Mach 4. Although air-breathing engines tend to be smaller and lighter than rocket engines that carry both fuel and oxidizer, that size advantage is lessened by the size of the rocket booster used for the initial acceleration.

Design Challenges for Hypersonic Flight

For a missile traveling at hypersonic speeds in the atmosphere, the intense effects of heating during much of the flight influence every aspect of the missile's design. Heating creates the need for special materials to handle extreme temperatures for many minutes. It also affects the stability of the flight as well as communications and targeting for the missile.

Heating and Thermal Shielding

Moving at very high speeds within the atmosphere creates thermal challenges for an object such as a hypersonic glide body. The leading edge (the nose cone or the front edge of a wing) compresses the air ahead of the object, heating the air. The more rapidly the air is compressed, the hotter it will become. Drag also generates heat along any exposed surfaces, but to a lesser degree than air compression does. As the object moves through the heated air, its surfaces will begin to warm.

Structural materials for objects that travel at hypersonic speeds must be able to withstand high temperatures for the duration of the flight inside the atmosphere without significantly deforming or melting. The lower the heat tolerance of the material used to build the object, the greater the need for heat shielding, such as the ceramic plates used on the space shuttles that were operated by the National Aeronautics and Space Administration (NASA).[538] The choice of materials for an object's structure and heat shielding has a significant effect on the object's weight and cost.

Air behaves differently at different temperatures, which has major implications for the materials needed at those temperatures. As air warms, its molecules increasingly vibrate. At temperatures over 2,000 Kelvin (3,140° Fahrenheit), the air molecules can be ripped apart. And at temperatures above 4,000 K (6,740° F), the atoms from those molecules will lose some of their electrons (become ionized), and the surrounding gas will be in a plasma state. Plasma effectively absorbs radio frequency signals, affecting the ability to communicate or use radar.

DoD has suggested that, according to its modeling of the first-generation hypersonic boost-glide missiles it is developing, temperatures around most of the body of a missile are likely to remain within a range of 1,000 K to 2,000 K, below the threshold for plasma formation.

Materials that have the potential to withstand those temperatures without melting or losing structural integrity include high-performance alloys known as superalloys.

For the parts of a missile exposed to higher temperatures, exotic materials such as ceramic composites and materials made from carbon fibers, which have melting temperatures around 3,500 K, can be used. Carbon-carbon composites, such as those used on the space shuttle, have been shown to withstand heat of more than 2,000 K. Those high-performance materials are especially difficult and expensive to manufacture—factors that could limit the speeds and ranges at which hypersonic missiles can operate.

Flying at hypersonic speeds through the atmosphere for more than a few minutes presents challenges greater than those of reentering the atmosphere. Materials routinely used on reentry vehicles (such as ballistic missiles and the space shuttle) work over the shorter time frame associated with reentry to absorb heat or dissipate it through chemical processes. Such materials are not sufficient for hypersonic flight because they will eventually conduct heat inward or be used up.

Stability and Maneuvering

At speeds greater than the speed of sound (over Mach 1), a shock wave is produced ahead of a moving body, resulting in what is known as a sonic boom. When speeds reach the hypersonic regime (over Mach 5), the boundary between the shock wave and the moving body—the shock layer—can become very thin. In addition, the air near the moving body is very hot and can be chemically altered. One of the biggest challenges that designers of hypersonic missiles face is predicting the behavior of the shock layer. The transition from smooth to turbulent air flow around a moving body can disrupt its stability and cause sudden, localized increases in temperature. High-speed computing has been fundamental to advancing understanding of that phenomenon in recent years.

Maneuvers such as turns increase drag, adding to a missile's thermal load (the heat it experiences during flight, which is a function of temperature and time). Increased drag slows

---

[538] The catastrophic effects that heating can have on an aircraft were dramatically illustrated in 2003, when an insulating tile on the Space Shuttle Columbia was damaged, causing the aircraft to overheat and break apart on reentry into the atmosphere.

the missile considerably, reducing its range. Those factors tend to limit how much a hypersonic missile can maneuver.

Communications and Targeting

DoD has stated that its hypersonic boost-glide missiles will be able to emit and receive radio signals, such as those from the Global Positioning System (GPS). That statement implies that portions of a missile's glide body where transmitters and receivers are located will remain below the temperature threshold for plasma formation in the surrounding air, as DoD's modeling suggested might be the case. Even so, the intersection of materials, temperature, and communications remains a challenge in designing hypersonic missiles, as ongoing research by organizations such as the Defense Advanced Research Projects Agency (DARPA) has shown. Although the sensitive electronics necessary for communications must be adequately shielded from a missile's exterior temperatures, the missile needs to have windows (called radomes) through which signals can be emitted or received. The materials for such radomes are inherently difficult to develop because they must allow radio frequency or infrared radiation to pass through them while also providing significant thermal shielding.

Scramjet Technology

An additional challenge in developing hypersonic cruise missiles is designing engine components that can maintain proper air–fuel mixing at the right temperature, air pressure, and density for a supersonic combustion ramjet to operate for the duration of a missile's flight. That challenge has been likened to the difficulty of keeping a match lit in a hurricane. Before 2022, the United States had successfully operated a scramjet for only a few minutes. But the recent successful demonstration of a scramjet implies that several design challenges have been overcome. Nonetheless, engineering hurdles remain in the development of hypersonic cruise missiles, which must meet size, range, and speed requirements while also managing the thermal and aerodynamic challenges of hypersonic flight described above for boost-glide missiles.

A Brief History of U.S. Hypersonic Research

Research to overcome the challenges of hypersonic flight has been going on since World War II, when Germany developed the V-2, a ballistic missile that traveled faster than Mach 5 for part of its flight.[539] Soon after the war, the United States built its first wind tunnel for aero-dynamic testing at hypersonic speeds and began developing its first intercontinental ballistic missiles. From 1959 to 1968, the United States test-flew a piloted hypersonic aircraft called the X-15, a research project that contributed to the development of NASA's piloted spacecraft.

---

[539] The V-2 is often considered to have made the first engineered flight at hypersonic speeds. But it was technically not a hyper- sonic missile, because it flew faster than Mach 5 within the atmosphere for only a small portion of its flight and because it did not use aerodynamic control surfaces to maneuver after its rocket stopped burning.

During that era, scientists made enough progress in aerodynamic design and advanced materials to address the challenges of extreme heating that ballistic missiles and spacecraft encounter briefly when they reenter the Earth's atmosphere. The greater challenges posed by extended flight and maneuvering through the atmosphere at hypersonic speeds and extreme temperatures remained the subject of active research programs.

Hypersonic flight continues to be extraordinarily challenging. But in recent years, computational science and materials science have advanced enough to transition to developing practical technologies for use in operational hypersonic missiles.

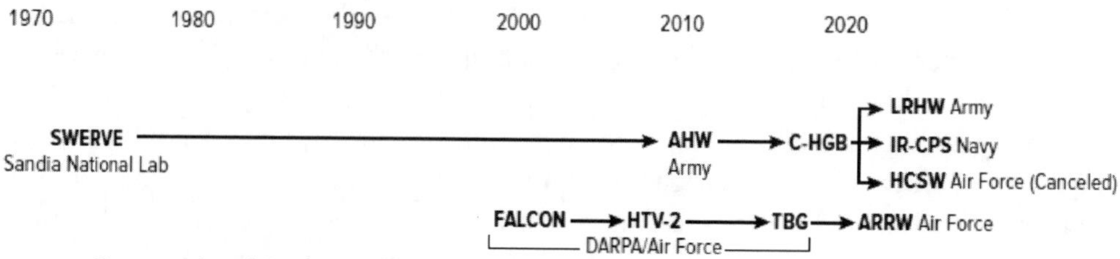

Data source: Congressional Budget Office.

AHW = Advanced Hypersonic Weapon; ARRW = Air-Launched Rapid Response Weapon; C-HGB = Common Hypersonic Glide Body; DARPA = Defense Advanced Research Projects Agency; DoD = Department of Defense; FALCON = Force Application and Launch From Continental United States; HCSW = Hypersonic Conventional Strike Weapon; HTV = Hypersonic Technology Vehicle; IR-CPS = Intermediate-Range Conventional Prompt Strike; LRHW = Long-Range Hypersonic Weapon; SWERVE = Sandia Winged Energetic Reentry Vehicle Experiment; TBG = Tactical Boost Glide.

Figure 1-1. Progression of DoD's Research Programs for Hypersonic Boost-Glide Missiles.

Technology for Hypersonic Boost-Glide Missiles

The modern era of DoD's involvement in research into hypersonic boost-glide missiles has roots in the Sandia Winged Energetic Reentry Vehicle Experiment (SWERVE) project, which began in the 1970s (see Figure 1-1). In 1985, the SWERVE project successfully tested a design for a missile glide body.

A second track of missile research began in 2003 when DARPA started the Force Application and Launch from Continental United States (FALCON) project. That project was designed to study technologies that would extend the reach of U.S. missiles so they could be launched from the continental United States rather than from locations closer to their targets.

In 2008, lawmakers established the Conventional Prompt Global Strike program to advance some of the technology originated by the FALCON project. The new program focused on a concept for an intercontinental-range hypersonic glider known as the Hypersonic Technology Vehicle-2. Reports at the time suggest that the program's rationale was to reduce the U.S. military's reliance on strategically located bases in multiple regions, for two reasons.[540] First, strategic locations change over time, and they are not always known or available far enough in advance to allow the U.S. military to establish bases there. Second, growth in the

---

[540] See National Research Council, *U.S. Conventional Prompt Global Strike: Issues for 2008 and Beyond* (National Academies Press, 2008), https://tinyurl.com/yjnfss59; and Department of Defense, Office of the Under Secretary of Defense for Acquisition, Technology, and Logistics, *Time Critical Conventional Strike From Strategic Standoff* (report of the Defense Science Board Task Force, March 2009), https://dsb.cto.mil/reports/2000s/ADA4 98403.pdf.

size of adversaries' antiaccess and area-denial regions created a desire for long-range weapons that could destroy an adversary's air defenses and strike its mobile offensive systems early (or, in some cases, later) in a conflict.

Other technology under consideration for that mission included putting conventional warheads on some of the Navy's Trident missiles—long-range, submarine-launched ballistic missiles that were previously armed with nuclear warheads. The Conventional Trident Modification (CTM) program envisioned a nonnuclear warhead carried by a maneuverable reentry vehicle that would provide the greater accuracy that a conventional warhead requires to compensate for its much lower explosive power. The CTM program was first funded by the Navy in 2007.

It was later canceled by lawmakers because of concerns that an adversary seeing a CTM launch might think the United States was launching a nuclear-armed Trident missile and might respond with a nuclear attack of its own.

To avoid such confusion, the United States has generally maintained different types of missiles to carry nuclear warheads and conventional warheads.

After the CTM program was canceled, the Conventional Prompt Global Strike program continued working on the intercontinental Hypersonic Technology Vehicle-2, conducting test flights in 2010 and 2011. The second test in 2011 failed when contact with the vehicle was lost. After that, the program's funding was sharply reduced. In 2012, DoD began to shift its focus to shorter-range hypersonic missiles, and "global" was dropped from the program's name, which became simply Conventional Prompt Strike.

During the same period, the Army began the Advanced Hypersonic Weapon program, which built on the glidebody design of the earlier SWERVE project. The Advanced Hypersonic Weapon was successfully tested in 2011 but failed during the boost stage of a flight test in 2014.

Despite reductions in their funding after test failures, those early 21st-century programs led to the two main efforts that DoD is pursuing today to develop hypersonic boost-glide missiles:

- The Advanced Hypersonic Weapon evolved into the Common Hypersonic Glide Body. That glide-body design forms the basis for missiles that the Army and Navy are developing, the Army's Long-Range Hypersonic Weapon and the Navy's Intermediate- Range Conventional Prompt Strike. (The Air Force also had a program based on the Advanced Hypersonic Weapon, known as the Hypersonic Conventional Strike Weapon, but it was canceled in 2020.)
- The Hypersonic Technology Vehicle-2 evolved into DARPA's Tactical Boost Glide vehicle, which forms the basis for the Air Force's current program to develop the medium-range Air-Launched Rapid Response Weapon.

Technology for Hypersonic Cruise Missiles

Research on supersonic combustion ramjet engines began in the 1950s. But the first successful flight test of a scramjet did not occur until 2002, when Australia's HyShot aircraft flew for a few seconds on a parabolic trajectory under scramjet power. Two years later, NASA's scramjet-powered X-43 successfully flew several tests in horizontal flight. The X-43 holds the record for the fastest jet-powered aircraft, having reached a speed of Mach 9.6 with the scramjet engine operating for 10 seconds during a flight test in 2004. Later development focused on the Air

Force's X-51 program. A scramjet-powered X-51 aircraft was successfully flown in 2013 at Mach 5.1 for 210 seconds.

The research from those programs is being applied to several current efforts to develop an operational hypersonic cruise missile. Although scramjet technology appears to be maturing, integrating the engine into a missile that is small enough to be tactically useful will require significant further efforts. Trade-offs exist between speed, size, and range that might be addressed with improvements to fuel efficiency. Such improvements are being explored in laboratory testing and with modeling and simulation.

Much of the research into hypersonic cruise missiles being conducted worldwide is exploring the use of hydrogen fuel. Hydrogen has many characteristics that make it desirable for use in scramjets, such as high flammability and high caloric value (a measure of the energy that can be extracted during combustion).

Hydrogen-fueled scramjets could theoretically operate at speeds higher than Mach 20.[541] And many researchers prefer hydrogen to hydrocarbon-based fossil fuels for environmental reasons, because it produces only water as a byproduct. But hydrogen fuels also have downsides. Their extreme volatility increases safety concerns during the storage and handling of a missile, and their lower density than hydrocarbon fuels makes it harder to design a missile with the sort of range that DoD wants.[542] For those reasons, DoD is likely to use hydrocarbon fuels in hypersonic cruise missiles, which would limit their maximum speed to about Mach 9.

DoD's Current Hypersonic Missile Programs

The Department of Defense has spent more than $8 billion since 2019 on programs to develop hypersonic missiles. Those programs include separate efforts by the Army, Navy, and Air Force to develop hypersonic boost-glide missiles; collaborations among DARPA, the Air Force, and the Navy to research technology for hypersonic cruise missiles; and various research programs for missile components (discussed in Appendix A). In its latest five-year budget plan, the 2023 Future Years Defense Program, DoD is requesting $13 billion over the 2023–2027 period for developing hypersonic missiles and almost $2 billion for procuring missiles. (The procurement request represents only Army and Air Force programs; the Navy has not yet requested procurement funding for the hypersonic missile it is developing.)

This section describes DoD's major hypersonic missile efforts and the research, development, test, and evaluation (RDT&E) funding provided or requested for them. Those funding numbers are in nominal dollars, meaning that they have not been adjusted to remove the effects of inflation.

DoD has stated that the urgency of hypersonic missile development outweighs the need for traditional oversight, so its three boost-glide missile programs are proceeding outside the usual acquisition process, under an approach called rapid prototyping. The RDT&E funding described below for those programs is for advanced component development and prototypes

---

[541] See P. Satyaprasad and others, "Missile Propulsion Systems," in N. Eswara Prasad and R. J. H. Wanhill, eds, Aerospace Materials and Material Technologies, vol. 2 (November 2016), pp. 305–330, https://doi.org/10.1007/978-981-10-2143-5_15.

[542] See Gautam Choubey and others, "Hydrogen Fuel in Scramjet Engines—A Brief Review," International Journal of Hydrogen Energy, vol. 45, no. 33 (June 24, 2020), pp. 16799–16815, https://doi.org/10.1016/j.ijhydene.2020.04.086.

(budget activity level 4) and, starting in 2023 for the Army's program, for system development and demonstration (budget activity level 5). In DoD's development strategy for hypersonic boost-glide missiles, the versions fielded initially will have the capabilities of the prototype missiles. Full-capability versions will be fielded later.

The Army and the Navy are developing separate missile systems, but they are collaborating on a joint glide body for those missiles—the Common Hypersonic Glide Body, or C-HGB (see Figure 1-2)—and a common two-stage booster. The Navy is responsible for designing the glide body and the booster, and the Army is responsible for producing them. (The Air Force participated in that collaboration until 2020, when it chose to focus on a different glide-body design for its boost-glide missile.) The Army and Navy will use identical glide bodies but plan to separately develop different launch platforms for the missiles.

Source: Army Rapid Capabilities and Critical Technologies Office (from a briefing to CBO on the hypersonic weapons budget, June 2020).

The Army and Navy are collaborating on the Common Hypersonic Glide Body (C-HGB). A glide body is the part of a hypersonic boost-glide missile that detaches from the spent rocket booster and glides to its target. Both services plan to use the C-HGB in the hypersonic boost-glide missiles they are developing: the Army's Long-Range Hypersonic Weapon and the Navy's Intermediate-Range Conventional Prompt Strike missile.

Figure 1-2. Depiction of the Common Hypersonic Glide Body.

The Army's Long-Range Hypersonic Weapon

The Army's boost-glide missile, the Long-Range Hypersonic Weapon (LRHW), is intended to be launched from the ground using mobile transporter- erector-launcher (TEL) vehicles. An LRHW battery will consist of four TELs, eight missiles, and one operations center for the battery. Most of the performance characteristics of the LRHW are classified, but its range has been publicly stated as exceeding 2,775 kilometers (km, or about 1,724 miles).

The LRHW program has been funded since 2020 and has received $1.7 billion through 2022. DoD requested $807 million for the missile in 2023 (see Table 1-1). In that budget request, the Army estimated that RDT&E for the program would conclude in 2027, at a total cost of $5.3 billion.

Table 1-1. U.S. Hypersonic weapons programs and their RDT&E funding

RDT&E Funding (Millions of nominal dollars)						
Service or Agency	Program	Description	Estimated Range (Kilometers)	Request for 2023	Total Past and Planned Funding[a]	Planned Initial Fielding Date
Army	Long-Range Hypersonic Weapon (LRHW)	Ground-launched boost-glide missile	3,000	807	5,269	2023
	Operational Fires	Ground-based launcher for hypersonic boost-glide missiles; uses the TBG and a tunable rocket motor for variable range	500	11	66	2023
Navy	Intermediate-Range Conventional Prompt Strike (IR-CPS)	Sea-launched boost-glide missile	3,000[b]	1,205	At least 8,902 (continuing)	2025
	Hypersonic Air-Launched Offensive Antisurface Warfare (HALO)	Air-launched cruise missile	N.A.	92	At least 444 (continuing)	2028
Air Force	Air-Launched Rapid Response Weapon (ARRW)	Air-launched boost-glide missile	1,000	115	1,315	2023
	Hypersonic Attack Cruise Missile (HACM)	Air-launched cruise missile	500[c]	462	1,812	N.A.
Defense Advanced Research Projects Agency	Hypersonic Air-Breathing Weapon Concept (HAWC)	Cruise missile technology	N.A.	60	At least 195 (continuing)	n.a. (Transitioned to the Air Force)
	Tactical Boost Glide (TBG)	Glide body prototype for ARRW	n.a.	30	At least 555 (continuing)	n.a. (Transitioned to the Air Force)
	Operational Fires	Ground-based launcher for hypersonic boost-glide missiles; uses the TBG and a tunable rocket motor for variable range	n.a.	N.A.	169	n.a. (Transitioned to the Army)

Data source: Congressional Budget Office, using Department of Defense data from the President's budget request. See www.cbo.gov/publication/58255#data.

n.a. = not applicable; N.A. = not available; RDT&E = research, development, test, and evaluation.

[a] Total RDT&E funding from the beginning of the program through the 2023–2027 Future Years Defense Program. Programs labeled "continuing" are expected to receive additional RDT&E funding after 2027.

[b] The range of the Navy's IR-CPS is expected to be about the same as that of the Army's LRHW, which will use the same glide body.

[c] The design details, including ranges, of HALO and HACM have not been made public. The range of a cruise missile depends largely on the amount of fuel it carries. Because of the Air Force's stated intent to design HACM to be smaller than ARRW, CBO anticipates that the range will be about 500 kilometers. CBO has not evaluated whether HACM can achieve that range and still have a smaller size and lower weight than ARRW.

The LRHW program has completed successful flight tests of a prototype missile, but it has also experienced test failures in recent years. Several more flight tests, to be conducted jointly with the Navy, are scheduled through 2028. The Army expects to field initial prototype LRHW missiles in 2023 and two additional batteries in 2025 and 2027, respectively. DoD's 2023 budget request includes $249 million in procurement funding to purchase LRHWs.

The Navy's Intermediate-Range Conventional Prompt Strike

The Navy's boost-glide missile, which will also carry the Common Hypersonic Glide Body, is a sea-launched missile called the Intermediate-Range Conventional Prompt Strike (IR-CPS). The Navy's original plan was to field the IR-CPS on Virginia class nuclear-powered submarines, using the Virginia payload module, a multimissile launcher that is planned to be built into newer versions of those submarines. To deploy the missiles sooner, the Navy now plans to field them on existing Zumwalt class destroyers first and then develop the launch capabilities for Virginia class submarines. The specifications of the IR-CPS are classified, but its range is expected to be about the same as that of the Army's LRHW (greater than 2,775 km).

The IR-CPS program has been funded by the Navy since 2019 and has received $2.6 billion through 2022. DoD requested $1.2 billion for the program in 2023. The Navy expects research and development on the IR-CPS to continue past 2027, at a total cost of more than $9 billion.

The Navy's version of the boost-glide missile is expected to take longer to develop than the Army's version for several reasons. First, the timing of the IR-CPS was initially linked to the timing of the Virginia payload module.

Second, launching a missile from a submarine requires a "cold launch," in which the missile is ejected from the launch tube by a gas generator before the missile ignites. That process is more complicated than the "hot launch" used for ground-launched missiles such as the Army's LRHW. In a hot launch, the missile ignites and is propelled out of the launcher by its own exhaust.

The Navy had intended to demonstrate a prototype-capability IR-CPS in 2022, but the missile failed its first flight test in 2022. Joint flight tests with the Army are scheduled to continue through 2028. The Navy plans to demonstrate a cold-launch prototype in 2024, field prototype missiles on Zumwalt class destroyers in 2025, and field full-capability missiles on Virginia class submarines in 2028.

The Air Force's Air-Launched Rapid Response Weapon

The hypersonic boost-glide missile that the Air Force is developing, the AGM-183A Air-Launched Rapid Response Weapon (ARRW), is based not on the Army and Navy's common glide body but on the Tactical Boost Glide (TBG) vehicle developed by DARPA (see Figure 1-3). The ARRW missile is designed to be launched from B-52 bombers, although there is discussion about using additional launch platforms, such as B-1 bombers. The Air Force has released some technical information about the missile: Its range will be at least 926 km (about 575 miles), and its booster will consist of a single-stage solid rocket motor.

The ARRW program received a total of nearly $1.2 billion in RDT&E funding from 2019 to 2022. For 2023, DoD requested $115 million for RDT&E and $46 million for missile procurement. Separately, DARPA has spent $555 million on the TBG program so far and requested $30 million for 2023.

The first flight test of the current development effort was not successful, but it was followed by several successful flight tests. The Air Force's 2022 budget request to purchase 12 ARRWs was cut by lawmakers, leaving questions about when and in what quantity the missiles will initially be fielded. The $46 million procurement request for 2023 is intended to fund one missile.

Source: Air Force Air-Launched Rapid Response Weapon program office (Lockheed Martin illustration).
The Tactical Boost Glide (TBG) vehicle developed by the Defense Advanced Research Projects Agency forms the basis for the hypersonic boost-glide missile that the Air Force is developing, the Air-Launched Rapid Response Weapon (ARRW). The TBG is shown here as a slender, pointed object on the front end of the missile. The cone in the upper left of the drawing is the protective shroud from the front of the missile, which has just been ejected in preparation for releasing the TBG.

Figure 1-3. Depiction of the Tactical Boost Glide Vehicle Being Carried by an ARRW Missile.

Other Hypersonic Programs

Besides the military's three programs to develop hypersonic boost-glide missiles, the Army and DARPA are developing ground-based launchers for boost-glide missiles. In addition, the Air Force, the Navy, and DARPA are working on technology for hypersonic cruise missiles.

OpFires Launchers

The Army is continuing development work on DARPA's Operational Fires (OpFires) program, which demonstrated a flexible ground-based launch system for short-range hypersonic boost-glide missiles, capable of launching multiple payloads at a variety of ranges. OpFires has been funded through DARPA since 2018 and has received $169 million through 2022. The Army spent a total of $55 million on OpFires in 2021 and 2022 and requested $11 million for 2023.

Budget documents indicate that the launch system will be fielded along with the LRHW in 2023.

Hypersonic Cruise Missiles

DoD is also developing hypersonic cruise missiles, which are based on scram- jet technology. DARPA has been funding research on scramjets under the Hypersonic Air-Breathing Weapon Concept (HAWC) program since 2014; the program has received $195 million through 2022. For 2023, DoD requested $60 million for the program, whose name has been changed to MoHAWC. The Air Force and the Navy have both been collaborating with DARPA on scramjet research, and each service now has its own program to develop a hypersonic cruise missile.

The Air Force's effort, the Hypersonic Attack Cruise Missile (HACM) program, received $190 million in 2022. An additional $462 million was requested for 2023. The program is described in budget documents as aiming to achieve a long-range, prompt strike capability, building on technology developed through DARPA's HAWC program. Three companies received funding to develop a prototype HACM; two of the prototypes were successfully demonstrated in 2021 and 2022. The prototype developed by Raytheon and Northrop Grumman reportedly flew more than 500 km, but it is unclear how long the scramjet was operating during that flight.

The Navy was also involved in flight tests during DARPA's HAWC program. It requested $92 million in 2023 for its own cruise missile development program: the Offensive Anti-Surface Warfare (OASuW) Increment 2, commonly known as the Hypersonic Air-Launched OASuW (HALO) program. Funding documents indicate that the Navy plans to field HALO missiles starting in 2028.

Design specifications for those various cruise missiles have not been publicly released. For a hypersonic (or any) cruise missile, design trade-offs exist between speed, size, and range.

The Path Forward for U.S. Hypersonic Missiles

In addition to the programs discussed above, DARPA and the laboratories of the military services are funding many research programs related to hypersonic flight. The topics of that research provide insights into the tech- nological challenges that remain in fielding hypersonic missiles. For example, many of those research programs seek to develop technical knowledge or components that will increase the range or payload of hypersonic missiles or enhance the ability to control them for maneuvering and targeting. (For more details about those programs, see Appendix A.)

DoD has stated that the hypersonic missiles currently in development will be developed and fielded in stages. The existence of ongoing research efforts to increase the range, maneuverability, and targeting capabilities of the next generation of hypersonic missiles suggests that the initial versions fielded by DoD will lack some of the performance and features often attributed to hypersonic missiles. As a result, it appears likely that DoD's first generation of hypersonic weapons will operate at or below the minimum ranges associated with the range categories for missiles (see Box 1-1). For instance, intermediate-range missiles are defined as reaching between 3,000 km and 5,500 km, but the range of the Army's and Navy's intermediate-range hypersonic boost-glide missiles (LRHW and IR-CPS) is expected to be about 3,000 km because heating will limit the speeds those missiles can sustain.

Thermal challenges also suggest that those missiles will be limited in what maneuvers they can perform, because their trajectories will be designed to minimize thermal stresses on the missiles. Furthermore, DoD's first- generation hypersonic missiles are not expected to have the accuracy or sensors needed to operate effectively in situations in which targets may be moving.

Policy Issues Related to Hypersonic Missiles

As significant progress has been made on the development of hypersonic missiles over the past decade, support for such weapons appears to have broadened in the United States. Potential reasons include the rise of China as a technological and military power, the pivot in U.S. defense policy toward a greater focus on China and Russia as near-peer competitors, and announcements by those two countries that they have overcome technological challenges and fielded hypersonic missiles. Nevertheless, hypersonic missiles could be seen as costly and unnecessary given the diverse arsenal of strike weapons that the U.S. military already possesses and given the immature state of potential adversaries' defenses against current U.S. missiles. Ultimately, whether hypersonic missiles are affordable and useful will depend on whether remaining technological hurdles can be overcome and at what cost.

Beyond general considerations of usefulness and cost— which would apply to any new weapon system—other policy issues are specifically associated with deploying hypersonic missiles. The most prominent of those issues involve hypersonic missiles' potential to confuse adversaries about the United States' nuclear intentions and their potential to escalate crises.

Nuclear Ambiguity

Nuclear ambiguity occurs when an adversary is uncertain about whether an incoming missile is armed with a nuclear or a conventional warhead. That uncertainty can lead the adversary to respond with nuclear weapons if it thinks it may be coming under a nuclear attack. The desire to avoid a nuclear response to a conventional strike is a significant, widely agreed upon principle of U.S. security policy. The Navy's Conventional Trident Modification program of the early 2000s was canceled largely because lawmakers were concerned that a Trident missile armed with a conventional warhead would be indistinguishable from one armed with a nuclear warhead.

Although the U.S. military has generally used different missiles for conventional and nuclear warheads, there is precedent for fielding nuclear and conventional versions of the same or very similar missiles. Examples include the Navy's Tomahawk Land Attack Missile and the Air Force's Air-Launched Cruise Missile. During the Conventional Trident Modification program, methods were proposed to ensure that adversaries could distinguish conventional from nuclear Tridents. Those proposals included having the conventional missiles use a different trajectory and maneuvering capabilities that would keep them from flying over countries other than the United States and the intended target.

Crisis Instability

Regardless of ambiguity about the nature of a missile's payload, the compressed time frames associated with a strike by a conventional hypersonic missile could be seen as fundamentally destabilizing during a conflict. A missile capable of flying 3,000 km in only about 15 minutes would allow little time for an adversary to make a well-thought-out response. The same issue would exist with conventionally armed ballistic missiles because their travel times would be similar to those of boost-glide missiles.

In addition, the maneuverability of hypersonic missiles creates uncertainty about their targets. When a ballistic missile is launched, despite its extreme speed, its general destination is roughly calculable—both because its high trajectory makes it observable to powerful terrestrial radar (as explained in Part 3) and because its ability to change its path after its rocket booster burns out is limited. A ballistic missile equipped with a maneuverable reentry vehicle probably could maneuver by no more than several tens of kilometers after reentering the atmosphere, whereas a hypersonic boost-glide missile could theoretically strike largely undetected anywhere within a radius of several thousand kilometers.

A hypersonic boost-glide missile would probably be detected at launch (if the adversary had space-based infrared sensors with the ability to do so and in the correct position). But the difficulties of tracking the boost-glide missile through its lower-altitude flight, combined with its much greater ability to change direction, mean that potential adversaries might not know whether they or another country in the region was the intended target.

That target uncertainty could increase the likelihood of a response (accidental or otherwise) before the actual nature of the attack was known.

Part 2: Scenarios That Define Potential Requirements for Hypersonic Missiles

The U.S. military's arsenal includes a wide variety of guided weapons that can be used to attack targets on land. Those weapons have ranges that vary from as short as a few kilometers (for the Joint Direct Attack Munition) to more than 1,500 kilometers (for the Tomahawk cruise missile) and speeds that vary from well below the speed of sound (for the AGM-154 glide bomb) to more than three times the speed of sound (for the Army Tactical Missile System). With that variety available, why is the Department of Defense interested in pursuing new and relatively expensive hypersonic missiles?

Longer ranges and higher speeds are the main attributes that distinguish the hypersonic weapons DoD is developing from the conventional strike weapons in today's inventory. Long range and high speed are nearly always desirable characteristics for a strike weapon, but they can be expensive. DoD's shift in focus, to potential conflicts with near-peer adversaries that have advanced militaries, has brought those characteristics to the fore. Of specific concern are potential conflicts with China or Russia, both of which have advanced air defenses, long-range bombers and missiles, and antiship missiles. With those systems, Chinese or Russian forces could establish large areas in which U.S. forces would be vulnerable to attack.

Such an antiaccess, area-denial strategy would reduce U.S. combat power and give an adversary greater freedom of action. In particular, the A2/AD weapon systems that China and Russia have deployed could force the U.S. military to use ground- or ship-launched weapons from as far as 4,000 km (about 2,500 miles) away and air-launched weapons from at least 1,000 km (about 600 miles) away. Launching such weapons from shorter distances would put the launchers and the units operating them at greater risk. Long-range weapons with high speeds could be launched from areas of lower risk but not take too much time to cover the longer distances to their targets.

This Part describes, in general terms, the planning scenarios for near-peer adversaries that DoD is currently using and their implications for the performance requirements of weapons designed to overcome the A2/AD threats found in those scenarios. In any conflict, combatants try to establish A2/AD zones to protect their own forces and territory and to limit where their opponents can safely operate. But near-peer adversaries pose particular A2/AD challenges because they are likely to possess longer-range weapons (and the ability to target them) in greater numbers than less advanced adversaries do. In particular, China and Russia both have many offensive and defensive weapon systems capable of establishing challenging A2/AD environments.[543] Beyond those challenges, other types of scenarios could also affect the U.S. military's requirements for hypersonic missiles.

---

[543] For more discussion of scenarios involving Russia and China, see Congressional Budget Office, Options for Fielding Ground- Launched Long-Range Missiles (February 2020), www.cbo.gov/ publication/56068.

China

Disputes over territory in the South China Sea have occurred in recent years. The People's Republic of China has threatened at times to take the island of Taiwan by force. And the United States has indicated in increasingly explicit ways its intent to come to Taiwan's aid should China attempt to do so. In addition, if the Chinese military tried to coerce other U.S. allies in the Pacific or shut down access to sea lanes there, the United States could choose to respond militarily.

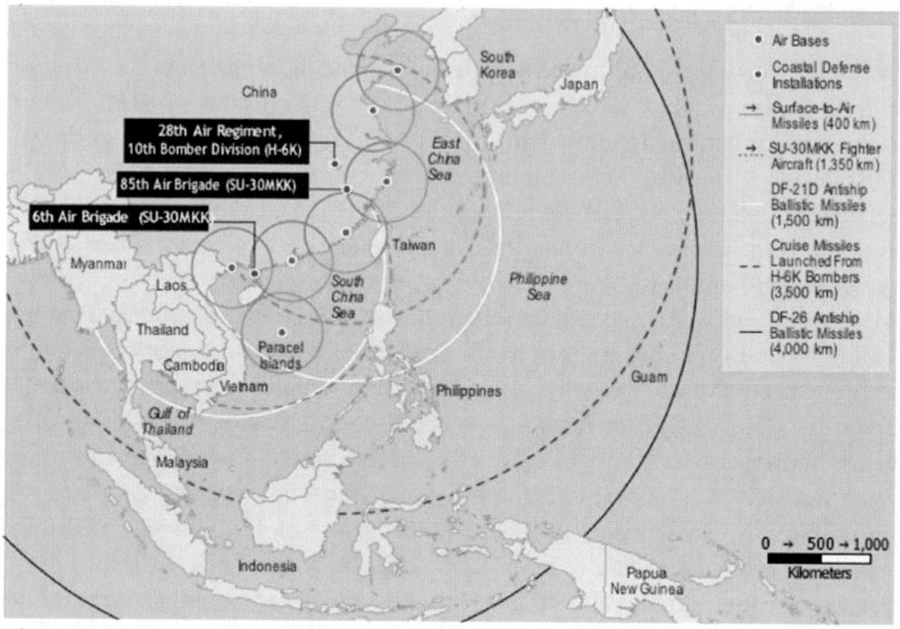

China has been developing a variety of weapon systems designed to keep U.S. and U.S. allies' forces far away from its coasts. In a potential conflict with China, those A2/AD systems could be targets for U.S. hypersonic missiles or alternatives.

Data source: Congressional Budget Office.

The weapon systems that make up China's A2/AD zone include air- and missile-defense systems that are assumed to cover China's coastline. Some air-defense sites might also be launch sites for antiship ballistic missiles; those launchers are potentially mobile. China's military airfields, where fighter aircraft and bombers that could launch cruise missiles are based, are generally located farther inland. Basing those A2/AD systems on Chinese-occupied islands in the South China Sea would extend their coverage areas.

A2/AD = antiaccess and area-denial; km = kilometers.

Figure 2-1. Approximate Coverage Areas of China's Potential A2/AD Defenses.

If a conflict erupted in the region, China has a variety of A2/AD weapons that could pose challenges to the United States and its allies. At present, China's antiship and air defenses have the potential to limit U.S. naval and air access to much of the South China Sea, particularly because China could extend the reach of those weapons by basing them on disputed islands in the South China Sea, such as those in the Paracel and Spratly chains.

Chinese surface-to-air missiles and antiship cruise missiles located on the mainland could cover an area with an estimated radius of 400 km, potentially including Taiwan (see Figure 2-

1). If those missiles were located on the Paracel Islands—which China has claimed sovereignty over, in violation of international law—China's reach into the South China Sea could be roughly doubled.

In addition, Chinese DF-21D antiship ballistic missiles have a range of 1,500 km. DF-26 antiship ballistic missiles may reach as far as 4,000 km, but their actual capabilities are not yet clear and would require that China have an intelligence and surveillance system that could detect and track targets at those long ranges with enough accuracy for the missiles to find their targets. Chinese H-6K bombers flying from an inland base could reach 3,500 km offshore, and SU-30MKK fighter aircraft have a range of 1,350 km.

China also employs modern air defenses, most of which were purchased from Russia or designed in Russia and licensed for production in China. The most notable are the S-300 and S-400 surface-to-air missile systems, which have maximum ranges of a few hundred kilometers. China has also fielded a variety of shorter-range air-defense systems that could engage naval aircraft in the region or land-based fighters and bombers. Those defenses would be less effective against cruise missiles, such as the Tomahawk, that fly a few hundred feet above the surface to avoid detection until they are close to their targets. But such defensive systems would still have some capabilities against low-flying missiles at short ranges.

To defend against ballistic missiles, China's S-300 and S-400 surface-to-air missile systems (like the U.S. Patriot missile system) have some capability as point defenses against shorter-range ballistic missiles as they approach their targets. (A point defense can protect a relatively small area, from a few kilometers to a few tens of kilometers in diameter.) China is working to develop interceptors to defend against longer-range ballistic missiles, but those systems are still in development and have not yet been deployed. Thus, China's ability to intercept very fast, high-flying missiles en route to their targets is likely to be limited for at least the near future.

Russia

As evinced by its invasion of Ukraine, Russia may harbor aspirations to restore the territory of the former Soviet Union. That territory includes the three Baltic Sea nations of Estonia, Latvia, and Lithuania, which are now members of the North Atlantic Treaty Organization (NATO).[544] If Russia invaded a NATO country, the United States would have a treaty-based obligation to respond. Such a response would be harder if Russian forces were able to establish a strong ground presence in one of those NATO countries and could protect themselves with long-range air defenses and missile attacks.

Like China, Russia has a variety of antiaccess and area-denial weapons that could pose challenges for the United States and its allies—particularly Russia's long- range missiles and air defenses. For attacking targets on land, Russia can use ground-launched Iskander ballistic missiles, which have a range of 300 km, and ground-, ship-, and air-launched land-attack cruise missiles, which have ranges up to several thousand kilometers. In addition, Russia's Kinzhal

[544] As of this writing, Russia is engaged in a protracted war in Ukraine. Depending on the outcome of that conflict and its effects on Russia and NATO, and the likely expansion of NATO to include Finland and Sweden, the set of Russia scenarios could expand to include direct attacks on other NATO countries. However, the basic characteristics of an A2/AD environment and the weapons that might be needed to address it are likely to remain roughly the same, although the likelihood of a conflict with Russia could rise or fall.

ballistic missiles, which have a range of 2,720 km when launched from fighter air- craft, can be used to strike targets on land or at sea and could limit naval access to the Baltic Sea. And Russia's air defenses, such as the S-300 and S-400 surface-to-air missile systems also deployed by China, could limit access by NATO air forces to areas that would include much of Poland, Lithuania, Latvia, Estonia, and Finland (see Figure 2-2).

However, the Russian military's failure to seize the Ukrainian capital of Kyiv in February 2022, which many analysts thought would be fairly easy for Russia, suggests some of the pitfalls of relying on technical assessments to measure military capability. Sensors, intelligence gathering, flexible command-and-control systems, and well-trained personnel are just as important. Thus, the risks to U.S. weapon systems in potential adversaries' A2/AD zones may be overstated.

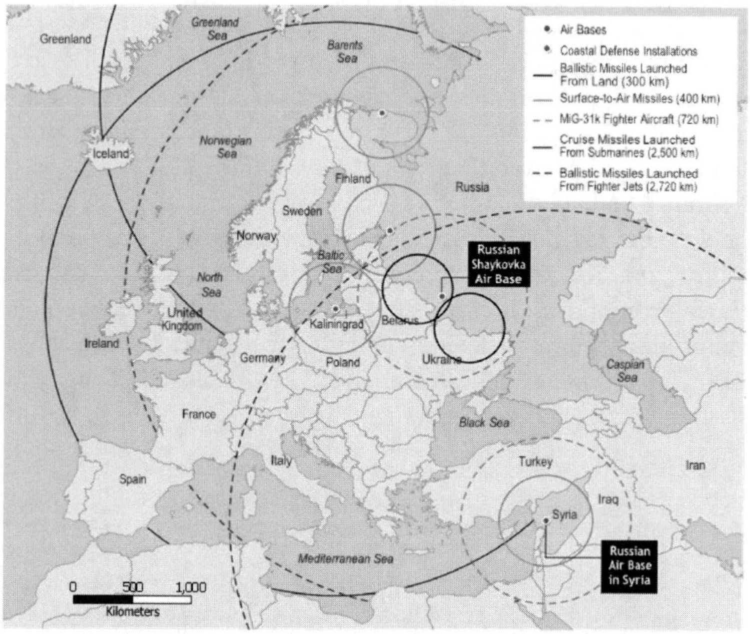

Russia has been developing weapon systems that could keep U.S. and NATO allies' forces far away from its borders. In a potential conflict with Russia, those A2/AD systems could be targets for U.S. hypersonic missiles or alternatives.

Data source: Congressional Budget Office.

Russian A2/AD systems are made up of air- and missile-defense systems strategically located in coastal and inland locations, covering much of Europe and nearby waterways. Basing those systems outside mainland Russia, such as in Syria and Kaliningrad, would extend their coverage areas.

A2/AD = antiaccess and area-denial; km = kilometers; NATO = North Atlantic Treaty Organization.

Figure 2-2. Approximate Coverage Areas of Russia's Potential A2/AD Defenses.

Other Scenarios

Hypersonic weapons could prove to be useful in ways other than thwarting the A2/AD defenses of near-peer adversaries. Notable examples would be scenarios in which the United States wanted to conduct a rapid, long-distance strike because it had intelligence indicating the location of a high-value target, such as the leader of a terrorist organization or a nuclear-armed missile about to be launched by a rogue nation. Such intelligence would have to be acted on quickly because of potential movement of the target. Although there might not be U.S. forces nearby, a hypersonic missile could potentially hit the target from a long distance in a time frame that would not be possible with a slower-flying missile, such as a subsonic cruise missile. That type of mission is what initially revived DoD's interest in hypersonic weapons in the early 2000s.

Part 3: Comparing the Capabilities of Hypersonic Missiles and Potential Alternatives

Considering the scenarios described in Part 2, the Congressional Budget Office analyzed how well different types of weapons could conduct rapid strikes with conventional warheads in an antiaccess and area-denial environment. In such an environment, three characteristics that are particularly important for strike weapons are range, responsiveness, and survivability: To be effective, a strike weapon must be able to reach the target (range) in a timely fashion (responsiveness) with a reasonable probability of not being intercepted en route (survivability). A strike weapon's range, responsiveness, and survivability for a given mission are a function not only of the weapon's performance but also of the type of platform that launches it. The performance of the opponent's weapons plays a role as well.

CBO used the metrics of range, responsiveness, survivability, and overall effectiveness in an A2/AD environment to characterize the potential performance of the following general types of weapon systems:

- Hypersonic boost-glide missiles;
- Ballistic missiles with various ranges and trajectories;
- Subsonic, supersonic, and hypersonic cruise missiles; and
- Long-range bomber aircraft.

Using the results of that general performance comparison, CBO created five illustrative options with specific quantities of particular weapon systems and compared the potential costs and effectiveness of those options (see Part 4).

The performance characteristics of the hypersonic missiles that the Department of Defense is developing are not publicly available, so CBO conducted its analysis by modeling generic versions of each type of weapon system using information that has been made public.

The technologies of the systems CBO analyzed exist in various states of maturity. But for the purposes of comparing performance, the systems were all assumed to be mature enough to deploy in the near future.

CBO concluded that although the hypersonic missiles now in development would theoretically have the energy to glide farther than a ballistic missile with the same initial speed, heating of the missile body would probably limit the range at which those hypersonic missiles could operate. Consequently, the hypersonic boost-glide missiles and cruise missiles currently being developed would have to be located near or within an A2/AD zone to conduct rapid strikes.

Hypersonic boost-glide missiles and ballistic missiles are considerably faster than the other weapons that CBO examined, including hypersonic cruise missiles. CBO concluded that over short to intermediate ranges, hypersonic boost-glide missiles might be slightly faster than ballistic missiles of the same range that were flying on a minimum-energy trajectory (a flight path that maximizes the range of a ballistic missile). But hypersonic boost-glide missiles would be no faster than ballistic missiles flying on a lower, or depressed, trajectory (which would require a little more power to achieve the same maximum range as a ballistic missile flying on a minimum-energy trajectory).

A purely ballistic missile with a conventional warhead would not be accurate enough for many targets, so the notional ballistic missile that CBO included in this comparison would be equipped with a maneuverable reentry vehicle capable of correcting its trajectory when it reentered the atmosphere. MaRVs would probably give ballistic missiles an accuracy similar to that envisioned for the hypersonic missiles DoD is developing (although those ballistic missiles would not be capable of flying the nonballistic trajectories used by hypersonic boost-glide missiles). Maneuverable reentry vehicles are a less challenging technology than hypersonic glide bodies; they have been studied since the 1960s and were deployed on the U.S. military's intermediate-range Pershing II nuclear-armed missiles in the 1980s.

In terms of performance against defenses, CBO's analysis found that subsonic cruise missiles are among the most difficult munitions to detect because they are small and fly close to the ground. But once detected and tracked, they are easier to intercept than missiles traveling at higher speeds. For all types of missiles, as speed increases from subsonic to hypersonic, the challenge of intercepting them increases.

The ability of hypersonic boost-glide missiles and ballistic missiles with MaRVs to avoid being detected and intercepted varies with the type of defense system—specifically, with the phase of a missile's flight in which it is targeted by the defense system. (For a description of those phases, see Box 1-1.) A hypersonic boost-glide missile may be detected by infrared sensors during portions of its boost phase as well as in its glide phase, when its temperature increases. The missile can also be detected by ground-based radar during early parts of its trajectory, when it is relatively high in the atmosphere, and at later stages, when it is closer to the radar.

Tracking and intercepting a missile in the middle of its flight would be considerably harder with hypersonic boost-glide missiles than with ballistic missiles. But the inherent challenges of ballistic missile defense—coupled with the fact that no potential U.S. adversaries have yet deployed midcourse defense systems—suggest that the advantages offered by hypersonic missiles against such long-range defenses may not be important at this time. Moreover, modifying a ballistic missile's trajectory (such as depressing its altitude) or using countermeasures (such as decoys to fool interceptor missiles) would make ballistic missiles difficult to counter even if potential adversaries deployed midcourse defenses.

Against defenses that target missiles in the terminal phase of their flight, ballistic missiles with MaRVs and hypersonic boost-glide missiles would probably be similarly effective because both would be able to maneuver as they got close to their targets.

Range

The threats that an A2/AD environment poses to missile launchers increase the importance of range for missile systems. Most of the missiles that the United States is deploying or developing for those purposes today have ranges of 5,500 kilometers (about 3,400 miles) or less, with the exception of long-range nuclear delivery systems, such as submarine-launched Trident missiles and ground-launched Minuteman III missiles (see Table 3-1). For any weapons with a range of less than 5,500 km, the ability to position launch platforms in proximity to a conflict but outside the range of an adversary's defenses is important. The defenses that China, Russia, or a regional power might use to threaten U.S. weapons and their launchers could have ranges from 400 km to 4,000 km (about 250 miles to 2,500 miles), depending on the type of defense.

Table 3-1. Ranges of various U.S. weapon systems

Type of Weapon System	Example	Range (Kilometers)
Existing Weapons		
Ground-launched tactical ballistic missile	Army Tactical Missile System (ATACMS)	300
Subsonic cruise missile	Tomahawk	1,000 to 2,400[a]
Submarine-launched ballistic missile	Trident	7,400
Ground-launched ballistic missile	Minuteman III	13,000
Supersonic stealth fighter aircraft	F-22 Raptor	1,500[b]
Subsonic stealth bomber	B-2 Bomber	4,800[b]
Weapons Under Development		
Intermediate-range hypersonic boost-glide missile (Ground- or sea-launched)	Long-Range Hypersonic Weapon (LRHW)/ Intermediate-Range Conventional Prompt Strike (IR-CPS) prototype	3,000[c]
Medium-range hypersonic boost-glide missile (Air-launched)	Air-Launched Rapid Response Weapon (ARRW) prototype	1,000[d]

Data source: Congressional Budget Office, using data from the Department of Defense. See www.cbo.gov/publication/58255#data. DoD = Department of Defense; km = kilometers.

[a] The standard variant of the Tomahawk missile has a range of roughly 1,000 km. The longest-range variant is reported to have an operational range of up to 2,400 km.

[b] This range is the unrefueled combat radius, which reflects an assumption that the aircraft has sufficient fuel to return to base. Range can be extended if the aircraft is refueled in flight.

[c] DoD has released information suggesting that the LRHW and IR-CPS missiles will have a range of at least 2,775 km. CBO used 3,000 km for its estimate.

[d] This range is an estimate based on the lower boundary of the range stated by DoD.

In a China scenario, for example, a U.S. missile with a range of 3,000 km (like the Army's planned Long- Range Hypersonic Weapon) would be able to reach China's populous and militarily important coastal regions if it was launched from the Pacific Island of Guam. But in doing so, it would be within range of China's medium-range bombers and ballistic missiles (see Figure 2-1). Whether those Chinese weapons could destroy the missile launcher would depend on how it was deployed (dispersed, camouflaged, or dug in), how well it was protected by air and missile defenses, and how quickly its missiles were launched at the start of a conflict.[545] The fate of the launcher would also depend on the accuracy and effectiveness of China's

[545] To protect the U.S. military's ability to launch a strike, missile launchers can be deployed in greater numbers than the minimum necessary for the strike and can be positioned in separate areas (dispersed), hidden by blending into the surroundings (camouflaged), or hidden under the ground (dug in).

weapons, the number of long-range missiles in China's inventory, and the ability of China's intelligence and command-and-control systems to pinpoint the launcher's location.

A U.S. Navy ship carrying a weapon similar to the planned Intermediate-Range Conventional Prompt Strike missile might have to stay as far as 4,000 km away from China's coast if China was able to track the ship and if its DF-21D or DF-26 antiship ballistic missiles proved effective at that range. In such a case, the IR-CPS missile might not be able to reach targets on China's coast, although it could still attack air-defense systems and other fixed targets deployed on artificial islands that China has built in the South China Sea. The IR-CPS might also be used against Chinese surface ships if they were not moving. However, a submarine carrying that missile could use its stealth capabilities to get closer to China's coast and strike targets there or perhaps farther inland. Subsonic cruise missiles like the Navy's Tomahawk might also be effective in that scenario.

The hypersonic boost-glide missile that CBO modeled on the Air Force's planned Air-Launched Rapid Response Weapon has a much shorter range than the Army's and Navy's boost-glide missiles—about 1,000 km—which would require an aircraft carrying it to get much closer to China's coast. China's air-defense systems have much shorter ranges than its bombers and ballistic missiles, about 400 km. But Chinese fighter aircraft (like the SU-30MKK) armed with air-to-air missiles might be able to attack U.S. aircraft as far as 1,000 km from the coast. ARRW missiles launched from that distance could probably still reach coastal targets.

In a Russia scenario focused on the Baltic region, ground-based missile launchers could be located at multiple spots in nearby North Atlantic Treaty Organization countries, but they might require some protection.

Submarines would be less likely to play a role in that scenario than they would in the waters around China because the Baltic Sea is smaller and space would be constrained. Aircraft, particularly ones with stealth characteristics, might be able to approach Russia's borders and launch missiles at inland targets.

In the face of A2/AD challenges from such potential adversaries, the United States could use a phased approach to destroying A2/AD systems. For example, if ships, submarines, and bombers were not yet in position, ground-based weapons that had already been deployed to a region could be fired the most quickly. The locations where such weapons could be deployed are limited in the Pacific region but are more numerous in the Baltic region. At present, U.S. allies in the Pacific have not agreed to allow the United States to base its planned hypersonic missiles in close proximity to China, leaving Guam as the most likely launch location. In Europe, by contrast, most if not all NATO countries might serve as launching points for U.S. missiles. Even with a range limited to 3,000 km, a missile like the Army's Long- Range Hypersonic Weapon might be able to degrade enough of the air and missile defenses in a small area early in a conflict to allow U.S. aircraft to operate in that area. In turn, those aircraft could use air-launched hypersonic missiles to eliminate additional A2/AD sites, further opening the area to U.S shorter-range forces.

Intermediate-range ballistic missiles equipped with maneuverable reentry vehicles would have the same, if not longer, range than the hypersonic boost-glide weapons that CBO modeled. The reason is that a MaRV is likely to be lighter than a hypersonic glide body. An air-launched ballistic missile with a MaRV could be boosted by a single-stage rocket, similar to that of the hypersonic ARRW. With a lighter payload, such a missile's range could exceed that of the ARRW. Alternatively, the payload capacity of a MaRV-equipped ballistic missile could be increased, making its range similar to the ARRW's.

Other weapons that CBO included in its comparison, such as subsonic cruise missiles (like the Navy's Tomahawk), have ranges longer than the air-launched hypersonic boost-glide missile (ARRW) but shorter than the ground- or sea-launched hypersonic boost-glide missiles (LRHW/IR-CPS). However, because those other weapons would be slower, their flight times would be much longer than the flight times of hypersonic missiles.

The U.S. military's interest in relatively long range weapons is driven by an assumption that the defensive systems in Chinese and Russian A2/AD areas would be effective. Those defensive systems require sophisticated intelligence collection and broad surveillance, which is almost certainly not constant or complete. Without detailed information about the tracking and targeting capabilities of Chinese and Russian air-defense and antiship systems, the severity of those A2/AD threats cannot be definitively determined.

Responsiveness

One of the main reasons that DoD is developing hypersonic missiles is that they can reach their targets very quickly from relatively long distances—a capability that is lacking in the current U.S. arsenal of longer-range conventional weapons. Quick response times also require that missiles' launchers already be located within range of their targets.

The U.S. arsenal currently lacks medium- and intermediate-range ballistic missiles primarily because the Intermediate-Range Nuclear Forces (INF) treaty between the United States and the Soviet Union prohibited either country from developing or possessing ground-launched missiles with ranges between 500 km and 5,500 km. The treaty did not limit air- or sea- launched missiles of those ranges, but the United States did not develop such ballistic missiles, presumably because they were considered too inaccurate and expensive to function as conventional weapons and were not considered necessary for nuclear deterrence.

One reason that DoD initially became interested in developing hypersonic boost-glide missiles is that their flight path (which involves spending less than half their flight in a ballistic trajectory) would make them compliant with the terms of the INF treaty. However, the United States withdrew from the treaty in 2019, citing Russian violations. That withdrawal widens the range of possible solutions to the problem of Russia's and China's efforts to develop longer-range weapons that increase the distance at which U.S. forces might have to operate.

The U.S. military has chosen to continue to focus on developing hypersonic weapons to meet its desire for fast medium- and intermediate-range missiles, rather than focusing on simpler alternatives, such as ballistic missiles with maneuverable reentry vehicles.

Estimating Average Speeds of Missiles

To measure the responsiveness of hypersonic missiles and other weapon systems, CBO estimated their flight times, taking into account such factors as range, initial speed, propulsion, trajectory, and loss of speed during flight because of drag. Different types of missiles have very different flight profiles, which can make comparing flight times difficult. For example, a cruise missile flies at a constant speed throughout most of its trajectory; a hypersonic boost-glide missile accelerates rapidly to a high speed and then slows down steadily as it glides along its trajectory; and a ballistic missile accelerates rapidly, slows for the first half of its trajectory, and then accelerates again for the second half of its trajectory.

Table 3-2. Average speeds of selected weapons at different ranges

Mach Number						
	Ballistic Missile					
Range (Kilometers)	Minimum-Energy Trajectory	Depressed Trajectory	Hypersonic Boost- Glide Missile	Hypersonic Cruise Missile	Long-Range Subsonic Bomber	Subsonic Cruise Missile
1,000	6	7–8	5–9	5–9	0.95	0.7
3,000	9	10–13	9–11	n.a.	0.85	0.7
10,000	16	18–20	9–14a	n.a.	0.7	n.a.

Data source: Congressional Budget Office, using data from Albert D. Wheelon, "Free Flight of a Ballistic Missile," American Rocket Society Journal, vol. 29, no. 12 (December 1959), pp. 915–926, https://doi.org/10.2514/8.4944; Lisbeth Gronlund and David C. Wright, "Depressed Trajectory SLBMs: A Technical Evaluation and Arms Control Possibilities," Science & Global Security, vol. 3, no. 1–2 (1992), pp. 101–159, https://tinyurl.com/2p8jnb48; James M. Acton, "Hypersonic Boost-Glide Weapons," Science & Global Security, vol. 23, no. 3 (2015), pp. 191–219, https://tinyurl.com/yc79ph7x; and Cameron L. Tracy and David Wright, "Modelling the Performance of Hypersonic Boost-Glide Missiles," Science & Global Security, vol. 28, no. 3 (2020), pp. 135–170, https://tinyurl.com/3upn3pk2. See www.cbo.gov/ publication/58255#data.

Average speed is calculated with respect to the ground—that is, velocity equals range divided by flight time. Where the table shows a range of average speeds, the lower number reflects conservative estimates of a weapon's trajectory and design considerations, and the higher number reflects optimistic estimates of trajectory and design considerations. "Mach number" refers to an object's speed relative to the speed of sound in the air at sea level; for example, Mach 5 is five times the speed of sound.

n.a. = not applicable.

CBO concluded that with current technology, hypersonic boost-glide weapons are unlikely to be able to reach a range of 10,000 kilometers because of heating. However, estimates for that range are included for purposes of comparison.

To facilitate comparison, CBO calculated a missile's average speed for each range, defined as the distance along the Earth's surface between the launcher and the target divided by the missile's time in flight (see Table 3-2). Although most of the missiles in CBO's comparison have ranges of 3,000 km or less, CBO analyzed weapons with ranges between 500 km and 10,000 km (about 300 miles to 6,200 miles) for completeness.

In its current development programs, DoD is designing hypersonic boost-glide missiles with likely ranges between 1,000 km and 3,000 km (about 600 miles to 1,900 miles). To estimate the flight time of such missiles, CBO created a model of a boost-glide trajectory that was based on an evaluation of publicly available data from an earlier flight test and potential improvements that DoD says it has made since then, allowing the missiles to glide faster and longer inside the atmosphere without damage from heating.[546] The actual performance of those missiles is probably somewhere between the previously demonstrated performance and the assumed improvements that CBO included in its model. (For more details about CBO's modeling of missiles, see Appendix B.)

For ballistic missiles, the longest flight time to a given target is the one associated with flying a minimum-energy trajectory (the path that allows the missile to have the longest possible range). The flight time of a ballistic missile can be shortened by altering the trajectory to a lower, but still arcing, path known as a depressed trajectory. The differences in flight times between a minimum-energy trajectory and a depressed trajectory become more pronounced over longer ranges (see Figure 3-1). CBO estimates that for ballistic missiles traveling 1,000

---

546 See James M. Acton, "Hypersonic Boost-Glide Weapons," Science & Global Security, vol. 23, no. 3 (2015), pp. 191–219, https:// tinyurl.com/yc79ph7x.

km to 3,000 km, depressed trajectories reduce flight times by 10 percent to 30 percent compared with minimum-energy trajectories.[547]

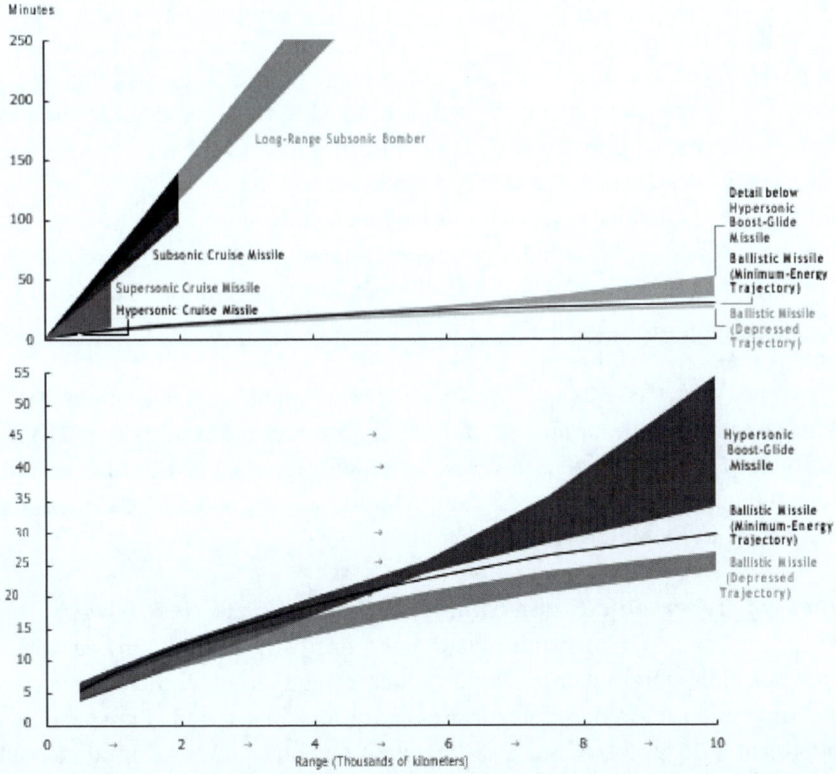

CBO modeled the flight times for various types of missiles and bombers. Ballistic and hypersonic missiles had the shortest flight times over all of the ranges in CBO's analysis.

Data source: Congressional Budget Office. See www.cbo.gov/publication/ 58255#data.

CBO modeled flight times for cruise missiles and subsonic long-range bombers using average speed along a flat trajectory. The range of flight times for those weapon systems reflects their range of speeds. For example, a supersonic cruise missile may travel 1,000 kilometers in as little as 10 minutes when flying at the fastest supersonic speed (just below Mach 5, or five times the speed of sound) or in as much as 48 minutes if flying at the slower end of the supersonic range (just above Mach 1). For ballistic missiles, CBO modeled flight times for a missile on a minimum-energy trajectory (which allows the missile to fly the longest distance) and on a range of depressed trajectories (with lower overall altitudes and modified booster properties that increase initial speeds). For hypersonic boost-glide missiles, the range of estimated flight times reflects a range of potential initial energies and trajectories that span the performance parameters suggested by current U.S. development programs.

For cruise missiles and subsonic bombers, the end of the colored band represents the estimated maximum range. CBO does not know the maximum range for hypersonic boost-glide missiles, but it extended the analysis to 10,000 kilometers to compare their potential flight times with those of ballistic missiles, which can currently achieve intercontinental ranges up to 12,000 kilometers.

Figure 3-1. Spans of Flight Times for Hypersonic and Other Weapon Systems.

[547] See Lisbeth Gronlund and David C. Wright, "Depressed Trajectory SLBMs: A Technical Evaluation and Arms Control Possibilities," *Science & Global Security*, vol. 3, no. 1–2 (1992), pp. 101–159, https://tinyurl.com/2p8jnb48.

A ballistic missile on a depressed trajectory generally spends more time in the atmosphere than a ballistic missile on a minimum-energy trajectory. As a result, it may also face greater thermal challenges, which can limit either its speed or its maximum range. However, a ballistic missile experiences those challenges for a much smaller portion of its flight than a hypersonic boost-glide missile of the same range.

Cruise missiles fly at low altitudes at roughly constant speeds. For its analysis, CBO used generic definitions of speed to group weapons into subsonic, supersonic, and hypersonic categories. On the basis of publicly available information, CBO limited the speed of the hypersonic cruise missiles currently under development to Mach 9.

In its comparison, CBO also considered long-range, subsonic stealth bombers, such as the current B-2 or the planned B-21, which could penetrate an adversary's A2/AD zone to deliver short-range guided weapons, such as the Joint Direct Attack Munition. To model that type of mission, CBO assumed the aircraft would fly at speeds between Mach 0.7 and Mach 0.85 and have a maximum range of 10,000 km.

The flight times that CBO calculated for a given distance between launcher and target would be increased if a missile or aircraft did not fly a direct route to its target. That increase could be significant for boost-glide and cruise missiles. Boost-glide missiles would suffer an additional speed penalty because they are unpowered, so they would lose energy during the maneuvers necessary to fly an indirect route to the target.

Responsiveness of Hypersonic Boost-Glide Missiles and Ballistic Missiles

Actual flight speeds for hypersonic missiles depend on many design considerations, such as their initial acceleration, aerodynamic shape (which affects drag), limits on maximum speed because of heating, and requirements for a maximum terminal speed (the speed at which the glide body approaches its target). CBO's estimates of response times take into account a range of potential design choices and improvements to drag conditions, but they do not account for any reductions in speed caused by maneuvering, which would increase flight times for hypersonic missiles. (For more information about how CBO estimated the upper and lower bounds on average speed for each type of missile, see Appendix B.)

Short Range

At ranges of less than about 1,000 km, there is significant overlap in the span of flight times for boost-glide and ballistic missiles, CBO estimates (see Figure 3-1, bottom panel). The hypersonic boost-glide missiles currently being developed by the United States are likely to have longer response times than ballistic missiles flying depressed trajectories. But if the United States can develop technologies that allow glide bodies to move faster without burning up, future boost-glide missiles have the potential to achieve roughly the same response times at short ranges as ballistic missiles flying depressed trajectories and slightly shorter response times than ballistic missiles flying minimum-energy trajectories (METs). How significant that small time advantage would be in combat is not clear.

Intermediate Range

At ranges greater than 1,000 km, hypersonic boost-glide missiles are likely to take longer to reach their targets—even under optimistic performance assumptions—than ballistic missiles on depressed trajectories. Boost-glide missiles may still take less time than MET missiles at ranges of less than about 5,000 km, but that advantage dissipates with range.

Long Range

Beyond 5,000 km, the shortest estimated response times of hypersonic boost-glide missiles exceed those of MET missiles. At those longer ranges, the time advantage that hypersonic boost-glide missiles gain by traveling a shorter path because of their low-altitude trajectory is offset by the reduction in speed that they experience because of drag as they fly through the atmosphere.

In addition, as range increases, estimates of the time difference between hypersonic boost-glide and ballistic missiles become increasingly sensitive to the hypersonic missiles' design parameters and initial speeds.

Responsiveness of Hypersonic Cruise Missiles

Hypersonic cruise missiles, which are powered by supersonic combustion ramjets, are likely to operate at speeds between Mach 5 and Mach 9. Because of their low, flat trajectory, hypersonic cruise missiles flying at those speeds would take less time to reach their targets than either boost-glide or ballistic missiles at very short ranges. Cruise missiles can end their flight at roughly any location between launch and their maximum range. By contrast, ballistic and boost-glide missiles designed for a certain range have a minimum distance they must fly as well as a maximum range.

CBO used a simple model of the flight time of cruise missiles, which does not take into account trade-offs between the burnout speed of the booster that provides initial acceleration and the ability of the scramjet to continue that acceleration (rather than just maintain the speed achieved by the booster). Using average speeds that do not depend on those design choices, CBO concluded that hypersonic cruise missiles would lose their time advantage over ballistic missiles between ranges of 700 km and 800 km.

Although scramjet technology has been demonstrated in both ground tests and flight tests, it would still require significant development to be able to be integrated into a useful weapon. DoD has stated that scramjet technology will allow for smaller, lighter cruise missiles that can be loaded on short-range tactical aircraft. Keeping the missiles' mass small enough, however, would require making concessions on both speed and range. Given those challenges, CBO limited the range for first-generation hypersonic cruise missiles in its analysis to about 500 km. The size of those missiles would limit the types of aircraft that could carry them and the number of missiles that would fit on an aircraft.

Responsiveness of Subsonic and Supersonic Cruise Missiles and Subsonic Bombers

Subsonic cruise missiles, such as the Navy's Tomahawk, are considerably slower than hypersonic missiles and ballistic missiles of the same range, flying at an average speed of Mach 0.7, CBO estimates. CBO compared subsonic cruise missiles with speeds up to Mach 1 over the ranges available with existing weapons—up to about 2,000 km—and found that their flight times would be 9 to 11 times as long, on average, as those of hypersonic boost-glide missiles and ballistic missiles of the same range (see Figure 3-1, top panel).

Supersonic cruise missiles would have flight times between those of subsonic and hypersonic cruise missiles. DoD does not currently field such weapons, nor is it developing them. But CBO included hypothetical supersonic cruise missiles in its analysis for comparison.

An Air Force B-2 bomber can fly at a maximum speed of Mach 0.95, offering similar response times but longer ranges than a subsonic cruise missile. The bomber's flight path, however, can increase or decrease the total response time for a strike, depending on the bomber's initial location and on whether it or its missile has to take a more circuitous path than the cruise missile to avoid defenses.

Importance of Positioning

The time needed to prepare a missile for launch includes positioning the launcher (whether it is located on land, at sea, or on an aircraft) so its weapons will be within range of the target. In a scenario in which the United States did not have time to move ships, submarines, and aircraft into position, ground-based launchers kept in the region would probably be able to launch missiles the most quickly and therefore be the most responsive. In scenarios in which the United States had more time to build up its forces in the region, it might be able to launch missiles from other platforms relatively quickly if they were in position.

Survivability

For a missile to be effective at destroying its target, its launcher must survive any attacks before launching the missile. And once launched, the missile must survive any attempts by air defenses or missile defenses to intercept it before it reaches its target.

Survivability of Launchers

The easiest way for a launcher to survive is to remain out of range of weapon systems that can attack it. Those systems include land-attack cruise missiles, ballistic missiles, and bombers for attacking ground-based launchers; antiship cruise and ballistic missiles and antisubmarine forces for attacking sea-based launchers; and air defenses and fighter aircraft for attacking airborne launchers. The need to avoid such weapon systems puts a premium on longer ranges for U.S. missiles.

Launchers that must operate within an antiaccess and area-denial zone can use several strategies to increase their chances of survival:

- They can operate near the outer edges of the A2/ AD zone, which boosts survival because adversaries typically have fewer weapon systems with long ranges and fewer sensors that can detect launches at those ranges. In general, the risk of attack tends to rise the closer a launch platform is to the adversary's territory.
- They can move so they are harder to track and hit.
- They can use stealth characteristics (as submarines and stealth bombers do) or camouflage (as ground-based launchers hidden among trees do) to avoid being detected. That ability means submarines are more likely to survive than surface ships, and stealth aircraft are more likely to survive than their nonstealth counterparts.
- They can be protected by active defenses, such as missile and air defenses.

There are limits to how long a launcher of shorter-range missiles can use those strategies to survive within an A2/AD zone. For example, stealth aircraft become more vulnerable if they must loiter in range of air defenses.

The same is true for ships moving at sea. But if those aircraft or ships must remain outside the A2/AD zone for protection until the last possible moment, the promptness of their missile strikes will be delayed by the extra time it takes them to get into position (within the range of the missiles they carry).

Survivability of Missiles

Once launched, strike missiles must survive any defenses that adversaries have deployed before they can hit their targets. One of the most advertised features of hypersonic missiles is their ability to penetrate defenses. The trajectories flown by hypersonic boost-glide and cruise missiles have inherent advantages for survival, but they also have unique vulnerabilities.

In general, strike missiles are difficult to defend against. Some (such as subsonic cruise missiles) fly low, staying below the horizon for much of their flight so they cannot be detected by radar until they are close to their targets.[548] Others (such as ballistic missiles and supersonic cruise missiles) use speed to reduce the effectiveness of defenses. Some missiles can also maneuver to make it harder for a defender to target or intercept them, and some can use countermeasures to help them penetrate defenses. For example, ballistic missiles can deploy decoys outside the atmosphere to make it more difficult for a defender to find and track the real warheads.

Hypersonic missiles are designed to use speed, altitude, and maneuvering to penetrate defenses. They fly almost as fast as ballistic missiles, but at lower altitudes. They also have the potential to maneuver midflight, either in altitude or in direction. Those features make it harder for them to be detected, tracked, and intercepted by air defenses (because they are too high to reach and too fast to catch) or by missile defenses (because they are too low to detect until late, too low for many midcourse interceptors to hit, and in many cases too maneuverable to track and intercept effectively with current interceptors). China and Russia are developing hypersonic weapons for just those reasons. They see advantages in having short- and medium-range missiles that could penetrate the missile defenses that protect U.S forces (particularly ships) in a theater of operations, as well as advantages in having long-range missiles that could penetrate the missile defenses that provide some protection for U.S. territory.

At the same time, the extreme heat that hypersonic missiles generate in the atmosphere makes them likely to be detectable by infrared sensors; the United States is developing space-based infrared sensors for that purpose. In addition, the trajectory of hypersonic boost-glide missiles near the end of flight is similar enough to that of aircraft that they may be vulnerable to air defenses. The effectiveness of air defenses would depend on the relative speeds and maneuverability of the interceptor and of the hypersonic missile near the end of its flight, when it had slowed considerably.

Benefits of Flying at Lower Altitudes

To illustrate the effects of geometry on the ability of a radar to detect a missile, CBO calculated the radar horizon distance, or the distance at which an object at a given height can be physically

[548] See Congressional Budget Office, National Cruise Missile Defense: Issues and Alternatives (February 2021), www.cbo.gov/ publication/56950.

seen by a radar. Because the Earth is spherical, a radar cannot see objects that are below the horizon.

Thus, a missile will remain out of view of a radar as long as it is below the radar's horizon (see Figure 3-2). Missiles can be detected earlier in their flight if the radar's altitude is increased—say, by giving the radar a taller antenna or putting it on an aircraft.

The first opportunity to detect a ballistic or hypersonic missile is during the boost phase, when its rocket motors emit a great deal of heat (infrared energy). Boost-phase detection systems are usually based in space, where factors such as the horizon are less relevant. The United States and the Soviet Union both initially deployed space-based infrared systems during the Cold War to detect long-range ballistic missile attacks by one another.

U.S. infrared systems have also proved useful for detecting launches of shorter-range ballistic missiles. Although those systems were developed for ballistic missiles, they would apply equally well to detecting launches of hypersonic boost-glide missiles, because both types of missiles get their initial boost from similar large rocket motors. (The boost phase is short, lasting just a few minutes, but an initial track of a ballistic missile can usually be calculated before the boosters burn out.) Current space-based systems have limited ability to detect launches of cruise missiles.

CBO's calculations indicate that a hypersonic boost-glide missile can avoid being detected by ground-based radar located near the target for much more of its flight than can a ballistic missile on a minimum-energy trajectory, particularly as ranges increase. For example, if a radar's antenna was 30 meters above the ground, it would detect a notional hypersonic boost-glide missile with a range of 1,000 km when the missile was just over 40 km above the Earth, about 850 km away from the radar (see Figure 3-3, left panel). A ballistic missile on a MET would be detectable only a little bit earlier.

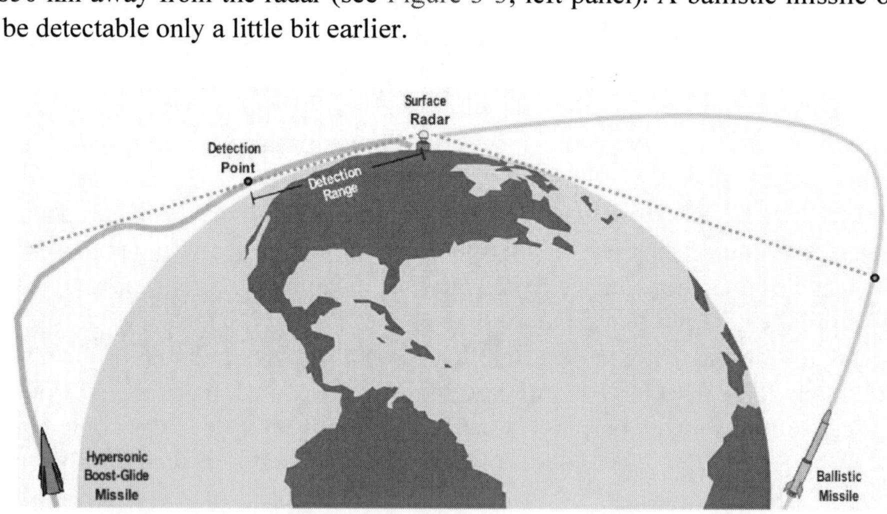

Because of the curvature of the Earth, a radar cannot detect a missile that is below the horizon. The distance to a radar's horizon increases with the height of the radar's antenna and the altitude of its target.

Data source: Congressional Budget Office.

Radars' actual horizon limits are much shorter than they appear in this figure because the figure is not to scale.

Figure 3-2. Effects of Altitude on a Radar's Ability to Detect a Missile.

On a longer flight, however, the relative advantage of the hypersonic boost-glide missile's lower altitude is much greater. For instance, if the missile was traveling 3,000 km, it would not be detectable when it reached its apogee (highest point), and thereafter it would continue to spend most of its flight below the radar's horizon.

Like the shorter-range boost-glide missile, it would come into continuous view about 850 km away from the radar, when it was gliding at an altitude of about 40 km (see Figure 3-3, right panel). The ballistic missile on a minimum-energy trajectory, by comparison, could be detected on its initial ascent and would remain in view of the radar for the remainder of its flight (assuming the radar was powerful enough to detect missiles 3,000 km away). A ballistic missile on a depressed trajectory would be visible for less time than one on a MET, but it would still be detectable for the majority of its flight.

The ability to detect and track a missile at any point does not automatically translate into the ability to intercept it. Effective defenses in the correct position are needed for that. Neither China nor Russia has yet deployed defenses that are likely to be very effective against medium- or intermediate-range ballistic missiles in the middle of their flight. But if those countries did, hypersonic missiles have features that would be advantageous against midcourse-phase missile defenses.

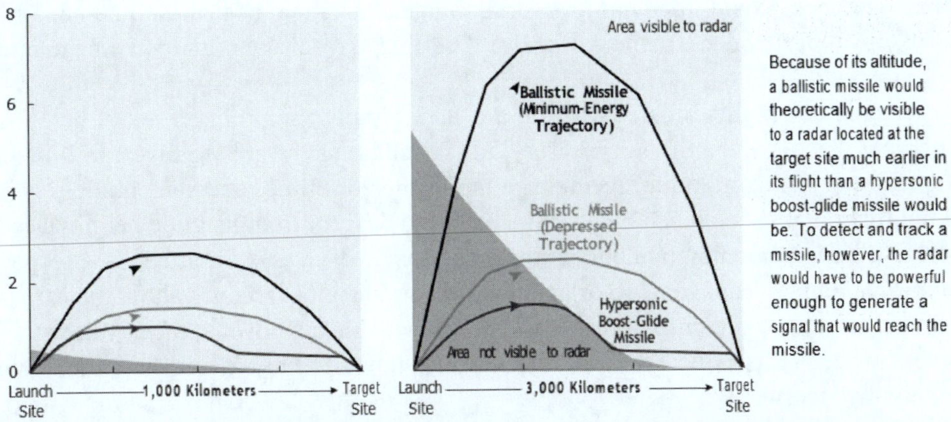

Data source: Congressional Budget Office.

In this example, the radar is located next to the target and has an antenna 30 meters high.

Ballistic missiles' trajectories are roughly parabolic, whether a missile is flying a minimum-energy trajectory (which allows it to fly the longest distance) or a depressed trajectory (with lower overall altitudes and modified booster properties that increase initial speeds). The trajectory of a boost-glide missile is modeled as a ballistic flight for the first half of its range, followed by a rapid descent to the glide altitude and a slowly descending glide path for the second half of its range.

In CBO's modeling, visibility to radar is a simple geometric calculation based on the curvature of the Earth, the altitude of the missile, the distance to the target, and the refraction of the radar caused by the atmosphere. That calculation indicates only that the missile could be visible above the horizon; the radar would need to have enough power and resolution to actually detect the missile.

Figure 3-3. Points at Which Hypersonic and Ballistic Missiles Could Be Visible to Ground-Based Radar at Ranges of 1,000 Kilometers and 3,000 Kilometers.

A missile defense system's interceptors must be designed to operate either inside the atmosphere (using control surfaces to maneuver and materials and sensors that can tolerate atmospheric heating) or outside the atmosphere (using thrusters to maneuver). Missile defenses that are designed to intercept ballistic missiles when they are outside the atmosphere in the midcourse phase of their flight cannot be used to intercept missiles inside the atmosphere, such as hypersonic boost-glide missiles during their glide phase.

Benefits of Flying Unpredictable Trajectories

The predictable path of a ballistic missile can be tracked with relative ease without continuous observation. Midcourse ballistic missile defense systems, such as the U.S. Ground-Based Midcourse Defense, rely on a combination of satellites to detect a launch, terrestrial radar and physics calculations to track the missile, and interceptors that are designed to engage the missile's warhead during the midcourse phase of its flight using the system's prediction of where the missile will be when the interceptor reaches it.

Hypersonic missiles pose challenges for such midcourse defense systems because they have the potential to make unpredictable changes to their flight path. That maneuverability has two implications for midcourse defenses. It requires that a hypersonic missile be tracked along much of its flight path. And it requires either that interceptors be located close enough to reach the hypersonic missile and intercept it before it maneuvers again or that the interceptors have enough energy to respond to a maneuver and still be able to reach the missile in time to intercept it.

Hypersonic boost-glide missiles fly on a coasting trajectory from an initial height in the atmosphere. To achieve that glide trajectory, the missile must be able to divert from its initial ballistic trajectory at some point, reenter the atmosphere, pull up, and start gliding. Near the end of the flight, it must also be able to initiate a rapid descent from its glide height. Designers hope to develop hypersonic glide bodies that can maneuver evasively, either as a response to detection or attempted engagement, or as a way to introduce uncertainty about their trajectory and ultimate target by executing a preset series of maneuvers unknown to the adversary. (Those maneuvers would reduce the speed of the missile, however, increasing its flight time and shortening its range.)

Responsive maneuvering is much harder to achieve than preplanned evasive maneuvering, but both are significant challenges for an object traveling through the atmosphere at hypersonic speeds. The Army, Navy, and Air Force are not pursuing responsive maneuvering in their boost-glide missile programs, but some research programs are looking into that technology.

Net Effect on Survivability

For all of the reasons noted above, hypersonic boost-glide missiles are likely to be detected on launch, but current defensive systems lack the ability to reliably intercept them. In the midcourse phase, acquiring tracking data with sufficient resolution to target the missiles with interceptors will be difficult unless potential adversaries develop space-based infrared sensors to track boost-glide missiles by their heat.

Developing interceptors with enough agility and maneuverability to engage those missiles will also be difficult.

Intercepting missiles in the midcourse phase is challenging, even with good tracking data and even for ballistic missiles (because they can employ countermeasures). The difficulty would be greater in some respects with hypersonic missiles, but the end effect might not be different, because neither China nor Russia can intercept long- range ballistic missiles today.

Terminal-phase defense systems may be better suited to intercepting hypersonic missiles, because the missiles will be close enough to track with ground-based radar and will be traveling at lower speeds by that time (possibly below Mach 5, in the supersonic range). However, terminal defenses have not been designed to counter hypersonic missiles. In addition, they are generally intended to cover a relatively small area, which makes them more appropriate for protecting high-value targets against attack than for protecting large portions of a country.

It is unclear how much advantage hypersonic missiles would have over ballistic missiles with maneuverable warheads against shorter-range, terminal defenses. By flying low, hypersonic missiles could force the ground- based radar associated with short-range defenses to detect a missile later in flight, shortening the time the defense has to intercept it. But because hypersonic missiles use up much of their energy during their glide, they are likely to be traveling more slowly than MaRVs as they near their targets. And both types of missiles can perform preprogrammed maneuvers near their targets to complicate the task of intercepting them.

Overall Effectiveness in an Antiaccess and Area-Denial Environment

The combination of range, responsiveness, and survivability determines the overall effectiveness of a weapon in an antiaccess and area-denial environment. Effectiveness is defined here as the ability to strike targets of interest given the constraints on distance, basing, and time imposed by an A2/AD environment and by U.S. strategy.

Trade-Offs Between Weapons' Characteristics

Significant trade-offs exist between range, speed, and platform survivability. Many existing weapon systems perform extremely well by some of those measures, but they do not combine all of the features needed be effective in an A2/AD zone, particularly during the early stages of a conflict. For example, Tomahawk cruise missiles have only a medium range (approximately 2,000 km), but they can potentially be launched from submarines operating well inside A2/AD defenses.

Because they fly at very low altitudes and can follow routes that avoid known defenses, they would be survivable on the way to their targets, but they might be vulnerable if there were sophisticated defenses around the targets. Tomahawk missiles are slow and could take up to two hours to reach their targets, depending on the location of the submarines from which they were launched. If submarines could not operate close to the target, faster missiles with longer range would be needed to provide short response times.

The specific capabilities of an adversary's defenses are also important. For example, an opponent might have sensors that could detect and track a high-flying strike missile but might lack weapons that could shoot down the missile. (Early detection might still be useful if it allowed people in the target area to be alerted in time to take cover or otherwise prepare for an attack.) Conversely, a surface-to-air missile system with long-range interceptors might not be able to take advantage of its interceptors' range if it could not detect a stealthy, low-flying cruise missile until the missile was very close. Likewise, China's long-range antiship missiles would only be effective if paired with the ability to detect and track ships at similarly long ranges.

An arsenal that included a variety of strike weapons with different ranges, speeds, launch modes, and flight trajectories would probably be the most successful against an adversary with defensive systems of varying capabilities.

Comparing Flight Distances Within Time Limits

A2/AD environments impose two important constraints: They increase the range at which U.S. forces would have to operate (known as the standoff distance), and that greater distance in turn lengthens the time it would take for a weapon to reach its target. U.S. doctrine for conventional warfare emphasizes striking early in a conflict and attacking some targets quickly enough that they do not have time to operate or respond effectively to the U.S. attack. Together, those factors place a premium on having very fast weapons that can operate from ranges as long as 3,000 km to 5,000 km in the case of surface-launched weapons and more than 1,000 km in the case of air-launched weapons.

Table 3-3. Range that a missile can travel in 15 minutes at various average speeds

Average Speed (Mach number)[a]	Range in 15 Minutes (Kilometers)	Missiles That Can Travel at That Average Speed	Defensive A2/AD Systems in That Range
0.7	200	Subsonic cruise missiles	Surface-to-air missiles
1	300	Supersonic cruise missiles	Surface-to-air missiles
5	1,500	Hypersonic cruise missiles, medium-range ballistic missiles	Fighter aircraft
10	3,000	Hypersonic boost-glide missiles, intermediate-range ballistic missiles	Air-launched cruise missiles
15	4,500	Intermediate-range ballistic missiles	Antiship ballistic missiles

Data source: Congressional Budget Office. See www.cbo.gov/publication/58255 #data.

The distance that the missiles could travel in 30 minutes would be roughly twice the ranges shown here. A2/AD = antiaccess and area-denial.

[a] "Mach number" refers to an object's speed relative to the speed of sound in the air at sea level; for example, Mach 5 is five times the speed of sound.

CBO compared hypersonic missiles and other weapon systems by evaluating the distance that each can fly within two-time constraints, 15 minutes and 30 minutes, by using the average speed over the entire trajectory. CBO's analysis indicated that two aspects of time-sensitive targeting are important, regardless of weapon type. First, average speeds of at least Mach 5 are essential for quick strikes against all but the closest targets. Second, as the time constraint is shortened, even the highest speeds are not adequate at longer distances; proximity to the target becomes important as well.

To evaluate the combined effects of standoff distance and speed, CBO calculated how far specific weapons might be able to travel in 15 minutes (see Table 3-3) and what those flight distances would mean in specific A2/AD scenarios. For example, in a South China Sea scenario, a subsonic cruise missile launched from Guam, flying at a speed slower than Mach 1, would travel roughly 200 km in 15 minutes, less than 10 percent of the way to China. A ballistic missile with an average speed of Mach 5 could travel about halfway to China (1,500 km) in that time (see Figure 3-4). Both of those ranges are too short to reach any targets of interest from Guam. A hypersonic boost-glide missile with an average speed of Mach 10 could just reach the coast of China (about 3,000 km) in 15 minutes, as could a ballistic missile with the same average speed.

In 30 minutes, higher-speed missiles (Mach 10 and above) could reach nearly all of China if they could sustain such speeds over very long ranges. That would be unlikely, however, for any type of missile other than ballistic missiles with intercontinental ranges.

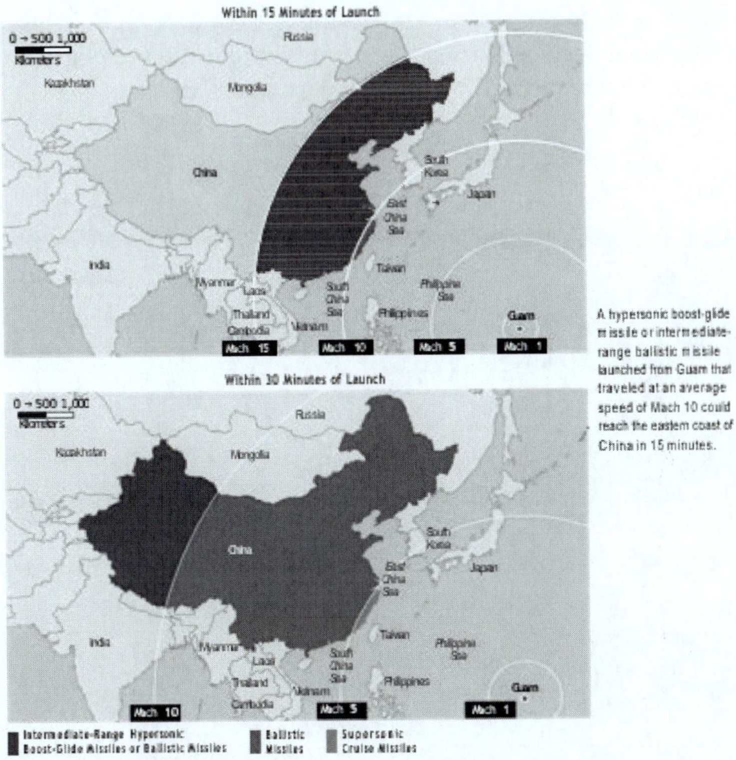

Data source: Congressional Budget Office.

The areas shown within the white circles represent the distance that a missile launched from the Pacific island of Guam and moving at an average speed of Mach 1, 5, 10, or 15 would travel in 15 minutes or 30 minutes. ("Mach" refers to an object's speed relative to the speed of sound in the air at sea level; for example, Mach 5 is five times the speed of sound.) The speed regimes correlate to subsonic cruise missiles (lower than Mach 1), supersonic cruise missiles (between Mach 1 and Mach 5), hypersonic cruise missiles (between Mach 5 and Mach 9), and hypersonic boost-glide missiles (average speeds between Mach 10 and Mach 15). Ballistic missiles could be designed with average speeds between Mach 5 (shorter range) and Mach 15 (longer range). Despite their speed, hypersonic cruise missiles would not be able to reach China from Guam, because their range will be limited to 1,000 kilometers or less by the amount of fuel they can carry.

Figure 3-4. Areas of China That a Missile Launched From Guam Could Reach Within 15 Minutes and 30 Minutes, by Average Speed.

In a Russia scenario focused on the Baltic region, conditions would be different because of the very different geography of that area. The shorter distances than in the Pacific suggest that shorter-range weapons could play a greater role early in a conflict, particularly if U.S. forces were stationed east of Germany. For example, a missile launched from the Germany–Poland border that had an average speed of Mach 5 could reach all of the Baltic Sea nations and into

Russia in 15 minutes (see Figure 3-5). A missile averaging Mach 10 could reach more than 2,000 km into Russia in that time. The same missiles based in Poland or Finland (which has applied to join NATO) could reach even farther into Russia.

In the Baltic scenario, missiles launched from ships or submarines would probably play a smaller role than missiles launched from aircraft or ground-based launchers. U.S. and allied forces would mostly be operating from land or in the air because Russia's antiship missiles could limit naval access to the Baltic Sea. The addition of Finland to NATO would also reduce the need for sea- launched options. Ground forces might play a particularly prominent role in that scenario because they are more difficult to detect and destroy than aircraft or ships.

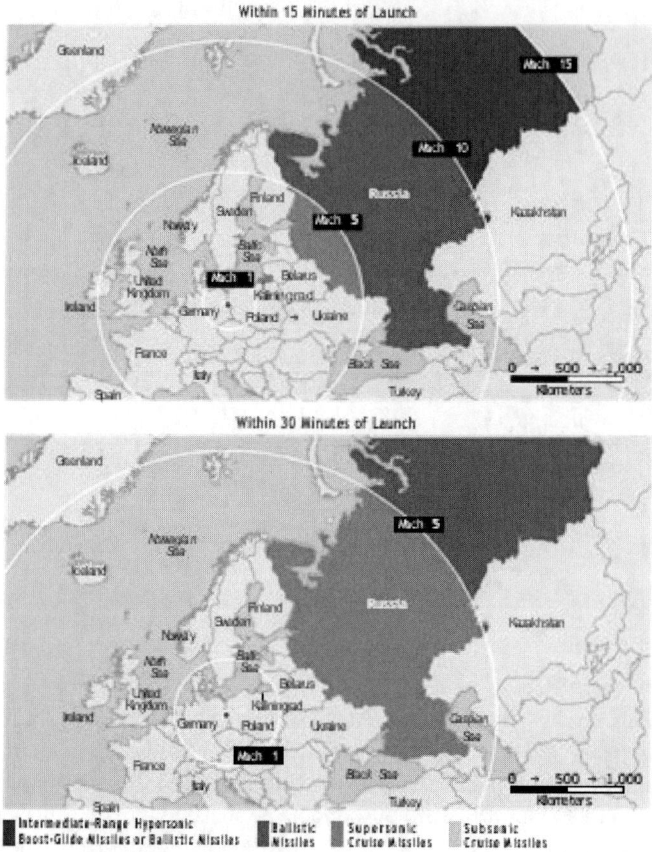

A hypersonic boost-glide missile or intermediate-range ballistic missile launched from the border of Germany and Poland that traveled at an average speed of Mach 10 could reach about 1,800 kilometers into Russia in 15 minutes.

Data source: Congressional Budget Office.

The areas shown within the white circles represent the distance that a missile launched from the eastern border of Germany and moving at an average speed of Mach 1, 5, 10, or 15 would travel in 15 minutes or 30 minutes. ("Mach" refers to an object's speed relative to the speed of sound in the air at sea level; for example, Mach 5 is five times the speed of sound.) The speed regimes correlate to subsonic cruise missiles (lower than Mach 1), supersonic cruise missiles (between Mach 1 and Mach 5), hypersonic cruise missiles (between Mach 5 and Mach 9), and hypersonic boost-glide missiles (average speeds between Mach 10 and Mach 15). Ballistic missiles could be designed with average speeds between Mach 5 (shorter range) and Mach 15 (longer range). Despite their speed, hypersonic cruise missiles would not be able to reach deep into mainland Russia from Germany, because their range will be limited to 1,000 kilometers or less by the amount of fuel they can carry.

Figure 3-5. Areas of Russia That a Missile Launched From Germany Could Reach Within 15 Minutes and 30 Minutes, by Average Speed.

Maturity of Weapon Systems

The analyses described in this Part compared weapon systems as though they were all technologically mature enough for deployment. But as another point of comparison, CBO also assessed the current development status of those systems (see Table 3-4).

Of the weapon systems in CBO's analysis, only intercontinental ballistic missiles and subsonic cruise missiles exist today in the U.S. inventory. Hypersonic boost-glide missiles and hypersonic cruise missiles are under development in the United States. But creating effective versions of those missiles is likely to pose greater technological challenges than would developing supersonic cruise missiles (which have been demonstrated by other countries) or medium- to intermediate-range ballistic missiles with MaRVs (which were part of the U.S. arsenal before the INF treaty and have some components, such as rocket boosters, roughly equivalent to those of hypersonic missiles).

Table 3-4. Characteristics of different missiles included in this analysis, by development and funding status

Key:
■ = Mature Technology
☑ = In Development as Pa t of DoD's Currer t Budget Plan
□ = Would Require Investr ent Not in DoD's Current Budget Pl n

					Able to Survive Defenses[a]		
Weapon Type	Exist Today in U.S. Inventory	Average Speed Greater Than Mach 5[b]	Range Greater Than 3,000 km[c]	Accurate Enough to Use a Conventional Warhead[d]	Air Defenses	Midcourse Ballistic Missile Defenses	Terminal Ballistic Missile Defenses
Intercontinental Ballistic Missiles	■	■	■	□	■	□	□
Medium- to Intermediate-Range Ballistic Missiles		□	□	□	□	□	□
Hypersonic Boost-Glide Missiles		☑	☑	☑	□	☑	☑
Subsonic Cruise Missiles	■			■	□	■	■
Supersonic Cruise Missiles				□	□	□	□
Hypersonic Cruise Missiles		☑	Unknown	☑	□	☑	☑

Data source: Congressional Budget Office.

DoD = Department of Defense; km = kilometers.

[a] All of the weapons shown here are designed to have some capability against missile defenses, including air defenses and midcourse and terminal ballistic missile defense systems. Ballistic missiles, because of their trajectories and high speed, are survivable against air defenses. By using countermeasures, such as decoys, they could become more survivable against potential midcourse ballistic missile defenses that adversaries might develop; and with the introduction of end-stage maneuvering, their survivability against terminal defenses would be similar to that of hypersonic boost-glide missiles. Hypersonic boost-glide missiles would be survivable against each stage of ballistic missile defenses because of their lower altitudes and maneuverability, but they would require the greatest investment.

[b] "Mach" refers to an object's speed relative to the speed of sound in the air at sea level; for example, Mach 5 is five times the speed of sound.

[c] CBO used 3,000 kilometers as a cutotf in range that represents the ability to operate from outside the reach of most antiaccess and area-denial systems. The final maximum range remains unknown.

[d] To use conventional warheads—which have a much smaller blast radius than nuclear warheads and thus require more accuracy—ballistic missiles would need to be fitted with maneuverable reentry vehicles. Hypersonic boost-glide missiles would require similar capability at the end of their flight. (Ballistic missiles and hypersonic boost-glide missiles have similar speeds and ranges.)

Part 4: Comparing Options for Planned Hypersonic Missiles and Alternatives

Building on its general comparison of the range, responsiveness, and survivability of different kinds of weapon systems, described in Part 3, the Congressional Budget Office created five illustrative options with specific missiles to show the array of approaches that the U.S. military could use to strike targets quickly from long distances. Three of the options include hypersonic missiles that the Department of Defense is currently developing, and the other two include alternative missiles that incorporate more conventional technology:

- Option 1 consists of the Army's Long-Range Hypersonic Weapon and the Navy's Intermediate- Range Conventional Prompt Strike missile (LRHW/ IR-CPS), which are intermediate-range hypersonic boost-glide missiles that would be launched from the ground or at sea, respectively. (For definitions of missiles' ranges and other characteristics, see Box 1-1.)
- Option 2 consists of intermediate-range ballistic missiles with maneuverable reentry vehicles that could be launched from the ground or at sea. Those missiles do not exist in the U.S. inventory today, but the technology for both the missiles and the MaRVs is well developed. The United States fielded similar missiles in the 1980s, before signing the Intermediate-Range Nuclear Forces treaty. In addition, the rocket boosters for the ballistic missiles would be the same ones that DoD is developing for the LRHW/IR-CPS missiles in Option 1.
- Option 3 consists of the Air Force's AGM-183A Air-Launched Rapid Response Weapon (ARRW), a medium-range hypersonic boost-glide missile.
- Option 4 consists of an air-launched, short- or medium-range hypersonic cruise missile like the ones that the Air Force and the Navy are in the early stages of developing, the Hypersonic Attack Cruise Missile (HACM) and the Hypersonic Air-Launched Offensive Antisurface Warfare (HALO) missile, respectively.
- Option 5 consists of a land-attack version of the Navy's existing Standard Missile-6, which would have extended range and speed and could be launched from the ground or at sea.

For those options, CBO estimated the costs to procure 300 missiles, integrate them with existing platforms, buy launchers and other associated equipment, and sustain the missile systems for 20 years (for more details, see Appendix C). In addition, CBO compared the effectiveness of the options and assessed how each could improve DoD's capability to rapidly strike targets from long distances, such as targets in adversaries' antiaccess and area-denial zones. For the hypersonic missiles now in development, CBO's estimates of capabilities are based on information publicly available from DoD.

CBO based its estimates of costs for the surface- and air-launched boost-glide missiles in Options 1 and 3 as closely as possible on public information about the characteristics of the LRHW/IR-CPS and ARRW missiles now in development and DoD's procurement goals for them. Although CBO refers to the missiles in those options as the LRHW/IR-CPS and ARRW for simplicity, CBO's estimates are not intended to predict the costs of the actual DoD programs, because considerable uncertainty remains about how those programs' schedules and inventory goals may evolve. Rather, CBO's options are meant to illustrate the relative costs of the high-speed missiles being developed by DoD and potential alternatives. CBO did not include in its estimates any calculation of the risks or consequences of cost and schedule growth. However, many complex technologies that have been fielded by DoD in the past have been subject to significant delays and cost overruns.

The five illustrative options are not mutually exclusive; DoD could field a mix of missile systems to use in specific situations. Therefore, CBO also analyzed the value of having redundancy in the U.S. inventory of counter-A2/AD weapons.

Comparing the options, CBO found that the intermediate-range ballistic missiles in Option 2 could provide the range, responsiveness, and most of the survivability of the LRHW/IR-CPS hypersonic boost- glide missiles in Option 1, at roughly three-quarters of the cost. CBO also found that ARRW hypersonic boost-glide missiles launched from aircraft (Option 3) would cost roughly one-third of what a similar number of LRHW/IR-CPS missiles would cost. But the ARRW missiles would have less range and would provide the same responsiveness only if aircraft were already in position to launch.

A smaller number of air-launched missiles could provide a worldwide availability similar to that of Option 1 or could complement intermediate-range missiles if an adversary's long-range air defenses could be degraded.

Thus, CBO estimated costs for two quantities of ARRWs in Option 3: the same number of missiles as in Options 1 and 2 (300) and a smaller quantity (100) that represents a complementary capability to either Option 1 or Option 2.

The ground- or sea-launched SM-6 missiles in Option 5 would be the least costly alternative, but they would offer the lowest performance in terms of both speed and range. (CBO did not estimate costs for the hypersonic cruise missiles in Option 4, because they are still early in their development and little information about them is available.)

CBO's illustrative options are not intended to cover all possible missiles that could be used in the A2/AD and other scenarios described in Part 2. For example, the Navy's Tomahawk medium-range subsonic cruise missile, though relatively slow, is likely to remain a significant part of DoD's arsenal in any conflict. Indeed, the Army plans to procure Tomahawks in coming years and deploy them with ground launchers. In addition, the Air Force is developing a longer-range version of its own air- launched, long-range cruise missile, the AGM-158.

To be effective, all of the options would depend on a network of sensors, tracking systems, and communications systems that would permit the type of over-the-horizon targeting envisioned by advocates of long-range standoff weapons. The response-time demands on those systems would be particularly great if DoD wanted to conduct rapid strikes on fleeting targets.

In general, such high-end sensor, tracking, and communications systems do not exist today. But DoD is developing a number of systems that could fulfill those roles if they proved successful. They include command- and-control systems such as the Air Force's Advanced Battlefield Management System and DoD's Joint All- Domain Command and Control System, as well as sensors and communications networks of the type that the Space Development Agency envisions for its Proliferated Warfighter Space Architecture. For the purposes of this analysis, CBO assumed that those systems had been deployed successfully, as had the aircraft and submarines that might carry the missiles in the options. However, the costs of those systems, which could be significant, are not included in CBO's analysis.

Characteristics of the Missile Options

The five notional missile systems that CBO analyzed run the gamut from intermediate- to short-range weapons and use all three launch modes (ground, sea, and air).

The options vary in the state of their development and technological maturity. Two of the options feature the first-generation hypersonic boost-glide missiles that the Army, Navy, and Air Force are developing and hope to deploy in the next several years. Two other options feature hypersonic cruise missiles that the Air Force and Navy are in the early stages of designing and short-range ballistic missiles that the Navy is developing by upgrading existing missiles. The remaining option features intermediate-range ballistic missiles that are not deployed or in development but that would incorporate well- established technology that has been used in past U.S. missiles.

Option 1: Intermediate-Range, Ground- or Sea-Launched Hypersonic Boost-Glide Missiles (Similar to LRHW/IR-CPS)

Under Option 1, the Army and Navy would complete and field the hypersonic boost-glide missile they are collaborating on, the LRHW and IR-CPS, which is expected to have a range of at least 3,000 kilometers (about 1,900 miles). The missile could be launched from vehicles on land as well as from ships and submarines at sea. It would use a two-stage rocket booster to reach an initial speed in the high hypersonic regime (Mach 15 or above), which would be necessary for it to glide to an intermediate range. The missile would reach an apogee of roughly 120 km before it reentered the atmosphere, after which the glide body would spend most of its flight at an altitude between 30 km and 40 km. The glide body would require thermal shielding to protect it and its sensitive electronics from temperatures exceeding 1,800 Kelvin (2,780° Fahrenheit).

The glide body would be enabled with the Global Positioning System, allowing the same sort of accurate and precise targeting as a Joint Direct Attack Munition guided bomb. But it would not have an onboard sensor to home in on its target during the final stage of flight. (Research to develop such a capability is ongoing, but it will not exist in the first-generation hypersonic missiles that DoD is building today.) The missile would also have a limited ability to maneuver outside of a preset course.

Missiles Deployed on Land

As reported by the Army, a hypersonic missile battery will include four transporter-erector-launchers that carry two missiles each. Reports have indicated that the Army plans to field six batteries and purchase 66 LRHW missiles. Potential launch locations in the Pacific region include the islands of Guam and Palau (and possibly countries closer to China if they were to allow the deployments). European locations could include U.S. military bases in Germany as well as bases of Eastern European and Nordic countries.

The Army's hypersonic missile units would be mobile, meaning that the exact location and time to set up and execute a missile launch might depend on the details of a particular conflict (as with an air or sea launch).

However, given the time pressures in a crisis, the time between receiving target coordinates and firing a missile would probably be relatively short for a ground-based system, provided that the system had already been deployed in the theater and was ready for launch. In such circumstances, ground launch could be faster than launches from sea or air if ships and aircraft were not in position to fire immediately.

Missiles Deployed at Sea

The Navy has indicated that it plans to buy 240 IR-CPS boost-glide missiles. They would be carried on Zumwalt class destroyers and on Virginia class submarines equipped with Virginia payload module (VPM) missile tubes. Current plans call for both Zumwalt destroyers and VPM-equipped submarines to have a maximum capacity of 12 IR-CPS missiles apiece, although it is likely that some of the missile tubes would be used for other weapons.

In the Pacific region, many launch locations at sea are possible. In the Baltic region, sea launch would probably be less likely, for two reasons: striking inland targets would be hard if Russian antiship and antisubmarine weapons limited U.S. operations in the Baltic Sea, and many alternative launch locations would be available on land. In either regional scenario, positioning a submarine or ship in an appropriate spot for launch might take significant time (hours or days). That time constraint might not be a problem in a slowly unfolding crisis, but it would create more difficulty in a sudden conflict. The coverage that a given IR-CPS missile would achieve would depend on its proximity to shore when it was launched.

For ease of comparison among the options, CBO estimated costs for a notional purchase of 300 missiles in each option. That amount is close to DoD's reported plan to purchase a total of 306 LRHW/IR-CPS missiles (66 LRHWs for the Army and 240 IR-CPSs for the Navy).

Option 2: Intermediate-Range, Ground- or Sea-Launched Ballistic Missiles Equipped with MaRVs

Under Option 2, DoD would develop and field a ballistic missile with a range of 3,000 km to 5,500 km (about 1,900 miles to 3,400 miles). The U.S. military fielded a medium-range ballistic missile armed with nuclear warheads (the Pershing II) until the late 1980s, when those missiles were destroyed in adherence to the INF treaty. The ballistic missiles in this option would have longer ranges than the Pershing II missiles and would carry conventional warheads.

The new ballistic missile would use the same two-stage rocket booster as the boost-glide missile in Option 1, but instead of a hypersonic glide body, it would be equipped with a maneuverable reentry vehicle for precise targeting. The MaRV would be able to change its flight path once it had reentered the atmosphere (in the terminal stage of flight), improving its accuracy.

CBO assumed that development of the new ballistic missile could be completed with the remaining funding planned for the corresponding ground- or sea-launched boost-glide missiles. The ballistic missile's boosters would be the same as those being developed for the LRHW/IR-CPS, and funding currently slated for the Common Hypersonic Glide Body could be used to complete past efforts to develop MaRVs. For an earlier project to arm Trident missiles with conventional warheads (the Conventional Trident Modification program), the Navy estimated that a MaRV could be developed and enter production in as little as two years.

To minimize flight time, the new missile would fly a depressed ballistic trajectory, reaching an apogee of between 300 km and 500 km. The reentry vehicle would need thermal protection similar to that of existing ballistic missile warheads, a technology that is well developed. The range of the new missile would be limited by the mass of its payload and its trajectory rather than by heating, because the missile would spend most of its flight outside the atmosphere.

Under this option, the Army would purchase 60 of the new ballistic missiles and the Navy would purchase 240, the same quantities as in Option 1.

Option 3: Medium-Range, Air-Launched Hypersonic Boost-Glide Missiles (Similar to ARRW)

Under Option 3, the Air Force would complete development of and deploy the Air-Launched Rapid Response Weapon, a hypersonic boost-glide missile with a range that CBO estimates at roughly 1,000 km (about 600 miles). That missile would have a shorter range and lower speed than the ground- and sea-launched hypersonic missiles in Option 1. It would also be smaller, enabling it to be carried on Air Force bombers and possibly on some larger fighter aircraft. CBO's analysis suggests that the missile's average speed over a 1,000- mile trajectory would be about Mach 7. Because of its smaller booster and lower intended speed and range, the ARRW would reach a lower apogee than the LRHW and IR-CPS: an altitude of perhaps 80 km to 90 km, still inside the atmosphere.

Other features of the boost-glide missile in this option, such as accuracy and maneuverability, would be similar to those of the boost-glide missiles in Option 1 and the ballistic missiles in Option 2. The ARRW currently being developed has a more complex aerodynamic design than the LRHW and IR-CPS, which indicates that it may have the potential for more complex maneuvering or ranges longer than 1,000 km. Whether those improvements can be achieved, however, is unclear.

The Air Force is considering several aircraft to carry the ARRW, including B-52 and B-1 bombers and F-15 fighters. Other aircraft are also possible. The B-52 is expected to be able to carry two ARRW missiles under each wing and might eventually be able to carry additional missiles internally. The F-15E and F-15EX might also be able to carry one ARRW apiece on their center- line weapon station.

The Air Force has not indicated how many ARRW missiles it plans to purchase. For this analysis, CBO estimated the costs of two ARRW procurement quantities: 300 missiles, to provide a direct comparison with the 300 LRHW/IR-CPS missiles in Option 1, and 100 missiles, to reflect the fact that fewer air-launched boost-glide missiles could provide a worldwide availability similar to that of a larger number of ground- and sea-launched boost-glide missiles.

Deploying ground units and moving submarines and ships long distances can take days or weeks. So ensuring that LRHW/IR-CPS missiles were in the right place for rapid launch would require dispersing missiles among ground batteries and ships deployed around the world. In contrast, a smaller number of ARRW missiles could be stored in just a few locations, ready to be flown to where they might be needed and launched on short notice.

Option 4: Short-Range, Air-Launched Hypersonic Cruise Missiles (Similar to HACM/HALO)

Under Option 4, the Air Force would develop and build a hypersonic cruise missile, powered by a supersonic combustion ramjet engine, with a range of less than 1,000 km. The missile would be similar to the ones envisioned in the Air Force's HACM program or the Navy's HALO program, although both of those missiles are still early in the process of development.

The Air Force missile in this option would be launched from B-52 bombers and perhaps other aircraft. It would require a small rocket booster to achieve supersonic speeds, at which point the scramjet engine would begin to operate. CBO assumed that the missile would travel at a roughly constant speed, which could be as low as Mach 5 or as high as Mach 9. It would maintain an altitude of about 30 km to 40 km for the duration of its flight, well within the atmosphere. As a result, the engine components and external housing would have to with- stand significant amounts of heating.

Estimates released by DoD indicate that its air-launched hypersonic cruise missiles would be much smaller than the air-launched hypersonic boost-glide missiles in Option 3, so up to four times as many of them could be carried on a B-52. However, the cruise missiles are in a much earlier stage of development than the boost-glide missiles, with much less mature technology, so it is not yet clear how large they would be. (CBO considers DoD's size estimate optimistic because the cruise missiles would need a rocket booster, they would have to carry fuel, and their efficiency would depend on the length of the engine.) The missiles in this option would probably also have a shorter range than the missiles in the other four options that CBO analyzed.

Because little is known about the characteristics of the hypersonic cruise missiles currently in development or the details of their underlying technology, CBO did not estimate the cost of the missile system in this option.

Option 5: Short-Range, Ground- or Sea- Launched Ballistic Missiles (Similar to SM-6 Block IB)

Under Option 5, the Army and the Navy would deploy the Standard Missile-6 Block IB, which the Navy is in the process of developing. The Block IB is a variant of the Navy's existing SM-6 Block IA, a short-range, supersonic guided missile that can be used for air defense, missile defense, and antiship warfare. Reports indicate that the Army is also interesting in fielding the Block IB missile.

The new missile is a significant modification of the Block IA version, with a range that may be as long as 1,000 km and an average speed in the low hypersonic regime (Mach 5 to Mach 6). That range and speed require the development of a new, larger booster as well as enhanced sensor technology. The SM-6 Block IB missile would be used for attacking surface targets and would be powered by a two-stage rocket booster. The SM-6 Block IB is not a hypersonic boost-glide missile, because it would fly a modified ballistic trajectory and would not use the air to produce lift.

To provide a direct comparison with the other options, CBO assumed that the Army and the Navy would purchase a combined total of 300 SM-6 Block IB missiles under this option, plus ground equipment for six Army batteries. The Army's missiles could be fired from modified ground-launch vehicles. The Navy's missiles could be fired from the same large tubes on Zumwalt class destroyers and Virginia class submarines as the IR-CPS missiles in Option 1. The Block IB missiles would also be compatible with the MK 41 vertical launchers on cruisers and Arleigh Burke class destroyers, greatly increasing the number of ships that could carry them. For that reason, it is possible that more SM-6 Block IB missiles would be purchased.

Costs of the Options

For its illustrative options, CBO estimated several types of costs: the costs to purchase the missiles; the costs to integrate the missiles with existing platforms and to buy associated equipment, such as launchers; and the costs to sustain the missile systems for 20 years. (Those estimates are described in more detail in Appendix C.) CBO did not include the costs of developing the missiles in its estimates. Also, CBO could not estimate the cost of the hypersonic cruise missiles in Option 4, because their technology remains immature and the ultimate cost to procure such a missile is unclear.

The intermediate-range hypersonic boost-glide missiles in Option 1 would be the most expensive of the options that CBO estimated, costing an average of $41 million apiece (in 2023 dollars) to procure—about 60 percent more than the intermediate-range ballistic missiles with MaRVs in Option 2, which would cost $26 million apiece (see Table 4-1). Including 300 missiles, five batteries of ground vehicles, launchers, ship modifications, and 20 years of sustainment would bring the total cost of Option 1 to $17.9 billion.[549] That total is about one-third more than the cost of the intermediate-range ballistic missile system in Option 2, which would total $13.4 billion, CBO estimates.

Of the medium- or short-range missiles in the options, the ARRW hypersonic boost-glide missiles in Option 3 would be the more expensive of the two types for which CBO estimated costs. Those air-launched boost-glide missiles would have an average procurement cost of $15 million apiece for 300 missiles and a total cost of $5.3 billion to procure and sustain them for 20 years.

The SM-6 Block IB ballistic missiles in Option 5 would cost the least per missile, $6 million, and the least overall, $3.5 billion.

Air-launched missiles, like those in Options 3 and 4, offer DoD the flexibility to move missiles quickly to where they are needed. Because of that flexibility, fewer air-launched missiles could potentially provide a world- wide availability similar to that of a larger number of ground- or sea-launched missiles, which are bound to the locations of their launch platforms. Smaller procurement quantities would increase the average cost per missile in Options 3 and 4 but would reduce the overall cost of the options. For example, reducing the number of hypersonic missiles in Option 3 to 100 would reduce the total cost of that option from $5.3 billion to $2.2 billion.

For the hypersonic missiles that are furthest along in development, the Army and Navy's LRHW/IR-CPS (Option 1) and the Air Force's ARRW (Option 3), CBO produced its own estimates based on analogies with existing missiles or on established cost-estimating relationships for missile components, such as boosters. As a check on those estimates, CBO compared its results with the limited cost information provided by DoD. CBO used a similar approach to estimate the cost of the intermediate-range ballistic missiles in Option 2. For the SM-6 Block IB missiles in Option 5, CBO relied on estimates by the Navy, because the Block IB is a modification of a missile that is currently in production and for which the Navy has provided cost estimates.

---

[549] Although reports have indicated that the Army plans to field six LRHW batteries, the first battery has reportedly already been equipped. Thus, the estimate for Option 1 only includes the costs of equipping the other five batteries.

Table 4-1. Costs of the missile options that CBO analyzed

			Costs of Option (Billions of 2023 dollars)			
	Number of Missiles Purchased	Average Procurement Cost per Missile (Millions of 2023 dollars)[a]	Missile Procure-ment[a]	Platform Integ-ration	20 Years of Sustainment	Total[b]
Intermediate-Range Missiles (Range 3,000-5,500 km)						
Option 1: Ground- or Sea-Launched Hypersonic Boost-Glide Missiles (Similar to LRHW/IR-CPS)	300	41	12.2	2.7[c]	3.0	17.9
Option 2: Ground- or Sea-Launched Ballistic Missiles Equipped with MaRVs	300	26	7.7	2.7	3.0	13.4
Medium-Range Missiles (Range 1,000-3,000 km)						
Option 3. Air-Launched Hypersonic Boost-Glide Missiles (Similar to ARRW)[d]	300	15	4.5	0.2[e]	0.6	5.3
	100	18	1.8	0.2[e]	0.2	2.2
Short-Range Missiles (Range less than 1,000 km)						
Option 4: Air-Launched Hypersonic Cruise Missiles (Similar to HACM/HALO)[f]	n.e.	n.e.	n.e.	n.e.	n.e.	n.e.
Option 5: Ground- or Sea-Launched Ballistic Missiles (Similar to SM-6 Block IB)	300	6	1.9	1.0	0.6	3.5

Data source: Congressional Budget Office. See www.cbo.gov/publication/58255#data.

ARRW = Air-Launched Rapid Response Weapon (being developed by the Air Force); HACM = Hypersonic Attack Cruise Missile (being developed by the Air Force); HALO = Hypersonic Air-Launched Offensive Antifsurface Warfare missile (being developed by the Navy); IR-CPS = Intermediate-Range Conventional Prompt Strike missile (being developed by the Navy); km = kilometers; LRHW = Long-Range Hypersonic Weapon (being developed by the Army); MaRVs = maneuverable reentry vehicles; n.e. = not estimated; SM = Standard Missile (the SM-6 Block IB is being developed by the Navy as a variant of the existing Block IA).

[a] Includes costs to procure the missiles and associated equipment, costs to assemble and integrate the individual missile components, and costs to test that the assembly and integration have been done correctly. (Costs to integrate the missiles onto their launch platforms are included under "platform integration.")

[b] Excludes costs to develop the missiles.

[c] Includes the costs of ground vehicles for five Army missile batteries and costs for the Navy to integrate the missiles onto its three Zumwalt class destroyers.

[d] CBO estimated the costs of two versions of this option: one with 300 missiles, the same as the number of hypersonic LRHW/IR-CPS missiles in Option 1, and one with 100 missiles to reflect the fact that the air-launched weapons in this option would not need to be spread among ships and submarines in ditferent parts of the world but could be stored in a few locations and flown to where they were needed on short notice. Thus, a smaller number of the air-launched hypersonic missiles in this option could provide a worldwide availability similar to that provided by the larger number of ground- or sea-launched hypersonic missiles in Option 1.

[e] Includes integration only with B-52 bombers.

[f] The Department of Defense is early in the process of developing this missile. Little is known about its characteristics, so CBO did not have a basis for estimating the cost of the missile.

Significant uncertainty exists about the costs of less mature technology and components related to hypersonic weapons, especially because developing them requires long-term investment in a wide variety of research areas that are associated with high risks of cost and schedule overruns. In particular, continued or increased amounts of research spending on aerothermo- dynamics and guidance systems might be necessary to achieve DoD's original goal of developing hypersonic missiles that can strike relocatable (or perhaps even moving) targets. For each option, the estimated costs described here represent one possible outcome based on the specific assumptions made in the option about the missiles' technologies and capabilities.

Effectiveness of the Options

Each of the missile systems in CBO's illustrative options would offer capabilities that would be relevant in the A2/AD and other scenarios described in Part 2.

The intermediate-range hypersonic boost-glide missiles and ballistic missiles in Options 1 and 2 would provide similar range, responsiveness, and survivability against current and emerging threats. The air-launched weapons—the medium-range hypersonic boost-glide missiles in Option 3 and the hypersonic cruise missiles in Option 4—would have shorter ranges and would rely on aircraft to get closer to their targets, which would increase both flexibility in positioning and the vulnerability of their launch platforms. Like the air-launched missiles, the ground- or sea-launched ballistic missiles in Option 5 would have short range and would complement the intermediate-range options.

Option 1: Intermediate-Range, Ground- or Sea-Launched Hypersonic Boost-Glide Missiles (Similar to LRHW/IR-CPS)

Launched from Guam, an intermediate-range boost- glide missile like the ones in Option 1 would be able to reach coastal regions of mainland China (a range of 3,000 km) in less than 15 minutes, CBO calculates. If the launchers were at the ready, it might take only about 5 minutes to launch the weapons once ordered to do so, for a total response time of 20 minutes. That rapid a response would require ground-based launchers to be deployed and ready to fire and submarines or surface ships to be within range and ready to launch.

Any additional steps that the launch platforms needed to take before being in range or ready to launch would add to the response time. For example, a ground-based launch system might take as much as 30 minutes to set up before it could fire a missile. A submarine located within 3,000 km of shore would be able to launch within a much shorter time, perhaps 5 minutes, provided that it was ready to fire. Thus, in a scenario in which mobile launchers were based on Guam and submarines were located about the same distance away from China, the intermediate-range boost-glide missiles in Option 1 could provide some of the fastest strike options from long distances early in a conflict. (If submarines were located closer to their targets than Guam, their missiles would arrive even sooner.)

The first-generation hypersonic boost-glide missiles that DoD is working on now will not be able to maneuver in response to movement by their targets. Therefore, the set of targets those missiles could attack would be limited to targets that were fixed or likely to remain fixed for at least 30 minutes. The targets would probably need to stay fixed for much longer than 30 minutes because it takes time (many minutes or longer) for U.S. intelligence, surveillance, and reconnaissance systems to find targets, confirm what they are, and transmit the information to regional commanders, as well as time for a commander's launch order to be disseminated to the crew operating the missile launcher. That set of targets would probably include coastal antiaircraft and antiship defensive systems (including missiles and sensors) and command or communications nodes. A successful strike on such targets could reduce, though not eliminate, the risks to U.S. aircraft and surface ships operating in the defended region. The ability to reach targets inland would depend on the maximum range of a missile and where it was launched from.

Given the high speed and low and unpredictable trajectory of an intermediate-range boost-glide missile, it would be difficult for an adversary to intercept the missile before it reached its target, according to CBO's analysis (unless the adversary developed hypersonic missile defenses at some future time). The launch platforms, however, might be vulnerable to attack by intermediate-range ballistic missiles and bombers, because Guam is within reach of those systems and because launching a missile from a ship or aircraft could reveal its location. The head of DoD's Indo-Pacific Command has stated that the missile defense system currently in place on Guam (a single Terminal High- Altitude Area Defense battery) leaves bases there vulnerable. DoD plans to explore adding a more capable missile defense system (the Aegis Ashore system) on Guam as part of its broader Pacific Deterrence Initiative.

Ground-based launchers would probably be more survivable in the Baltic region than on Guam, mostly because they would have many areas from which to operate.

Aircraft based relatively near the conflict zone could also have many airfields to operate from. Those conditions would greatly increase the effectiveness of ground- or air-launched boost-glide missiles in a Baltic scenario. Hypersonic missiles launched from submarines and surface ships would probably not play as large a role in the Baltic region, given the alternative land- and air-basing options.

Option 2: Intermediate-Range, Ground- or Sea-Launched Ballistic Missiles Equipped with MaRVs

The ballistic missiles equipped with maneuverable reentry vehicles in Option 2 would offer roughly the same effectiveness as the hypersonic boost-glide missiles in Option 1 in both the China and Baltic scenarios.

The two types of missiles would have the same range and reach targets just as quickly. The main difference is that the ballistic missiles would need to be able to defeat midcourse-phase ballistic missile defenses—if potential U.S. adversaries developed such systems in the coming decades that could contend with U.S. countermeasures. (Midcourse defenses designed to counter ballistic missiles would pose little threat to hypersonic boost-glide missiles because of their lower trajectories.) It is not clear whether there would be much difference in the ability of MaRVs or hypersonic glide bodies to penetrate short-range (terminal-phase) defenses located close to a target.

Because MaRVs are likely to be lighter than hypersonic glide bodies, the ballistic missiles in Option 2 could be designed to have longer ranges or larger payloads than the boost-glide missiles in Option 1. They could also be designed to fly alternative ballistic trajectories, such as depressed trajectories, that would increase their survivability. MaRVs would allow the missiles to be just as accurate as hypersonic weapons, permit them to execute maneuvers to avoid terminal-phase defenses, and let them be retargeted during flight. In addition, a MaRV would enter the terminal phase of flight at a higher speed than a boost-glide missile, which would shorten the time a terminal defense system had to intercept it. However, as with hypersonic missiles, any maneuvers that a MaRV made would reduce its speed.

Option 3: Medium-Range, Air-Launched Hypersonic Boost-Glide Missiles (Similar to ARRW)

The shorter range of the hypersonic-boost glide missiles in Option 3 means that aircraft would have to fly well within the range of China's or Russia's A2/AD defenses during a conflict to launch them. In addition, because of the travel time necessary to position the aircraft, the missiles' effectiveness would depend on the number of aircraft in the general region early in the conflict.

The medium-range hypersonic boost-glide missiles would probably also travel more slowly than their intermediate-range counterparts in Options 1 and 2, though still within the hypersonic regime (about Mach 7, on average, over the course of their flight, CBO estimates). CBO's analysis indicates that from launch, a medium-range hypersonic boost-glide missile could reach its maximum range of roughly 1,000 km in less than 15 minutes. It would probably also maintain a lower altitude early in flight than an intermediate-range hypersonic missile and would reach a lower apogee, making it even harder for ground-based radar to detect.

In a China scenario, medium-range, hypersonic boost-glide missiles would not be effective at reaching deep inland targets, because of their limited range. In a Baltic scenario, depending on the basing location, the ability to launch multiple missiles from a relatively short standoff distance could offer advantages to the United States in a conflict with Russia.

Option 4: Short-Range, Air-Launched Hypersonic Cruise Missiles (Similar to HACM/HALO)

In both the Pacific and Baltic scenarios, the effectiveness of the short-range hypersonic cruise missiles in Option 4 would be similar to that of the medium-range hypersonic boost-glide missiles in Option 3. The cruise missiles' even shorter range would increase the time that aircraft carrying the missiles would have to spend in defended areas. But if the cruise missiles were smaller, aircraft could carry more of them, increasing coverage of a given area.

CBO projects that hypersonic cruise missiles' range would be less than 1,000 km (probably closer to 500 km) and that their speed would be in the low hypersonic regime (between Mach 5 and Mach 9). Hypersonic cruise missiles would travel at lower altitudes than hypersonic boost-glide missiles for almost all of their flight, making them more difficult to track.

Option 5: Short-Range, Ground- or Sea- Launched Ballistic Missiles (Similar to SM-6 Block IB)

The short-range ballistic missiles in Option 5 would not be able to reach China if launched from Guam.

Thus, in the Pacific scenario, they would not play a large role if U.S. forces had to operate from long distances during the initial phase of a conflict, unless they could be launched from Okinawa, the Japanese mainland, or Taiwan. However, previous analysis by CBO concluded that ground-launched SM-6 missiles might be effective for defending U.S. bases and the territory of U.S. allies in the Pacific.[550] For that mission, they might operate as antiship or air-defense weapons.

Launched from a surface ship, the upgraded SM-6 missiles in Option 5 would only reach the coast of China if the ship was located within the A2/AD zone, thus requiring protection.

---

[550] See Congressional Budget Office, Options for Fielding Ground- Launched Long-Range Missiles (February 2020), www.cbo.gov/publication/56068.

Alternatively, as a sea-launched antiship missile, the SM-6 Block IB might be effective at destroying China's ship-based A2/AD assets.

In a Baltic scenario, a short-range, ground-launched ballistic missile like the upgraded SM-6 would have more applications. The ship-launched variant would be less useful, however, given the likely contested zone in the Baltic Sea.[551]

Pursuing Multiple Options

None of the options that CBO examined are mutually exclusive; lawmakers could opt to pursue several of them at once. DoD's current multipronged approach to developing hypersonic missiles has some advantages—because of the broad range of potential scenarios in which different weapons could be used—despite its higher cost than a single-pronged approach.

For example, ground-launched, intermediate-range hypersonic boost-glide missiles (Option 1) or ballistic missiles equipped with MaRVs (Option 2) would probably provide the quickest response times. But in a Pacific scenario, they would provide less coverage of targets in China than other types of missiles would. That means ground-launched missiles could be most relevant in the first phase of battle and could open access for sea- and air-launched missiles by destroying some defenses. Submarines could also provide initial strikes with hypersonic missiles if they were within range of their targets.

Air-launched medium-range weapons (like the boost- glide missiles in Option 3) would become more useful after an adversary's A2/AD systems had been significantly degraded. Those air-launched weapons would also provide shorter response times once the aircraft could loiter safely within range of their targets.

Appendix A: Other Military Research Programs Related to Hypersonic Flight

Besides the missile development programs discussed earlier in this report, the Department of Defense (DoD) funds a number of programs that focus on more basic or applied research involving hypersonic flight. Those programs are being pursued by the Defense Advanced Research Projects Agency (DARPA) and in both military and academic laboratories. The technical areas that those programs focus on provide insight into the remaining challenges for fielding hypersonic missiles with advanced capabilities.

Between them, DARPA's and the military services' budgets include dozens of program elements that involve basic research, applied research, or advanced technology development relevant to hypersonic weapons. Some illustrative examples are described below. Basic research includes projects at military laboratories and universities to develop a better understanding of the performance of advanced materials and the complex air-flow phenomena of hypersonic flight. Applied research includes efforts to bring various materials and components of hypersonic missiles to maturity and to develop methods for producing those materials and components cost-effectively. Overarching goals for such research include

551 Ibid.

increasing missiles' range, payload capacity, speed, and maneuverability and reducing costs and processing time for components.

The services and DoD are also funding research into technological advances necessary for the complex command-and-control systems and the surveillance and reconnaissance capabilities that would be essential for defending against adversaries' hypersonic missiles. Those research efforts are not included here.

Defense Advanced Research Projects Agency

DARPA continues to contribute to the Air Force's boostglide missile program through its research on the Tactical Boost Glide vehicle described in Part 1. In addition, DARPA has several research programs exploring materials that can perform at high temperatures—specifically, materials that could be used for the structures of hypersonic missiles or for radomes (transparent areas through which radio signals can travel). Those programs are investigating materials that can provide thermal protection as well as structural components that can withstand the thermal loads that hypersonic boost-glide missiles are subject to.

As part of its efforts to support development of hypersonic cruise missiles, DARPA is also investigating propulsion techniques and cost-effective system designs under the Hypersonic Air-Breathing Weapon Concept (HAWC) program, now called MoHAWC. The program received additional appropriations in 2022 to complete flight tests.

Service Programs

Like DARPA, the Air Force and Navy are investing in research to develop materials that have a combination of properties that could make them suitable for use in hypersonic weapons. Those research efforts include development of processing methods to generate potential materials and computational methods to simulate the materials' performance.

Several services are conducting research into aerodynamics at hypersonic speeds. For example, the Navy is developing simulation tools to model the aerodynamics of hypersonic bodies (including a program exploring their interactions with cloud formations). The Army is investigating the aerodynamics of various shapes at hypersonic speeds and developing models to predict the effects of turbulence.

In addition, the Army's Foundational Hypersonics Weapons Research program focuses on the fundamental physics of air flows, the mechanics of maneuver control, vehicle shapes, and computational and experimental techniques, as well as ways to reduce processing costs for carbon-carbon composites. The Navy's overarching Hypersonic Technologies program is intended to fund novel ideas for advancing the performance of hypersonic weapons in areas such as advanced materials, propulsion, stability and control, seekers and sensors, guidance, navigation, and payloads.

At a higher level of technological development, the Army's Advanced Hypersonic Technology program continues to support development of the Long-Range Hypersonic Weapon and advanced technologies to incorporate into that missile. The Army is also investing in test infrastructure (such as wind tunnels), high-temperature electronics, and high-performance computing in service of hypersonic applications. The Air Force is likewise focusing on inserting advanced technology into high-speed systems through a research program to develop simulation and test-bed technology to make technology insertion less risky for hypersonic missile programs. In addition, the Air Force is developing hardware-in-the- loop test beds for novel guidance concepts as part of its research into advanced guidance technology applicable to hypersonic missiles.

For its part, the Air Force is conducting computational research to fully characterize the challenges that conventional warheads face during hypersonic flight, including thermal and vibration issues. And the Marine Corps is funding research to develop a design for a hypersonic munition compatible with the M142 High Mobility Artillery Rocket System (HIMARS).

In the area of hypersonic cruise missiles, the Air Force continues to fund research exploring engine components and aerodynamic design to improve the efficiency and operability of scramjets, in addition to its program to develop the Hypersonic Attack Cruise Missile. (The Navy's applied research program for scramjet technology, the Hypersonic Booster program, has been terminated for 2023 and funding redirected to the service's Hypersonic Air-Launched Offensive Antisurface Warfare program.) Further, both the Air Force and the Navy are conducting research into advanced and alternative fuels that have the potential to improve scramjets' speed and performance.

Rapid and affordable manufacturing of hypersonic missile components also remains a challenge across the services. The Air Force is funding a program to develop manufacturing technology for materials, command-and-control communications components, and radio-frequency and digital components for hypersonic missiles.

Appendix B: CBO's Approach to Modeling Missile Trajectories

To estimate the flight times of different types of missiles and their visibility to radar, the Congressional Budget Office used analytic models of the missiles' trajectories. In modeling those flight paths, CBO had two goals. The first was to estimate flight time as a function of range and of parameters such as the speed of a rocket when its booster stops burning (known as the burnout speed).

The second was to model the visibility of a missile to ground-based radar as a function of altitude. This appendix provides details about the models CBO used.

Flight Time

The time a missile spends in flight depends on its trajectory, the range to its target, and its speed. A missile's speed in turn depends on the initial burnout speed and the deceleration associated with reentry, maneuvering, and atmospheric drag when the missile flies through the air.

Trajectory Considerations

CBO considered four types of missile trajectories: the minimum-energy trajectory and a depressed trajectory for ballistic missiles (see the figure in Box 1-1), the boost-glide trajectory for hypersonic missiles, and the flat, low-altitude trajectory for cruise missiles. The minimum-energy trajectory (MET) is the flight path that maximizes the total range of a missile on a ballistic trajectory for a given amount of energy input. CBO used the MET as a baseline for ballistic flights of a given range. For comparison, CBO also considered a lower-altitude ballistic trajectory. That depressed trajectory requires more energy than the MET for the same range, but a missile on such a path flies a shorter distance and thus can arrive at its target faster than a missile on the MET.

For hypersonic boost-glide missiles, common depictions of trajectories by the Department of Defense (DoD) and others suggest that a missile's glide body reenters the atmosphere very shortly after the booster rocket stops burning and spends almost all of its trajectory inside the atmosphere. During the time spent inside the atmosphere, the glide body will be difficult to detect by ground-based radar because of its altitude and will be able to maneuver using control surfaces to interact with the air. (Missiles that are outside the atmosphere must rely on thrusters to change their flight path.)

Despite those advantages, CBO's analysis suggests that for the foreseeable future, boost-glide missiles that travel more than 1,000 kilometers (km) are unlikely to spend more than half of their flight inside the atmosphere because of the intense thermal stresses of hypersonic flight through air. Specifically, the longer a glide body spends interacting with the atmosphere at hypersonic speeds, the more thermal stress it will endure. Trajectory designs that balance those competing incentives (difficulty of detection and maneuverability on the one hand and thermal stress on the other hand) will result in hypersonic boost-glide missiles that spend roughly half of their flight gliding within the atmosphere. Such a trajectory also avoids constraints on missiles that were imposed by the Intermediate-Range Nuclear Forces (INF) treaty, which prohibited missiles that spent more than 50 percent of their flight on a ballistic trajectory. The INF treaty is no longer in effect, but some of the programs that DoD is working on today were under development when the treaty was still in force.

Given the thermal challenges of hypersonic flight, it is unclear exactly how much of a boost-glide missile's trajectory is likely to be on a ballistic path and how much will be spent on a glide path inside the atmosphere. The longer the range of a boost-glide missile, the more likely it is to spend a larger fraction of its flight outside the atmosphere. To address that uncertainty, CBO modeled the boost-glide trajectory using two different values for the time spent on a ballistic path. In the lower case, the missile spends half of its flight on a ballistic path and the other half inside the atmosphere in the glide phase. Potential improvements in thermal design could enable longer glide times, so CBO also explored the effects of decreasing the amount of the flight spent on a ballistic path to one-quarter of the range. That trajectory would allow the missile to remain at low altitudes longer, but at the cost of a longer total flight time than a trajectory in which the glide phase was limited to only half of the flight.

Calculating Flight Times for Minimum-Energy and Depressed-Trajectory Ballistic Missiles

CBO estimated flight times for a ballistic missile on a minimum-energy trajectory using Python code shared with CBO by David Wright of the Massachusetts Institute of Technology. The code is a physics-based model for flight on a round, nonrotating Earth with atmospheric drag.[552]

[552] For more information about the physics of ballistic missiles, see Albert D. Wheelon, "Free Flight of a Ballistic Missile," American Rocket Society Journal, vol. 29, no. 12 (December 1959), pp. 915–926, https://doi.org/10.2514/8.4944.

In the model, the equations of motion—including thrust, gravitational, and atmospheric drag forces—are integrated to calculate the position and speed of a missile throughout its flight. The output of the model is the flight time after burnout and the burnout speed. CBO estimated the time until the booster burns out by using data for known missiles of similar ranges and linearly interpolating between those data.

To estimate flight times for the depressed trajectory, CBO considered the effect of the shorter path that a missile on a depressed trajectory travels as well as the potential for using a larger rocket booster with a higher burnout speed. CBO estimated that flight times for a ballistic missile on the depressed trajectory could be 10 percent to 30 percent shorter than flight times on the MET.[553]

Calculating Flight Times for Hypersonic Boost-Glide Missiles

The dynamics of flight through the atmosphere at hypersonic speeds require complex physical models to simulate. The aerodynamic properties of DoD's prototype hypersonic missiles are not publicly available. Thus, for the glide phase of the trajectory, CBO started with the results from an earlier analysis of a test that the United States conducted on the Hypersonic Technology Vehicle-2 (HTV-2) in 2008 and varied several parameters to reflect potential design choices and improvements in glide bodies' aerothermodynamic properties made since the HTV-2 was tested.[554] CBO did not use a detailed model of the aerodynamics of hypersonic flight.

Burnout Speed, Apogee, and Deceleration

The speed at which the glide body enters the ballistic (midcourse) portion of the flight is a function of the burnout speed of the rocket booster. CBO does not know the burnout speeds of the rocket boosters intended to be used for the Army's Long-Range Hypersonic Weapon (LRHW), the Navy's Intermediate-Range Conventional Prompt Strike (IR-CPS), or the Air Force's Air-Launched Rapid Response Weapon (ARRW), so CBO used two values for the burnout speed in its modeling:

- The burnout speed associated with a ballistic missile on a minimum-energy trajectory of the same range as a given boost-glide missile (called the MET- equivalent boost), and
- The burnout speed associated with the HTV-2 (6 km per second) regardless of a missile's range.

CBO chose those values to cover a wide range of potential booster choices. Competing demands on booster design include size and cost, thermal challenges associated with the glide speed, and flight time. However, factors such as the diameter and depth of the missile tubes in the new payload module for Virginia class submarines can limit options for using larger boosters.

---

[553] Those estimates are based on Lisbeth Gronlund and David C. Wright, "Depressed Trajectory SLBMs: A Technical Evaluation and Arms Control Possibilities," *Science & Global Security*, vol. 3, no. 1–2 (1992), pp. 101–159, https://tinyurl.com/2p8jnb48.

[554] See James M. Acton, "Hypersonic Boost-Glide Weapons," *Science & Global Security*, vol. 23, no. 3 (2015), pp. 191–219, https://tinyurl.com/yc79ph7x; and Cameron L. Tracy and David Wright, "Modelling the Performance of Hypersonic Boost- Glide Missiles," *Science & Global Security*, vol. 28, no. 3 (2020), pp. 135–170, https://tinyurl.com/3upn3pk2.

The glide bodies under development in the LRHW, IR-CPS, and ARRW programs have been reported to reach significantly lower apogees (maximum altitudes) than the apogees associated with ballistic trajectories.[555] To divert from a ballistic path, reenter the atmosphere, and begin gliding at hypersonic speeds, a missile must undergo a maneuver to reorient it toward Earth. In addition, the glide body will lose speed as it reenters the atmosphere, and once inside the atmosphere, it will continue to slow because of atmospheric drag as it travels along its flight path.

CBO used the deceleration observed in the model for a ballistic missile to estimate the deceleration of a hypersonic glide body when it performs the maneuver to reorient itself and reenters the atmosphere. CBO estimates that speed decreases by roughly 20 percent when a glide body goes below an altitude of 100 km and reenters the atmosphere. For simplicity, CBO considers that decrease in speed as occurring instantaneously. For the remainder of a flight, CBO modeled two glide phases: one in the upper atmosphere and one in the lower, denser atmosphere, where drag has a more significant impact. The altitude of the glide body is assumed to follow a linear descent in each phase, from 100 km to 40 km in the upper atmosphere and from 40 km to 30 km in the lower atmosphere. To estimate the deceleration profile of a glide body in each phase, CBO performed linear fits on published model results of the glide phase for a variety of initial speeds.[556]

CBO's model for a glide body's velocity in the direction of motion reflects the net effect of lift, drag, and gravity by using a consistent reduction in velocity of 0.2 km/ sec for every 1,000 km of range in the upper atmosphere and 0.8 km/sec for every 1,000 km of range in the lower atmosphere. Those values are consistent with other models of hypersonic boost-glide missiles.[557] To reflect potential improvements in glide-body design that would reduce drag, CBO also considered a "low-drag" design in which deceleration is half that of the baseline values: a consistent reduction in velocity of 0.1 km/sec for every 1,000 km of range in the upper atmosphere and 0.4 km/sec percent for every 1,000 km of range in the lower atmosphere.

How Different Values Affect Flight Times

Altering properties such as the trajectory, drag, and burnout speed of a boost-glide missile affects the missile's time to reach its target. Those effects are greater the longer the range to the target. At a range of 10,000 km, for example, the flight time of a missile with a MET-equivalent boost and baseline drag increases from 35.8 minutes to 42.0 minutes when the glide phase increases from 50 percent to75 percent of the trajectory (see Figure B-1, left-hand panels). For a missile with a 6 km/sec boost and baseline drag, the same change increases the time to fly 10,000 km by more than 10 minutes, from 44.3 minutes to 54.4 minutes.

Reducing the effective drag (moving from the baseline-drag scenario to the low-drag scenario) has a relatively small effect on the flight time of a boost-glide missile. That difference is more pronounced—but still small—when the glide phase increases, whether because the

---

[555] Officials from the Office of the Under Secretary of Defense for Research and Engineering, personal communication with CBO staff (March 11, 2021).

[556] See James M. Acton, "Hypersonic Boost-Glide Weapons," *Science & Global Security*, vol. 23, no. 3 (2015), pp. 191–219, https://tinyurl.com/yc79ph7x; and Cameron L. Tracy and David Wright, "Modelling the Performance of Hypersonic Boost- Glide Missiles," *Science & Global Security*, vol. 28, no. 3 (2020), pp. 135–170, https://tinyurl.com/3upn3pk2.

[557] See Cameron L. Tracy and David Wright, "Modelling the Performance of Hypersonic Boost-Glide Missiles," Science & Global Security, vol. 28, no. 3 (2020), pp. 135–170, https://tinyurl.com/3upn3pk2.

missile's overall range is longer or because the glide phase is lengthened to 75 percent of the missile's trajectory.

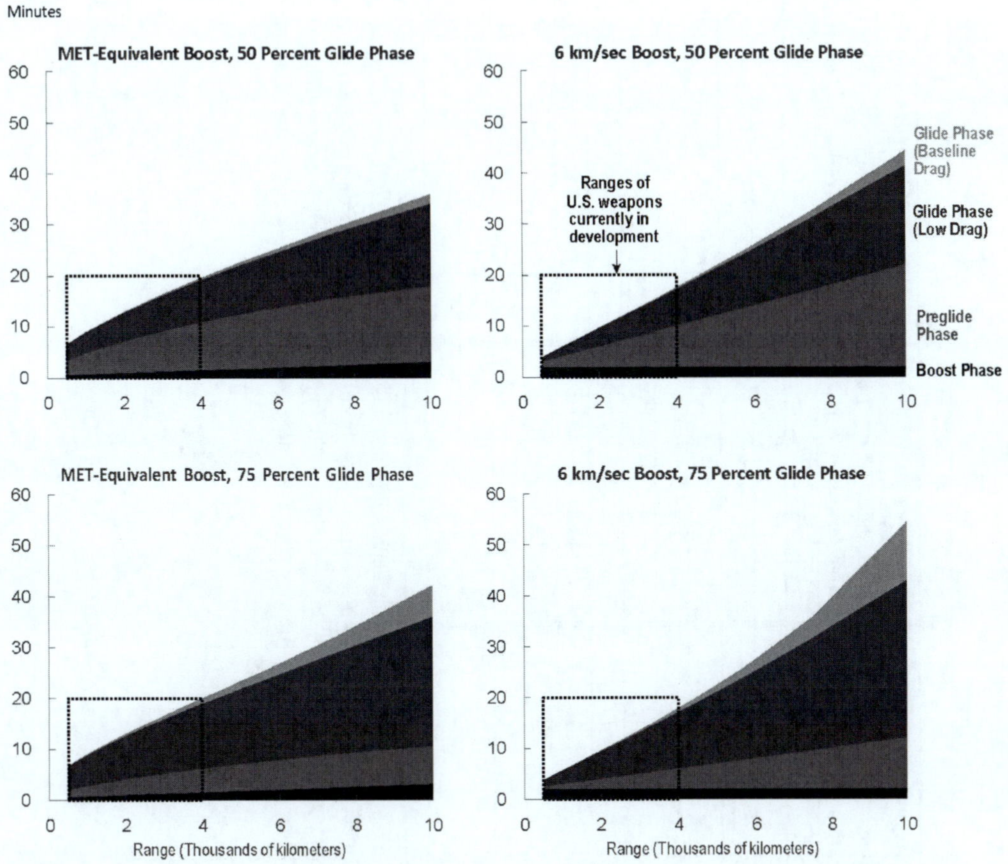

The effect of initial speed (a MET-equivalent or 6 km/sec burnout speed) is the most significant difference between the options that CBO modeled. At ranges greater than 6,500 km, flight times are longer with a 6 km/sec boost than with a MET-equivalent boost. The more of a flight spent in the glide phase—because of either trajectory design or longer overall range—the greater the impact of drag conditions.

Data source: Congressional Budget Office. See www.cbo.gov/publication/58255#data.

Total flight time for a hypersonic boost-glide missile is the sum of the three phases of the missile's flight. For a missile with an initial boost equivalent to that of a ballistic missile on a minimum-energy trajectory of the same range, the boost phase varies as a function of range. For a missile with a baseline boost of

6 km/sec, the boost phase is constant across all ranges. The preglide phase (the period after boost and before reentering the atmosphere) is a function of the trajectory design—specifically, whether 50 percent or 75 percent of the distance traveled is spent in the glide phase. The glide phase is represented by either the dark brown area (for the low-drag design) or by the orange and dark brown areas together (for the baseline-drag design).

The black dashed box represents the anticipated ranges of the U.S. hypersonic boost-glide missiles currently in development: the Air Force's Air-Launched Rapid Response Weapon, the Navy's Intermediate-Range Conventional Prompt Strike missile, and the Army's Long-Range Hypersonic Weapon.

km/sec = kilometers per second; MET = minimum-energy trajectory.

Figure B-1. Flight Times for a Hypersonic Boost-Glide Missile, by Phase of Flight, Range, and Design Parameters

Table B-1. Estimated Flight Times for Hypersonic Boost-Glide Missiles Under Various Conditions

Minutes

	MET-Equivalent Boost				6 km/sec Boost			
	50 Percent Glide Phase		75 Percent Glide Phase		50 Percent Glide Phase		75 Percent Glide Phase	
Range (km)	Baseline Deceleration	Low-Drag Deceleration	Baseline Deceleration	Low-Drag Deceleration	Baseline Deceleration	Low-Drag Deceleration	Baseline Deceleration	Low-Drag Deceleration
500	6.5	6.5	6.7	6.6	3.8	3.8	3.7	3.7
1,000	8.9	8.8	8.9	8.7	5.7	5.7	5.6	5.6
1,500	10.9	10.8	11.0	10.7	7.6	7.6	7.5	7.4
2,000	12.9	12.7	12.9	12.5	9.6	9.5	9.5	9.3
2,500	14.6	14.4	14.7	14.1	11.5	11.4	11.4	11.2
3,000	16.3	16.0	16.4	15.7	13.5	13.4	13.5	13.1
3,500	17.9	17.5	18.2	17.3	15.5	15.2	15.6	15.1
4,000	19.5	19.1	20.0	18.9	17.5	17.2	17.8	17.0
4,500	21.0	20.5	21.6	20.3	19.5	19.1	20.0	19.0
5,000	22.5	22.0	23.3	21.8	21.7	21.1	22.4	21.0
7,500	29.4	28.3	32.1	28.9	32.4	31.0	35.7	31.5
10,000	35.8	33.9	42.0	35.8	44.3	41.3	54.4	42.8

Data source: Congressional Budget Office. See www.cbo.gov/publication/58255#data.

CBO calculated the flight times for hypersonic boost-glide missiles under a wide range of possible conditions and missile properties. The missiles' range varied from 500 km to 10,000 km. The burnout speed of the rocket was modeled under two conditions: a boost equivalent to that of a ballistic missile on a minimum- energy trajectory of the same range, and a baseline boost of 6 km/sec over all the ranges. Under each of those conditions, the trajectory was modeled using a glide phase that lasted either 50 percent or 75 percent of the total range and using a baseline deceleration profile as well as a low-drag deceleration profile that slowed the glide body less.

km/sec = kilometers per second; MET = minimum-energy trajectory.

The hypersonic boost-glide missiles that DoD is currently developing are likely to have ranges between roughly 500 km and 4,000 km. At those ranges, CBO's model predicts that differences in the percentage of the glide phase and in drag would have minimal effects on flight times (see the area inside the black dashed lines in Figure B-1). For a missile with a MET-equivalent boost, increasing the glide phase from 50 percent to 75 percent of the trajectory does not increase the flight time at 1,000 km and increases it by roughly 30 seconds at 4,000 km. Reducing the drag factor by half shortens that missile's flight time by just a few seconds at short ranges and by about one minute at 4,000 km if the missile's glide phase lasts for 75 percent of the trajectory.

Differences in initial speed (a MET-equivalent or 6 km/ sec burnout speed) have the greatest effect on flight time of the variables included in CBO's modeling. A burnout speed of 6 km/sec is the speed associated with a MET flight with a range of about 5,700 km. Below that range, the 6 km/sec burnout speed is faster than the MET- equivalent boost; above that range, it is slower. Using a 6 km/sec booster at ranges greater than 5,700 km results in flight times that are significantly longer than for ballistic trajectories. CBO believes it is unlikely that such ranges are being considered for hypersonic boostglide missiles, because of the thermodynamic challenges of such long flights and the slow speed at which a glide body would be traveling as it neared its target. Even so, a single rocket booster with a 6 km/sec burnout speed would not be appropriate to span the full set of ranges considered here.

At a range of 1,000 km and with baseline deceleration because of drag, increasing the burnout speed from the MET equivalent to 6 km/sec would reduce a missile's flight time by more than 3 minutes, about half the total resulting flight time with a 50 percent glide phase (see Table B-1). At 4,000 km, the reduction in time would be similar, 2 minutes, but that savings represents a much smaller percentage reduction in the total flight time. At 10,000 km, the MET-equivalent burnout speed is higher than 6 km/sec, so reducing the speed to 6 km/sec would increase the flight time by more than 5 minutes.

Comparing Flight Times for Ballistic Missiles and Hypersonic Boost-Glide Missiles

CBO compared its estimates of flight times for ballistic missiles on minimum-energy or depressed trajectories and hypersonic boost-glide missiles over different distances (see Figure 3-1, lower panel). Both depressed-trajectory ballistic missiles and hypersonic boost-glide missiles have a variety of possible parameters, so CBO calculated a range of potential flight times for both types of missiles. For depressed-trajectory missiles, the top of the range of flight times for a given distance represents the MET burnout speed but with a shallower burnout angle (the angle above the horizon at which the missile is moving at burnout), and the bottom of the range represents a higher burnout speed and an even shallower burnout angle. For hypersonic boost-glide missiles, at distances of less than 5,700 km, the top of the range of flight times represents glide bodies with baseline drag that are launched with a boost equivalent to that of a ballistic missile on a MET, and the bottom of the range represents glide bodies with half the baseline drag that are launched with a 6 km/sec boost. At distances of more than 5,700 km, the relative positions of the two burnout speeds are reversed.

CBO found that depressed-trajectory ballistic missiles generally have the shortest flight times over all distances, although hypersonic boost-glide missiles could have comparable flight times at distances of less than 1,000 km if they have a low-drag design and are launched by a 6 km/sec booster. CBO also found that ballistic missiles on a minimum-energy trajectory have shorter flight times than hypersonic boost-glide missiles at distances greater than about 5,000 km. At shorter distances, their flight times fall within the range of flight times estimated for hypersonic boost-glide missiles. Only when glide bodies have less drag than is likely for the current generation of U.S. hypersonic missiles, and are launched on very fast boosters over short ranges, do hypersonic boost-glide missiles have a clear time advantage over ballistic missiles on a minimum-energy trajectory.

Calculating Flight Times for Cruise Missiles

CBO modeled the flight times of cruise missiles using three bands of speed: Mach 0.5 to 1 for subsonic, Mach 1 to 5 for supersonic, and Mach 5 to 9 for hypersonic. CBO assumed a completely flat trajectory for cruise missiles to calculate flight time as the ratio of range to velocity. The maximum range of a cruise missile is limited by the amount of fuel it can carry. CBO estimated those maximum ranges at 2,000 km for a subsonic cruise missile, 1,000 km for a supersonic cruise missile, and 500 km for a hypersonic cruise missile.

Radar Visibility

CBO also analyzed how different types of trajectories can affect how soon in its flight a missile is visible to ground- based radar. The horizon of a radar is a simple geometric calculation that takes into account the curvature of the Earth in order to estimate the distance at which a radar can detect an object at a certain height above the Earth. The basic calculation, which includes an effective radius of the Earth to account for the fact that the atmosphere tends to bend (refract) radio waves toward the Earth, is:

$$D = \sqrt{2xhx\frac{4R_E}{3}}$$

where D is the distance from the radar to the object and h is the height of the radar above the surface of the Earth, which has a radius R_E. To be detectable, any object beyond that distance must be above the Earth's surface at a minimum height of:

$$H_T = \frac{\left(R_T - \sqrt{2xhx\frac{4R_E}{3}}\right)}{2x\frac{4R_E}{3}}$$

where R_T is the range to the target.

To estimate when a missile is above the radar horizon, CBO populated a matrix of heights and ranges for a 30-meter-high ground-based antenna and compared the trajectories modeled above with the condition for H_T to determine whether a missile at each point along its trajectory would be above the radar horizon and thus in view of the radar. CBO varied the height of the radar up to 100 meters (m) to determine the calculations' sensitivity to that parameter and found that the ranges at which various types of missiles could be detected did not change significantly. For example, a radar 100 m high could detect a missile about 20 km earlier in its flight than a radar 30 m high. For a missile traveling at Mach 10, that difference would translate to an additional 6 seconds of warning.

CBO used simple geometric calculations to illustrate the challenge of detecting low-flying missiles and aircraft from ground-based radar. Much more detailed modeling would be required to predict the performance of any given radar system. In general, air-defense radars lack the power to detect hypersonic glide bodies at the ranges considered in this analysis. More powerful radar systems may have that ability, but they still face a variety of challenges that constrain their performance, including clutter in the environment (such as radar returns from trees and other objects on the surface that are close to the horizon). Also, a radar usually has a minimum elevation angle that is at least a few degrees above the horizon, which limits how low it can see. CBO did not include that effect in its simple calculations, but the basic result would be the same: Objects that fly low remain below the radar horizon longer.

Appendix C: How CBO Estimated Costs for Weapons Included in This Analysis

The Congressional Budget Office examined several types of missiles that could help the U.S. military counter the antiaccess and area-denial environments being created by potential U.S. adversaries, particularly China and Russia. Those weapons include hypersonic boost-glide missiles based closely on the ones being developed by the Army, Navy, and Air Force; intermediate-range, ground- or sea-launched ballistic missiles (which would have a range similar to that of the Army's and Navy's planned boost-glide missiles); short-range, air- or sea-launched hypersonic cruise missiles; and short-range, ground- or sea-launched ballistic missiles like the Standard Missile-6 (SM-6) Block IB, an upgraded version of the Navy's existing SM-6 Block IA missile.

CBO's Approach

CBO estimated several types of costs for those weapon systems: costs to procure the missiles and associated equipment, costs to integrate a weapon system onto existing platforms (such as ships and aircraft), and costs to sustain the system after its deployment. Those costs are presented in 2023 dollars.

CBO's estimates do not include costs to finish developing the missiles, for two reasons. First, the services are using nontraditional funding mechanisms to develop hypersonic missiles. That approach makes it difficult to determine how much development funding is being used for hypersonic technology in general, how much is being used to develop specific operational systems, and how much is being used for what would be considered procurement in a more traditional program. Second, given the relative immaturity of hypersonic weapons technology, the Department of Defense's (DoD's) estimates may be optimistic, but CBO does not have a good way to estimate likely development costs.

CBO based its estimates for surface- and air-launched hypersonic boost-glide missiles as closely as possible on publicly available information about the characteristics of, and procurement goals for, the three versions of those missiles currently in development—the

Army's Long-Range Hypersonic Weapon (LRHW), the Navy's Intermediate-Range Conventional Prompt Strike (IR-CPS) missile, and the Air Force's AGM-183A Air-Launched Rapid Response Weapon (ARRW). Although CBO refers to those weapons as the "LRHW/IR-CPS" and "ARRW" missiles in this appendix for simplicity, CBO's estimates are not intended to be predictions of the actual costs of those programs. (The full details of those systems are not publicly known, and the technology is still being developed, so CBO could not estimate costs with confidence.) Rather, CBO's estimates are meant to illustrate the relative costs of various types of high-speed weapons.

For the hypersonic missiles that are furthest along in development—the LRHW, IR-CPS, and ARRW missiles—CBO based its estimates of costs on analogies with existing missiles or on established cost-estimating relationships for missile components, such as boosters. To check those estimates, CBO compared them with the limited cost information provided by DoD. CBO used a similar approach to estimate the costs of notional intermediate-range ballistic missiles with maneuverable reentry vehicles. CBO could not estimate the costs of hypersonic cruise missiles, because the technology for those missiles is at an early stage, and the ultimate cost to procure such a missile is unclear. For the SM-6 Block IB missile, CBO started its estimates with cost data provided by the Navy for the SM-6 Block IA, which is currently in production, and adjusted for the different characteristics of the IB variant.

CBO's estimates represent one set of possible outcomes that reflect the specific assumptions made about missiles' technologies and capabilities. Significant uncertainty exists about the costs of less mature technologies or missile components, which represent long-term investments in a wide variety of research areas that are associated with high risks of cost and schedule overruns.

Costs of Hypersonic Boost-Glide Missiles for the Army and Navy

The Army's LRHW system and the Navy's IR-CPS system will use essentially the same missiles but with different canisters to accommodate the missiles' different launch platforms—ground vehicles, submarines, and surface ships. (Canisters are the protective cases in which missiles are transported and sometimes launched.) The two services are jointly developing the LRHW/IR-CPS missile but are separately developing their own canisters and launchers.

Procurement Costs

To estimate the procurement cost of an LRHW/IR-CPS missile—including the booster, glide body, and canister—CBO used the costs of similar components for existing weapon systems, as well as published estimates from the LRHW and IR-CPS program offices for components of their systems. (Additional costs for the service-specific launchers and other equipment are discussed below with platform costs.)

Rocket boosters are a well-understood technology. Widely used cost models exist for boosters that are based on the performance parameters of existing systems. CBO identified a rocket similar in size to the one shown in drawings released by the LRHW and IR-CPS program offices: the Orion 32 family of boosters being developed by Northrop Grumman. The first stage of the LRHW appears to be about the same size as a first-stage Orion 32XL, and the second stage is roughly half the size of a second-stage Orion 32. To estimate the cost of the two-stage LRHW/IR-CPS booster, CBO applied cost-estimating relationships developed by Technomics (a defense-analytics consulting firm) to the general characteristics of the Orion 32 boosters, adjusting for known differences with the LRHW/IR-CPS.

For the nose cone, which sits atop the missile and protects the glide body during the first part of the missile's flight, CBO based its estimate on the nose cone for the SM-3 Block IIA missile, scaled up to reflect the greater size of the LRHW/IR-CPS nose cone.

For the Common Hypersonic Glide Body (C-HGB), which will be used in both the LRHW and IR-CPS missiles, CBO started with estimated costs for the Mk21A ballistic reentry vehicle that is being developed for the Ground-Based Strategic Deterrent, an intercontinental ballistic missile slated to replace the Minuteman III.

CBO adjusted the costs of the Mk21A to account for the larger size of the C-HGB, which CBO estimated from images released by the LRHW and IR-CPS program offices. CBO further adjusted those costs to account for the additional thermal protection that the C-HGB will need (because it will spend a much longer time than the Mk21A moving at high speed through the atmosphere) and for a guidance and control system to allow the C-HGB to follow its planned trajectory and accurately hit its target. (A ballistic reentry vehicle such as the Mk21A does not have a guidance system; instead, it relies on the missile that carries it to release it at a point in space where gravity will pull it toward its target.)

Both the LRHW and the IR-CPS will be launched from a canister that will also protect the missile during storage and transportation. The canister will be mounted in or on the launcher platform that will fire the missile. CBO based its estimate of the procurement cost for an LRHW/IR-CPS canister on the cost of canisters for the SM-3 missile (which is fired from the vertical launch system on Navy surface ships), with adjustments to account for the much larger size of the LRHW/IR-CPS missiles.

Like the SM-3, the LRHW is a "hot-launch" missile that leaves the canister under its own rocket power. In contrast, the IR-CPS can be hot-launched from surface ships but will be "cold launched" from submarines. In a cold launch, high-pressure gas forces the missile out of the canister before the missile's booster ignites. CBO had no basis to estimate how the costs of the two canisters might differ, so it used the same cost estimate for both. On the one hand, the IR-CPS canisters will have to deal with the added complexity of cold launch, but the canisters and missiles will be in the controlled environment of a submarine or ship. On the other hand, hot launch is less complex, but the LRHW canisters will need to be hardened to protect the missiles during travel on rough roads in all types of weather.

CBO's estimate of the procurement cost per missile also includes costs for assembling and integrating the missile's components and for testing to ensure that the components are not defective and that the assembly and integration have been done correctly. Testing is important for systems like the LRHW/IP-CPS, whose complexity is on a par with that of systems that launch satellites into space. For its estimate, CBO used information about assembly, integration, and testing costs provided by the Navy.

Under the assumptions described above, CBO estimated that the procurement cost for the first LRHW/IR-CPS missile and canister (T-1) would amount to $81 million (see Table C-1). Averaged over a total purchase of 300 missiles, the procurement cost per missile would decline to $41 million, CBO estimates. For that and the other estimates of procurement costs in this analysis, CBO applied a learning curve of 90 percent, meaning that the per-missile cost declines by 10 percent each time the number of missiles produced has doubled.

Table C-1. Estimated average procurement cost per missile for an LRHW/IR-CPS

Millions of 2023 Dollars

For First Missile (T-1)		Over Total Purchase of 300 Missiles
Booster and Missile Body	28.1	14.0
Glide Body	38.5	19.2
Canister	2.1	1.0
Missile Assembly, Integration, and Testing	12.8	6.4
Total for All-Up Round[a]	81.3	40.7

Data source: Congressional Budget Office. See www.cbo.gov/ publication/58255#data.

CBO estimated values for each component on the basis of similar programs. Items such as the glide body and nose cone were scaled by the estimated size of the component as shown in images released by the Department of Defense. A 90 percent learning curve (in which the per-missile cost declines by 10 percent every time the number produced has doubled) was applied to the estimated procurement quantities.

IR-CPS = Intermediate-Range Conventional Prompt Strike missile (being developed by the Navy); LRHW = Long-Range Hypersonic Weapon (being developed by the Army).

[a] An all-up round, or AUR, is a fully assembled missile as delivered to its launch site.

Platform and Platform Integration Costs

CBO's estimates of the costs of integrating LRHW/ IP-CPS missiles onto their launch platforms are based on the platforms suggested in the services' current plans. The estimates do not include the costs of building launch platforms, such as submarines or ships, that are already in the force or that are planned to enter the force. However, CBO's estimates do include the costs of launcher equipment peculiar to the LRHW/IR-CPS system and other costs associated directly with integrating the missiles onto their launch platforms, including a portion of the fire control systems and software used to communicate target information to the missile system (though not the services' broader command-and-control infrastructure).

The Army plans to launch LRHW missiles from M870 trailers towed by Oshkosh M983A4 tractor trucks, the same family of vehicles that tow current Patriot surface- to-air missile launchers. The Army has indicated that it plans to field six LRHW batteries. Each battery will have four launchers (which will carry two missiles apiece), a truck-mounted battery-operations center, and various support vehicles.

The Army has requested \$144 million of research, development, test, and evaluation funding for ground service equipment for the LRHW from 2023 to 2025. Using its previous analysis of options for fielding long- range ground-launched missiles, CBO estimates that purchasing new ground equipment for the five remaining LRHW batteries would cost about \$200 million, or \$40 million per battery.[558] Those costs might be lower if the Army could modify some existing equipment for use in LRHW batteries.

The Navy plans to launch IR-CPS missiles from Zumwalt class guided missile destroyers and later from new Virginia class attack submarines equipped with the Virginia payload module (VPM). In estimating platform costs for the IR-CPS system, CBO used the Navy's estimate of \$900 million to replace the 155 millimeter guns on the three existing Zumwalt class destroyers with four vertical launch tubes per ship. CBO did not include the cost of VPM launchers in its estimate because the Navy already planned to install those launchers—beginning with the Block V variant of Virginia class submarines—to replace missile capacity that will be lost when

[558] See Congressional Budget Office, Options for Fielding Ground- Launched Long-Range Missiles (February 2020), www.cbo.gov/ publication/56068.

the Navy's four Ohio class guided missile submarines are retired. The VPM on a Block V Virginia class submarine will include four vertical launch tubes.

Each Zumwalt and VPM launch tube will be able to accommodate three IR-CPS missiles, although the Navy may opt to load some of the tubes with different weapons. IR-CPS canisters will not be directly mounted in the Zumwalt or VPM tubes. Rather, the three IR-CPS missiles will be contained in an advanced payload mod-ule (APM) that will be mounted in the destroyers' and submarines' tubes.

The Navy received nearly $900 million for IR-CPS platform integration from 2019 to 2022 and requested $364 million for 2023. CBO assumed that platform integration costs would continue at the average annual amount provided during the 2019–2022 period until 2028, when the missiles are due to be fielded on Block V submarines, for a total of $1.0 billion from 2024 to 2028.

CBO estimated an additional cost of $1.2 billion to procure 58 APMs and associated shipboard systems. CBO based its estimate for an APM and combat system on the cost of the launcher that the Navy purchased to convert tubes for Trident ballistic missiles into launchers for Tomahawk cruise missiles, with an adjustment to that cost to account for the greater size of IR-CPS missiles.

The 58 APMs would be enough for a full load of IR-CPS missiles on the Navy's 3 Zumwalt class destroyers (36 missiles in 12 launch tubes) and 11 submarines with Virginia payload modules (132 missiles in 44 launch tubes), with two spare APMs left over. In practice, however, the Zumwalt and Virginia launch tubes would probably not all carry APMs with IR-CPS missiles.

Sustainment, Support, and Other Costs

The last category of costs that CBO estimated for the missile systems in this analysis is costs to keep the missiles in the military's inventory for 20 years. Those costs cover such activities as maintaining the missiles (including repair and overhaul as they age); modifying, improving, and replacing outdated components; maintaining the facilities that handle the missiles; and other costs.

Annual sustainment and support costs can vary widely, from a few thousand dollars per missile for newer, smaller systems to nearly $3 million per missile for older, larger systems, such as Trident nuclear ballistic missiles.

CBO estimates that annual sustainment costs for LRHW/IR-CPS missiles would total about $0.5 million per missile. That estimate is based on the maintenance portion of sustainment costs for Trident missiles, as estimated by DoD. CBO did not base its estimate on the full sustainment costs for Trident missiles because those missiles are much older (they have been deployed for more than 30 years) and carry nuclear warheads—two factors that can substantially increase support costs.

The services could incur additional costs if they needed to establish new units and increase the number of personnel in each service to operate the LRHW/IR-CPS missiles. For example, on the basis of its earlier analysis of options for fielding long-range ground-launched missiles, CBO estimates that operation and support costs for an LRHW battery would total about $25 million per year, including costs for military personnel.[559] The Army would avoid that marginal

[559] Ibid.

cost if it converted existing units into LRHW batteries. The Navy, for its part, does not plan to purchase additional ships or submarines to accommodate IR-CPS missiles.

Costs of Hypersonic Boost-Glide Missiles for the Air Force

LRHW/IR-CPS missiles might be able to be launched from the cargo bay of a large transport aircraft, but the Air Force prefers to use bomber and fighter aircraft to launch strike missiles.[560] Thus, the Air Force is developing a smaller hypersonic boost-glide missile, the AGM-183A ARRW, which is expected to have a much shorter range and a smaller warhead than the LRHW/IR-CPS missile. In the Air Force's plans, ARRW missiles will be launched from under the wings of B-52 bombers, but they may eventually be carried by other bombers or by large fighters, such as the F-15E or F-15EX. The potential ability of aircraft to get closer to their targets than surface launchers reduces the disadvantage of having missiles with a shorter range.

Procurement Costs

To estimate the procurement costs of an ARRW missile, CBO used an approach similar to the one it used for the LRHW/IR-CPS. The ARRW's single-stage booster has been described as a modified version of the booster used on the MGM-140 Army Tactical Missile System (ATACMS). Thus, CBO based the cost of the ARRW booster and missile body on actual costs for the ATACMS, scaled to account for the ARRW's significantly longer length and slightly larger diameter.

For the glide body, CBO adjusted its estimate for the Common Hypersonic Glide Body used in the LRHW/ IR-CPS to account for the ARRW's smaller size and the smaller amount of thermal protection that will be necessary because the ARRW will glide for a shorter distance in the atmosphere.

For costs to assemble and integrate the ARRW's components, CBO scaled its estimate for the LRHW/IR-CPS to account for differences in the costs of the missiles' physical components and their relative complexity. (For example, the LRHW/IR-CPS uses a two-stage booster, and the ARRW uses a single-stage booster.)

Those adjustments result in an average procurement cost of $15 million per missile for a purchase of 300 missiles, CBO estimates (see Table C-2). The Air Force has not indicated how many ARRW missiles it plans to purchase, so for comparison purposes, CBO used the same procurement quantity as for the LRHW/IR-CPS. Accounting for testing and spare missiles, 300 ARRWs would allow for more than 65 B-52 missions with four missiles each. Alternatively, the Air Force may choose to procure fewer missiles. If it purchased 100 missiles, for example, the average procurement cost per missile would increase to $18 million.

By comparison, in its 2022 budget, the Air Force requested $161 million to procure "an estimated 12 missiles," implying a cost per missile of about $13 million. In CBO's estimates, procurement costs for the first 12 missiles average $18 million each, about one-third more than the Air Force's budget request. That difference reflects uncertainty about the physical and performance characteristics of ARRW missiles, which underlie CBO's bottom-up estimate, as well as uncertainty about the number of missiles the Air Force could have actually acquired for

[560] In 1974, the Air Force demonstrated the ability to launch a Minuteman intercontinental ballistic missile from the cargo bay of a C-5A transport plane, but the service did not end up pursuing that operational capability.

$161 million. (Several test failures have caused initial procurement funding for ARRW missiles to be delayed until after 2023, a change that is likely to add to the program's cost.)

Table C-2. Estimated average procurement cost per missile for an ARRW

Millions of 2023 Dollars

	For First Missile (T-1)	Over Total Purchase of 100 Missiles	Over Total Purchase of 300 Missiles
Booster and Missile Body	8.1	4.8	4.0
Glide Body	15.9	9.4	7.9
Shipping Container	0.5	0.5	0.5
Missile Assembly, Integration, and Testing	3.0	2.7	2.3
Total for All-Up Round[a]	27.5	17.5	14.9

Data source: Congressional Budget Office. See www.cbo.gov/ publication/58255#data.

CBO estimated values for each component on the basis of similar programs. Items such as the glide body and nose cone were scaled by the estimated size of the component as shown in images released by the Department of Defense. A 90 percent learning curve (in which the per-missile cost declines by 10 percent every time the number produced has doubled) was applied to the estimated procurement quantities.

ARRW = Air-Launched Rapid Response Weapon (being developed by the Air Force).

[a] An all-up round, or AUR, is a fully assembled missile as delivered to its launch site.

Platform and Platform Integration Costs

The Air Force has not indicated a need to buy additional aircraft to accommodate ARRW missiles. B-52s are being used as launch platforms in the ARRW test program, so CBO assumed that most costs to integrate the missiles onto those aircraft would be covered by development and testing funds. However, CBO estimates that fully integrating ARRWs into the combat system of operational B-52s would cost the Air Force an additional $200 million.

Further costs would be incurred if the Air Force opted to integrate ARRW missiles onto its other bombers or the F-15E/EX. In particular, it can be more difficult to get safety authorization for a new weapon to be carried internally on aircraft such as the B-1, the B-2, or eventually the B-21(as opposed to being carried under the wing on a B-52). CBO did not include those potential further costs in its estimates.

Sustainment, Support, and Other Costs

CBO estimates that sustaining and supporting the ARRW system would cost about $0.1 million per missile each year. That number is based on CBO's estimate for the LRHW/IR-CPS system, scaled down to account for differences in the missiles' assembly, integration, and testing costs. In its estimate of sustainment costs for the ARRW, CBO did not include additional operation and support costs or military construction costs to store and maintain an inventory of missiles. Although not zero, those costs are likely to be small relative to other parts of the ARRW program and relative to what the Air Force spends to maintain its inventory of other conventional strike weapons.

Costs of Intermediate-Range, Ground- or Sea-Launched Ballistic Missiles Equipped with MaRVs

As an alternative to the hypersonic boost-glide missiles that DoD is developing, CBO examined notional ballistic missiles that would use boosters similar to those of the LRHW/IR-CPS but would carry ballistic reentry vehicles rather than more sophisticated glide bodies.

A purely ballistic reentry vehicle with a conventional (nonnuclear) warhead would not be accurate enough for many targets, so CBO's notional ballistic missile would be equipped with a maneuverable reentry vehicle (MaRV). Maneuverable reentry vehicles have been studied since the 1960s and were deployed on the United States' intermediate-range Pershing II nuclear-armed missiles in the 1980s.

Procurement Costs

To estimate the procurement cost of an intermediate range, ground- or sea-launched ballistic missile, CBO started with its estimate for the LRHW/IR-CPS and substituted the cost of a MaRV for the cost of those missiles' hypersonic glide body. CBO based the cost of a MaRV on the Navy's estimates of the cost of the Conventional Trident Modification reentry vehicle proposed in 2006, adjusted by a growth factor representing cost growth in past missile programs.[561]

Table C-3. Estimated average procurement cost per missile for an intermediate-range, ground- or sea-launched ballistic missile equipped with a MaRV

Millions of 2023 Dollars

	For First Missile (T-1)	Over Total Purchase of 300 Missiles
Booster and Missile Body	26.0	13.0
Reentry Vehicle	15.5	7.7
Canister and Material Handling Equipment	2.1	1.0
Missile Assembly, Integration, and Testing	8.1	4.1
Total for All-Up Rounda	51.6	25.8

Data source: Congressional Budget Office. See www.cbo.gov/ publication/58255 #data.

CBO estimated values for each component on the basis of similar programs. A 90 percent learning curve (in which the per-missile cost declines by
10 percent every time the number produced has doubled) was applied to the estimated procurement quantities.

MaRV = maneuverable reentry vehicle.

[a]. An all-up round, or AUR, is a fully assembled missile as delivered to its launch site.

Those adjustments result in an average procurement cost of $26 million per missile for a purchase of 300 ground- or sea-launched ballistic missiles equipped with MaRVs, CBO estimates (see Table C-3). That figure is 37 percent less than CBO's estimate of the cost to procure an LRHW/IR-CPS hypersonic boost-glide missile with the same range.

---

[561] For a description of CBO's approach to analyzing growth in acquisition costs, see Congressional Budget Office, An Analysis of the Obama Administration's Final Future Years Defense Program (April 2017), pp. 47–50, www.cbo.gov/publication/52450.

Platform and Platform Integration Costs

In CBO's estimation, platform integration costs would be roughly the same for the intermediate-range ballistic missiles as for the LRHW/IR-CPS missiles.

Sustainment, Support, and Other Costs

In CBO's estimation, sustainment costs would be roughly the same for the ballistic missile system as for the LRHW/IR-CPS systems.

Costs of Standard Missile-6 Block IB Missiles

The Navy's effort to develop a Block IB variant of the SM-6 short-range ballistic missile is relatively recent, and the exact specifications of the missile are still being worked out. Production of Standard missiles is long established, and the Navy has had success modifying the missiles for different purposes. Compared with most earlier variants of Standard missiles, the Block IB will use a larger-diameter (21-inch) rocket motor to achieve longer range and higher speed. Standard missiles have primarily been used on the Navy's surface combat ships, but air-launched versions have been used in the past.

Modifications would be necessary to launch the missiles from submarines. In addition, the Army is considering purchasing the SM-6 for ground launch.

Procurement Costs

To provide a direct comparison with the other weapon systems, CBO assumed that the Army and the Navy would purchase a combined total of 300 Block IB missiles. On the basis of information provided by the Navy, CBO estimates that 300 missiles would have an average procurement cost of $6 million per missile (including the canister). Production quantities for the Block IB could be larger if the Navy deployed the missile on more surface combatants than the three Zumwalt class destroyers planned for the IR-CPS and if the Army purchased Block IBs to equip its Mid-Range Capability (MRC) batteries. (Current plans call for MRC batteries to be equipped with SM-6 missiles and Tomahawk cruise missiles.)

Platform and Platform Integration Costs

CBO estimates about $1 billion for platform and platform integration costs, including the same number of launchers (58) as for the Navy's IR-CPS and ground equipment for six Army batteries. Costs would be lower than for the IR-CPS for two reasons. First, there would be no need to modify destroyers to use the Block IB missile; it would be compatible with those ships' existing MK 41 vertical launchers. And second, the launchers to be used in submarines' Virginia payload module tubes would be smaller (and less expensive) than the advanced payload module used to fire the IR-CPS missile. CBO's estimates do not include the costs of building launch platforms, such as submarines or ships, that are already in the force or that are planned to enter the force.

Sustainment, Support, and Other Costs

In CBO's estimation, sustainment costs would be roughly the same for the SM-6 missile as for the ARRW.

About This Document

This report was prepared at the request of the Ranking Member of the Senate Committee on Armed Services. In keeping with the Congressional Budget Office's mandate to provide objective, impartial analysis, the report makes no recommendations.

Corinne Kramer prepared the report with guidance from David Mosher and Edward G. Keating. David Arthur helped develop the cost analysis and assisted with other elements of the report.

Michael Bennett, Charles Pineles-Mark, and Robert Shackleton (formerly of CBO) provided additional assistance. Eric J. Labs fact-checked the report.

John Hong of the Institute for Defense Analyses, Dean Wilkening of the Johns Hopkins University Applied Physics Laboratory, Robert S. Wilson of the Aerospace Corporation, and David Wright of the Massachusetts Institute of Technology commented on an earlier draft. The assistance of external reviewers implies no responsibility for the final product; that responsibility rests solely with CBO.

Jeffrey Kling and Robert Sunshine reviewed the report. Christian Howlett edited it, and R. L. Rebach created the graphics and prepared the text for publication.

CBO seeks feedback to make its work as useful as possible. Please send comments to communications@cbo.gov.

Phillip L. Swagel Director
January 2023

Chapter 9

Hypersonic Weapons: DOD Could Reduce Cost and Schedule Risks by Following Leading Practices*

United States Government Accountability Office

Abbreviations

ARRW	Air-launched Rapid Response Weapon
CPS	Conventional Prompt Strike
DOD	Department of Defense
HACM	Hypersonic Attack Cruise Missile
HALO	Hypersonic Air-Launched Offensive anti-surface warfare
LRHW	Long-Range Hypersonic Weapon
MTA	middle tier of acquisition
OD	outcome determination
Q	quarter
SM-6	Standard Missile 6
TBD	to be determined

Why GAO Did This Study

Offensive hypersonic weapons move at least five times the speed of sound and have unpredictable flight paths. DOD has begun multiple efforts to develop offensive hypersonic weapons that can be launched on the ground, in the air, and at sea. High costs and failed tests are a concern for some of these efforts.

A joint explanatory statement includes a provision for GAO to review DOD's efforts to develop offensive hypersonic weapons. This report addresses the extent to which DOD's hypersonic efforts are: (1) employing leading practices for product development, (2) identifying and analyzing cost risks, and (3) implementing effective enterprise risk management, among other objectives.

GAO reviewed documentation for the six hypersonic efforts currently being developed by the Army, Air Force and Navy, and interviewed DOD officials.

* This is an edited, reformatted and augmented version of the United States Government Accountability Office Report, Publication No. GAO-24-106792, dated July 2024.

In: Emerging Military Technologies
Editor: Jerry Shields
ISBN: 979-8-89530-613-0

GAO assessed these efforts using its leading practices for product development, cost estimating, and risk management, as appropriate.

What GAO Recommends

GAO is making 10 recommendations to DOD, including increasing the incorporation of feedback from users into system designs, expanding the use of modern digital engineering tools, improving the Conventional Prompt Strike cost estimate, and expanding enterprise-level reporting activities. DOD concurred with these recommendations.

What GAO Found

The Department of Defense (DOD) is increasingly investing in the development of hypersonic weapons. These weapons' unique characteristics—such as the ability to maneuver at very high speeds—could allow the U.S. to strike heavily defended targets from a distance.

All six offensive hypersonic weapon efforts GAO identified have placed a high priority on delivering quickly, with all intending to deliver a "minimum viable product"—one with the initial capabilities needed for users to recognize value. Four of the efforts, however, are not soliciting user feedback to determine what capabilities to include in their minimum viable product, a leading practice for product development identified by GAO in July 2023.

In addition, four efforts have not adopted leading practices for using digital engineering tools, another leading practice for product development. These tools include virtual representations of physical products. Employing modern digital engineering tools and directly soliciting user feedback both have the potential to speed up the design process, reduce costs, and develop a more usable product.

Comparison of Ballistic and Hypersonic Missile Trajectories

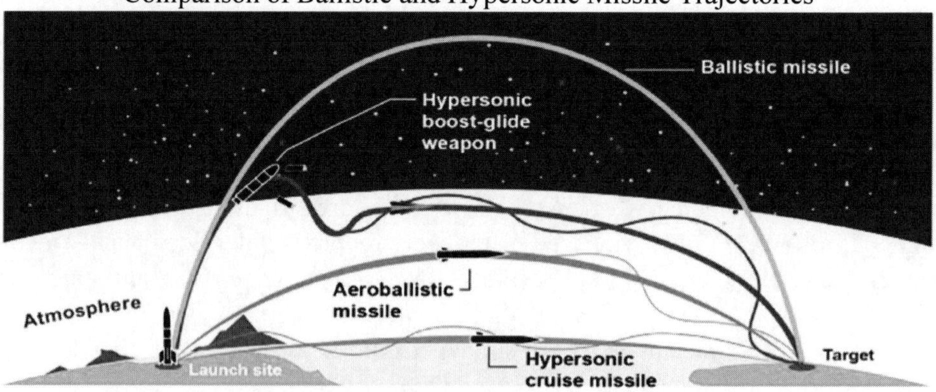

Source: GAO analysis of Department of Defense data. |GAO-24-106792.

While DOD has identified and analyzed cost risks, the cost of these weapons is difficult to estimate. This is in part due to DOD's limited experience developing and fielding hypersonic weapons. For example, the Navy's estimate for Conventional Prompt Strike—among the most mature cost estimates available— compensates for the lack of quality historical data by relying heavily on the views of subject matter experts. Expert views are best used sparingly, as they can be prone to bias, unless estimators analyze and account for that bias. Addressing this and other issues in accordance with GAO leading practices for cost estimates could provide Navy decision-makers a more accurate estimate.

DOD implemented most elements of an effective risk management framework, which allows agencies to assess and monitor threats to achieving their goals. DOD is not, however, comprehensively reporting to Congress about progress against DOD-wide risks to fielding hypersonic systems. Reporting this information at an enterprise level provides a more complete picture of DOD's efforts and progress, while promoting transparency.

July 29, 2024

Congressional Committees,

Hypersonic weapons, which move at least five times the speed of sound and have unpredictable flight paths, could allow the United States to strike valuable, heavily defended targets from a distance in the early phases of a conflict. Department of Defense (DOD) research into these weapons dates back decades, but these research efforts did not culminate in the fielding of new systems. However, DOD has now made a variety of investments to develop operational, offensive hypersonic weapons, in part due to the continued growth of anti-access and area denial capabilities by potential adversaries.

Explanatory materials accompanying the James M. Inhofe National Defense Authorization Act for Fiscal Year 2023 include a provision for GAO to review DOD's efforts to develop offensive hypersonic weapons.[562] This report addresses (1) the acquisition approaches DOD is using for offensive hypersonic weapons, (2) the extent to which DOD is employing leading practices for product development to address risks associated with these weapons, (3) the extent to which DOD has identified and analyzed cost risks for these weapons, and (4) the extent to which DOD is managing enterprise-level risks for the development of hypersonic systems.

For the purposes of this report, we defined an offensive hypersonic weapon as a system that (a) travels at speeds in excess of Mach 5, (b) spends a majority of its flight path inside the atmosphere, (c) is capable of maneuvering in flight, and (d) is intended to provide an operational strike capability. Applying these criteria, we identified six offensive hypersonic weapons efforts.

For all our objectives, we reviewed documentation, such as acquisition strategies, and interviewed officials from the Army, Navy, and Air Force, as well as other DOD stakeholders. For the first objective we used this information to provide descriptions of the offensive hypersonic weapons programs and to determine their acquisition pathways and activities. To assess whether DOD is employing leading practices for product development, we reviewed program documentation that addressed schedules, technology issues, and requirements, and compared this information with our leading practices for product development. We focused on practices related to the development of a minimum viable product, as we found these were most applicable given the efforts' stage of development. Specifically, we assessed whether the efforts prioritized schedule by off-ramping capabilities when necessary, directly solicited and

[562] 168 Cong. Rec. H9425, H9648 (Dec 8, 2022) to accompany the James M. Inhofe National Defense Authorization Act for Fiscal Year 2023.

incorporated user feedback to inform the minimum viable product, and used digital engineering tools, such as digital twinning.[563]

To assess cost risks, we interviewed cost estimators and reviewed related documentation. We assessed the most recent cost estimate created by the Conventional Prompt Strike effort — which was among the most advanced of the efforts we reviewed and approaching a major acquisition decision — based on our leading practices for the "accurate" characteristic of a quality cost estimate.[564] To assess enterprise-level risks for the development of hypersonic systems, we interviewed DOD officials who focused on these risk areas, reviewed relevant documentation, and compared these findings with our leading practices for managing enterprise risks.[565] For additional details on our objectives, scope, and methodology, see appendix I.

We conducted this performance audit from April 2023 to July 2024 in accordance with generally accepted government auditing standards. Those standards require that we plan and perform the audit to obtain sufficient, appropriate evidence to provide a reasonable basis for our findings and conclusions based on our audit objectives. We believe that the evidence obtained provides a reasonable basis for our findings and conclusions based on our audit objectives.

Background

Hypersonic Weapons

The technical and operational concepts that underpin hypersonic weapons are not new. DOD explored various concepts for hypersonic flight from the 1950s to the 1980s, but none of these concepts resulted in an operational capability. Between 2001 and 2014, the U.S. government developed and tested several experimental hypersonic vehicles with a mix of success and failure, including the Hypersonic Technology Vehicle, the Advanced Hypersonic Weapon, and the X-43 test vehicle.

U.S. Strategic Command has had a requirement for a high-speed, long- range, conventional strike weapon since the early 2000s. Efforts to develop hypersonic weapons accelerated starting in 2018 and 2019, in response to potential adversaries such as China and Russia continuing to improve their anti-access and area denial capabilities. Hypersonic weapons are one possible means to counter these capabilities.

High-level DOD strategy documents have articulated DOD's commitment to hypersonic technologies. For example, the 2022 National Defense Strategy identified hypersonic weapons as one capability that will contribute to DOD's strategy and identified it as one of the technology areas in which the United States would seek to build an enduring advantage.[566] In

[563] GAO, Leading Practices: Iterative Cycles Enable Rapid Delivery of Complex, Innovative Products, GAO-23-106222 (Washington, D.C.: July 27, 2023); and Leading Practices: Agency Acquisition Policies Could Better Implement Key Product Development Principles, GAO-22-104513 (Washington, D.C.: Mar. 10, 2022).

[564] GAO, Cost Estimating and Assessment Guide: Best Practices for Developing and Managing Program Costs, GAO-20-195G (Washington, D.C.: Mar. 12, 2020).

[565] GAO, Enterprise Risk Management: Selected Agencies' Experiences Illustrate Good Practices in Managing Risk, GAO-17-63 (Washington, D.C.: Dec. 1, 2016).

[566] Department of Defense, *2022 National Defense Strategy* (Washington, D.C.: Oct. 27, 2022).

2022, the Under Secretary of Defense for Research and Engineering identified hypersonic technologies as one of 14 critical technology areas that would guide DOD's investments.[567]

As a result of identifying hypersonic technologies as a top priority, DOD established a Principal Director for Hypersonics within the Office of the Under Secretary of Defense for Research and Engineering. The role of the Principal Director is to establish a plan for hypersonic technology development and act as a coordinator for these efforts across DOD. While the Principal Director serves as a coordinator, the military departments remain the primary developers of offensive hypersonic weapon systems.

There are multiple approaches to developing hypersonic weapons.

- Hypersonic cruise missiles work by using a rocket motor to boost the missile to supersonic speeds before switching to a special type of engine known as a scramjet. The scramjet combines supersonic airflow with fuel, which it then ignites to produce thrust.
- Boost-glide systems work by using a missile to propel a payload to the edge of space at which point a glide body separates from the missile and, using aerodynamic lift to extend its range, maneuvers itself toward the target.
- Aeroballistic missiles represent an additional approach to achieving hypersonic speeds. This design is rocket-powered and combines aspects of ballistic and aerodynamic flight. While it does not adhere to a boost-glide trajectory, it still travels at hypersonic speeds and maneuvers within the atmosphere.

Figure 1 provides a comparison of the flight paths of some hypersonic weapon types as well as ballistic weapons.

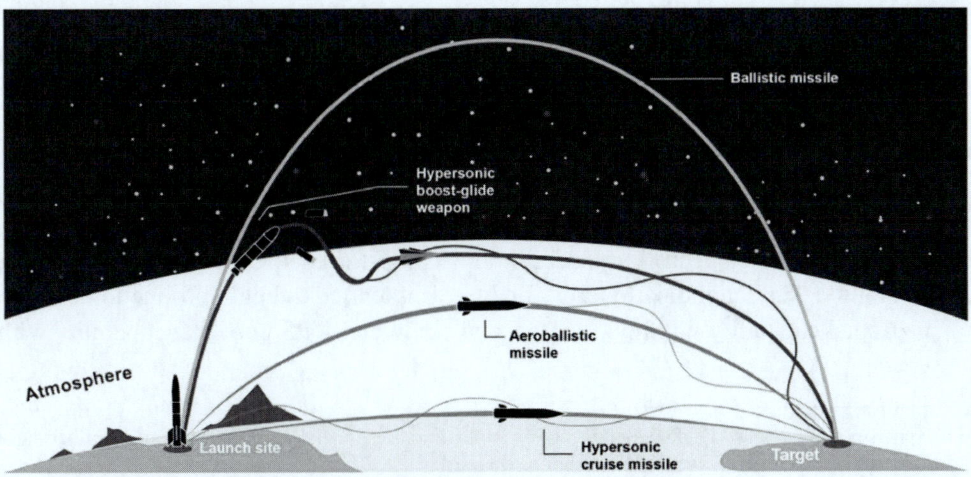

Source: GAO analysis of Department of Defense data. | GAO-24-106792.
Note: Figure not drawn to scale.

Figure 1. Comparison of Ballistic and Hypersonic Flight Paths.

[567] Department of Defense, Under Secretary of Defense for Research and Engineering, *USD(R&E) Technology Vision for an Era of Competition* (Washington, D.C.: Feb. 1, 2022).

We previously reported on several challenges in developing hypersonic weapons.[568] For example, the high-temperature environment around a hypersonic system in flight presents complex physical and chemical challenges that are typically not encountered in other weapon systems. To compensate, offensive hypersonic weapons need specialized components to achieve sustained maneuvering during high-speed flight. The technologies used to deal with these challenges have effects on the system's capability, cost, and manufacturability. In addition, developing and manufacturing offensive hypersonic weapon systems requires a workforce with specialized knowledge.

Finally, developing hypersonic systems requires several types of specialized test facilities. Wind tunnels are used to obtain aerodynamic data for hypersonic vehicles. Arc-heated facilities test material performance at high temperatures. Test tracks and arenas provide data on the lethality of a weapon system. Flight tests are also needed to expose technologies and systems to realistic environments. Flight tests, conducted at flight test ranges, are difficult and expensive to conduct, partly because of the long distances involved and the need to place sensors along that distance to collect test data.

DOD Acquisition Pathways

In January 2020, DOD established the Adaptive Acquisition Framework.[569] The framework emphasizes several principles that include simplifying acquisition policy, tailoring acquisition approaches, and conducting data- driven analysis. The Adaptive Acquisition Framework contains six acquisition pathways. The two that are most relevant for this report are:

- Major capability acquisition: This pathway leads major defense acquisition programs, major systems, and other complex acquisitions through phases, such as technology development, system development, and production, separated by major reviews known as milestone decisions. Programs on this pathway generate knowledge through structured analysis.
- Middle tier of acquisition (MTA): This pathway includes two expedited paths. The first path, rapid prototyping, is intended to quickly develop and demonstrate a capability in an operational environment within 5 years. Rapid prototyping also results in prototypes that a military department can field to the warfighter as an interim capability. The second path, rapid fielding, is intended to begin production of a new or upgraded capability within 6 months, and complete fielding of that capability within 5 years.[570]

---

[568] GAO, Hypersonic Weapons: DOD Should Clarify Roles and Responsibilities to Ensure Coordination across Development Efforts, GAO-21-378 (Washington, D.C.: Mar. 22, 2021).

[569] Department of Defense, The Defense Acquisition System, DOD Directive 5000.01 (July 28, 2022); and Operation of the Adaptive Acquisition Framework, DOD Instruction 5000.02 (Jan. 23, 2020). For more information on how DOD is using the Adaptive Acquisition Framework, see GAO, *Weapon Systems Annual Assessment: Challenges to Fielding Capabilities Faster Persist*, GAO-22-105230 (Washington, D.C.: June 8, 2022).

[570] For programs using the MTA pathway, the program start date (and thus the date from which the 5-year timeframe is measured) for programs designated on or after December 30, 2019, is generally the date that the program was designated, which is the date that an acquisition decision memorandum was signed initiating an MTA rapid prototyping or rapid fielding program. MTA programs designated before December 30, 2019, and certain programs designated after this date, generally maintain their MTA program start date as the date funds were first obligated.

Each pathway has requirements for reviews, cost and schedule goals, and reporting to decisionmakers. Acquisitions using the MTA pathway are generally not subject to the same acquisition and requirements processes as those on the major capability acquisition pathway.

The military departments can also transition efforts from one pathway to another. For example, DOD's MTA policy states that military departments will develop a process for transitioning successful prototypes from the rapid prototyping pathway to new or existing programs for production, fielding, and operations and sustainment under the rapid fielding pathway or another acquisition pathway.[571] In addition, the military departments could choose to terminate a program that has not progressed as expected. Figure 2 shows the major capability acquisition and MTA pathways.

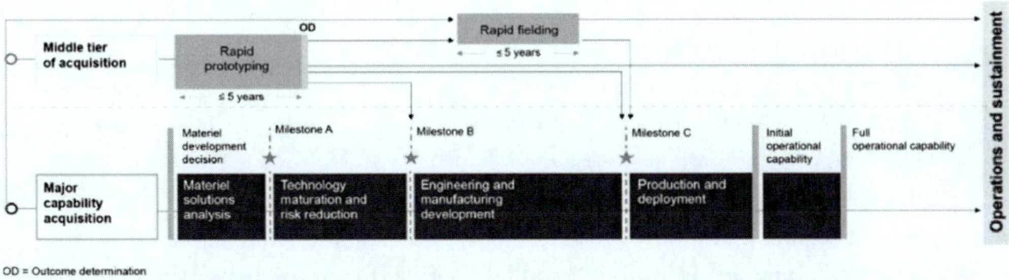

Source: GAO analysis of Department of Defense (DOD) data. | GAO-24-106792.

Figure 2. Selected Department of Defense Adaptive Acquisition Framework Pathways.

GAO Leading Practices

We have identified leading practices relating to the effective management and oversight of government programs and acquisitions.

Our *leading practices for product development* identify several practices used by leading companies to develop complex, innovative products. Leading companies employ an iterative process when developing complex products. The iterative process involves a continuous cycle, through which companies rapidly develop and deploy products.

This process prioritizes schedule by focusing on the development of a minimum viable product—one with the initial capabilities needed for users to recognize value. Three practices are key for developing and improving a minimum viable product:

- Leading companies prioritize schedule by off-ramping capabilities when necessary. To achieve speed to market, these companies will prioritize developing a minimum viable product by removing capabilities that pose a risk to delivering the product on schedule. The off-ramped capabilities can be deferred to a later release or terminated.
- Leading companies seek and obtain continuous user feedback— feedback from the actual operators of the product—throughout the iterative cycles. These companies capture this feedback to determine the minimum viable product and to inform improvements to the minimum viable product.

[571] Department of Defense, Operation of the Middle Tier of Acquisition, DOD Instruction 5000.80 (Dec. 30, 2019).

- Leading companies also use digital engineering tools, including digital twins—virtual representations of physical products. Digital twins incorporate dynamic data of a physical object or a system—meaning the model changes and updates in real-time as new information becomes available. Digital twins differ from 3D models, which are static visualizations that are updated manually, and are essentially paper design drawings in digital form.[572]

Figure 3 illustrates the relationships between the different phases of the iterative design cycle.

Our *Cost Estimating and Assessment Guide*, states that reliable cost estimates must among other things, be "accurate". We define the characteristic of accurate as:

- using the best methodology from the data available,
- properly adjusting for inflation,
- validating underlying formulas and inputs, with the resulting estimate containing few, if any, minor mathematical mistakes,
- regularly updating the estimate to ensure it reflects program changes and actual costs,
- documenting and explaining variances between estimated and actual costs, and
- being based, to the extent possible, on a historical record of cost estimating and actual experiences from other comparable programs.[573]

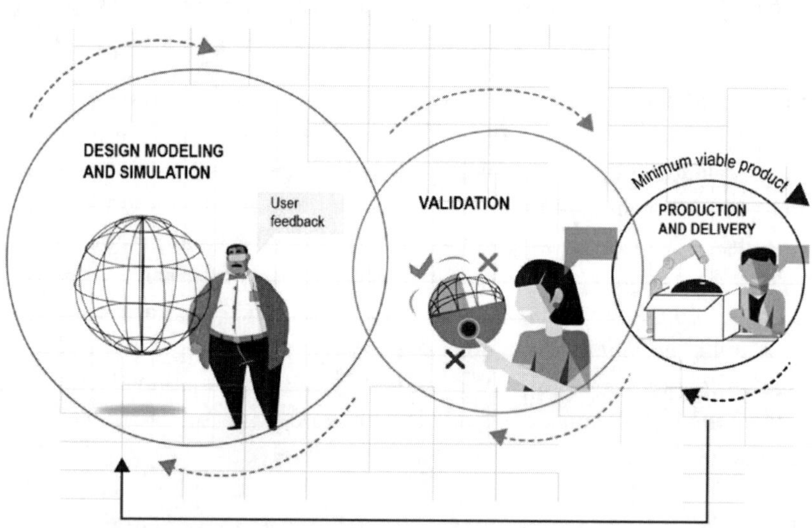

Source: GAO analysis of leading company iformation, GAO (illustration). | GAO-24-106792.

Figure 3. Leading Companies Progress through Iterative Design, Validation, and Production Cycles to Develop a Minimum Viable Product.

---

[572] GAO-23-106222; and GAO-22-104513.

[573] GAO-20-195G. The *Cost Estimating and Assessment Guide* also articulates three other characteristics of a reliable cost estimate: comprehensive (meaning it reflects the current schedule and technical baselines), well-documented (meaning all parts can easily be repeated or updated and can be traced to original sources), and credible (meaning it discusses and documents any limitations, including the possibility of uncertainty or bias). We chose to focus on the accurate element because it was most relevant to the scope of this report.

Enterprise Risk Management
describes a forward-looking management approach that allows agencies to assess risks (that is, both threats and opportunities) that could affect the achievement of their goals. This approach employs six elements, including:

- align process to goals and objectives,
- identify risks,
- assess risks,
- select risk response,
- monitor risks, and
- communicate and report on risks.[574]

Figure 4 below illustrates the elements of the enterprise risk management framework.

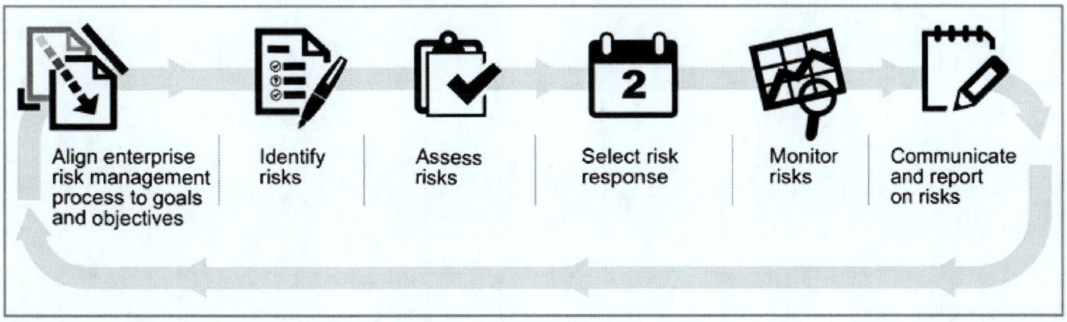

Source: GAO. | GAO-24-106792.

Figure 4. Essential Elements of Enterprise Risk Management.

Acquisition Approaches for Hypersonic Weapons Emphasize Rapid Development

Each of the six offensive hypersonic weapons efforts we identified follow an acquisition strategy that places a high priority on quickly delivering capabilities. The efforts' plans for achieving this differ and progress for some efforts have been delayed due to the need to correct issues identified during initial flight testing. Table 1 provides some details on the six offensive hypersonic weapon efforts.

All the efforts we assessed eventually intend to use the major capability acquisition pathway should they continue to production but differ in terms of what stage they will enter it. Alternatively, decision makers may choose to terminate an effort if it does not develop as expected.

[574] GAO-17-63.

Table 1. Acquisition approaches of offensive hypersonic weapon efforts

Weapon	Military department	Initiation date	Expected fielding date (quarter and fiscal year)	Current acquisition pathway	Type	Platform
Long-Range Hypersonic Weapon	Army	August 2019	4Q 2024	MTA rapid fielding	Boost-glide	Truck-based launcher
Conventional Prompt Strike	Navy	October 2019	4Q 2025	MTA rapid prototyping	Boost-glide	Zumwalt-class surface ships and Virginia-class submarines
Air-launched Rapid Response Weapon	Air Force	August 2018	To be determined	MTA rapid prototyping	Boost-glide	B-52 bomber aircraft
Hypersonic Attack Cruise Missile	Air Force	September 2022	2Q 2027	MTA rapid prototyping	Hypersonic cruise missile	F-15E fighter aircraft
Hypersonic Air-Launched Offensive anti-surface warfare	Navy	March 2023	2029	MTA rapid prototyping	To be determined	F/A-18 (E/F) fighter aircraft
Standard Missile-6 Block IB	Navy	March 2018	4Q 2029	Major Capability Acquisition	Aeroballistic missile	Surface ships

MTA = middle tier of acquisition Q = Quarter.

Source: GAO analysis of DOD documentation. | GAO-24-106792.

Note: For the purposes of this report, we use "initiation" to refer to the MTA effort's start date. For programs using the MTA pathway, the program start date for programs designated on or after December 30, 2019, is generally the date that the program was designated, which is the date that an acquisition decision memorandum was signed initiating an MTA rapid prototyping or rapid fielding program. MTA programs designated before December 30, 2019, and certain programs designated after this date, generally maintain their MTA program start date as the date funds were first obligated.

Long-Range Hypersonic Weapon

The Long-Range Hypersonic Weapon effort seeks to deliver a truck-based long-range missile system capable of launching a hypersonic payload, known as the Common Hypersonic Glide Body, on a boost-glide trajectory.

Source: U.S. Army. | GAO-24-106792.

Long-Range Hypersonic Weapon (LRHW)

The LRHW effort began in 2019 under the Army Rapid Capabilities and Critical Technologies Office's accelerated prototyping authority.[575] The Army delivered the first LRHW battery in 2021, albeit without missiles. A battery is composed of one operations center and four transporter-erector-launchers. Each transporter-erector-launcher consists of a truck and trailer, which carry a launcher that raises to a near-vertical orientation to launch the missile. According to DOD officials, since the first battery was delivered, soldiers have been familiarizing themselves with the launchers and offering feedback. To procure and field the second and third batteries, the Army initiated an MTA rapid fielding effort for LRHW in August 2023, managed by the Army Program Executive Office for Missiles and Space.

The Army delayed delivery—planned for fiscal year 2023—to at least late fiscal year 2024 due to difficulties discovered in flight testing. During the first flight test in June 2022, the missile launched but failed mid-flight. The next three flight tests failed during the initial launch sequence. According to program officials, integration issues between the launcher and missile

[575] According to the Army, the Rapid Capabilities and Critical Technologies Office's charter at the time permitted it to initiate such projects on its own authority without using the middle tier of acquisition rapid prototyping pathway. Though not a formal middle tier of acquisition effort, the LRHW effort's acquisition strategy still generally resembled a rapid prototyping effort, with a 5-year timeframe, and plans to demonstrate the capability while also creating a residual or limited operational capability at the conclusion of the effort.

caused the test failures. As a result, the Army is conducting a risk review before conducting any further tests. The Army has also paused production of LRHW missiles and does not intend to resume it until the review is complete and there has been a successful flight test of the missile.

According to officials, once the effort conducts a successful flight test, the first missile configured for operational use can be delivered in 6 weeks, with an 11-month timeline to complete all eight such missiles needed for Battery 1. A successful flight test is also required before the Army will declare the system operational. Officials said that the earliest date such a test could occur is the fourth quarter of fiscal year 2024.

LRHW is closely aligned with the Navy's Conventional Prompt Strike (CPS) effort as both use the same missile. The Army procures the hypersonic missile and associated canister through the Navy's CPS effort, while design and production responsibilities for certain subsystems are divided between the Army and Navy. Figure 5 shows this division of responsibilities.

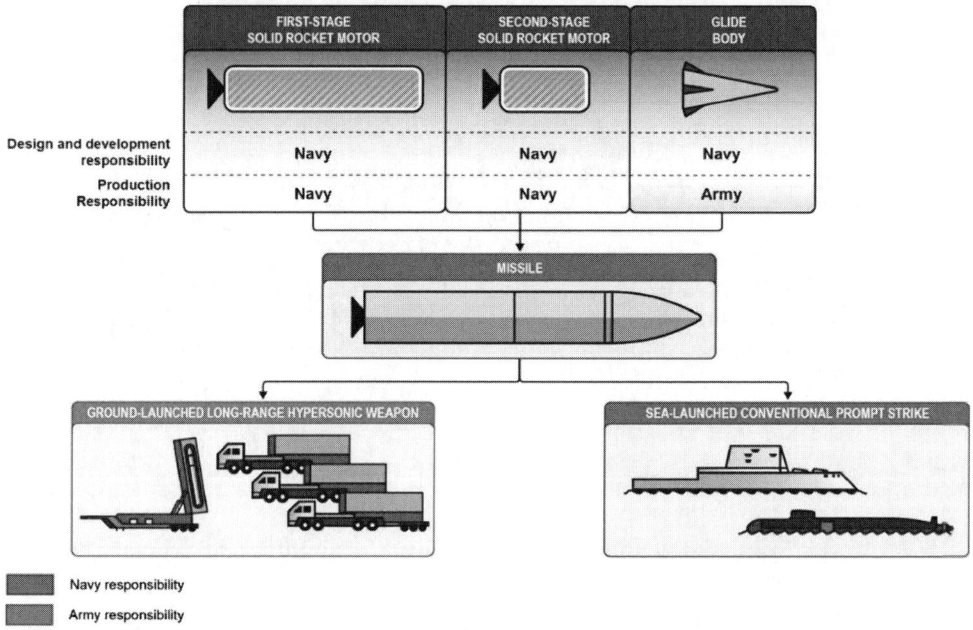

Source: GAO analysis of Department of Defense and contractor information. | GAO-24-106792.
Note: Figure not drawn to scale.

Figure 5. Long-Range Hypersonic Weapon and Conventional Prompt Strike Service Responsibilities.

Conventional Prompt Strike (CPS). The Navy initiated the CPS effort in 2019 as a rapid prototyping effort with an initial phase focused on the development of the missiles for use by both the Navy and Army. The Navy plans to field the CPS system on *Zumwalt-class* destroyers starting in the mid-2020s, and on *Virginia-class* submarines in the early 2030s.

The same test failures that delayed LRHW have also delayed CPS. CPS officials adjusted their testing plan in response to the flight test challenges, creating multiple incremental test events prior to the next flight test.

CPS officials plan to initiate a parallel MTA rapid fielding effort after conducting a successful flight test. The rapid fielding effort will focus on integrating CPS onto the first *Zumwalt-class* surface ship. The overall CPS effort will then later transition to the major capability acquisition pathway at low-rate initial production for fielding on Virginia-class submarines and the remaining *Zumwalt-class* ships, if testing is successful. For both classes of vessels, the Navy is developing a cold- gas eject system that will be used to conduct the initial launch of the missile before the rocket motors ignite. According to officials, this technology has been previously used by the Navy for submarines, but fielding on the *Zumwalt-class* will be a first for the surface fleet.

Air-launched Rapid Response Weapon (ARRW). The Air Force initiated the ARRW effort in 2018 as an MTA rapid prototyping effort. This effort built on the Defense Advanced Research Project Agency's Tactical Boost-Glide demonstrator, combining it with a new operational rocket motor.

Flight test results for the ARRW effort were initially mixed. The first flight test in late 2022 was successful but gathered limited data. The second flight test in early 2023 failed due to faults with the missile shroud and heat shielding. The effort conducted two additional flight tests in September and October 2023. The need for additional tests delayed ARRW and required Air Force officials to apply for and receive a waiver to the 5-year time frame on rapid prototyping efforts. The waiver extended the effort by one year into 2024 to allow completion of the remaining flight tests. The Air Force plans to conduct a final test in March 2024.

Currently, ARRW and Air Force officials have not decided on plans for transitioning to a new pathway at the conclusion of the rapid prototyping and testing phase. According to officials, the Air Force will review the effort and make decisions on its future in 2024 after the last flight test.

Conventional Prompt Strike

The Conventional Prompt Strike effort seeks to deliver a long-range missile system capable of launching the same missile used in The Long-Range Hypersonic Weapon effort from surface ships and submarines.

Source: U.S. Navy. | GAO-24-106792.

Air-launched Rapid Response Weapon

The Air-launched Rapid Response Weapon effort seeks to develop a boost-glide air-to-surface capability launched from a B-52 bomber aircraft.

Source: U.S. Air Force. | GAO-24-106792.

There is currently no planned funding for future phases of development or procurement for ARRW in DOD's budget.

Hypersonic Attack Cruise Missile (HACM). HACM effort builds on the Defense Advanced Research Project Agency's Hypersonic Air-breathing Weapon Concept, a scramjet demonstrator. According to HACM officials, the effort evaluated and selected the most technically mature design from the demonstrator program and used it as the basis for HACM's scramjet. In addition to the scramjet, officials said the Air Force will also develop a new solid fueled rocket motor, a section to connect the scramjet and rocket, an explosive payload, and guidance systems.

The Air Force initiated HACM as an MTA rapid prototyping effort in 2022. According to officials, the effort is engaging in an incremental development process that seeks to build and demonstrate the systems through a series of increasingly capable configurations, rather than designing for and demonstrating the final configuration from the start. In this process, the effort conducts an incremental critical design review for the initial, basic configuration planned for the first test flight, rather than a full critical design review for a fully capable configuration. Air Force officials plan a full critical design review for the final design in spring 2025. Air Force officials plan to conduct at least one test flight of the final configuration prior to the conclusion of the rapid prototyping effort in 2027. HACM plans to conduct several flight tests for the prototyping effort, both in Australia from Australian Air Force F-18s, and in the United States from F-15Es. HACM officials said that they added more opportunities for flight testing than prior efforts, based on the assumption that some tests might fail.

If the tests are successful, the Air Force plans to transition HACM to the major capability acquisition pathway at either development start or production start in 2027, depending on what capabilities the Air Force is willing to accept and whether production facilities are ready.

Hypersonic Attack Cruise Missile

The Hypersonic Attack Cruise Missile effort seeks to develop an air-launched hypersonic cruise missile for use with the F-15E fighter aircraft.

Source: U.S. Air Force. | GAO-24-106792.

Hypersonic Air-Launched Offensive anti-surface warfare (HALO). The HALO effort initiated under the MTA rapid prototyping pathway in March 2023. The Navy placed orders with two contractors with competing designs, according to officials. The Navy intends to take both designs through a preliminary design review, then select a single contractor to continue system development. The HALO effort intends to deliver an initial capability by fiscal year 2029.

HALO officials said that they do not plan to keep the effort within the rapid prototyping pathway through the creation and demonstration of a successful prototype. Instead, officials said that they only plan to stay within this pathway through the conclusion of the current concept development contracts and preliminary design reviews, currently scheduled to conclude in fiscal year 2025. At that point, the effort plans to transition to the major capability acquisition pathway at system development.

Standard Missile 6 (SM-6) Block IB. The Navy initiated the SM-6 Block IB effort in 2018 with an MTA rapid prototyping effort to develop the new, more capable second-stage rocket motor, designed by the Navy. The Navy planned to follow this with a second rapid prototyping effort to incorporate the new motor into an improved SM-6 missile.

While developing the new rocket motor, Navy officials discovered that an additional upgrade to the entire SM-6 family was required due to a component obsolescence issue. According to officials, the approach the effort took to upgrading the second stage rocket motor was not feasible. As a result, the Navy ended the prototyping effort and transitioned the SM-6 Block IB to a major capability acquisition pathway in 2021. The Navy has awarded two other transaction agreements to industry for the second stage rocket motor, with plans to later select a single design. As of September 2023, Navy officials said the program office has identified dates to meet certain milestones, however these dates are subject to change.

Hypersonic Air-Launched Offensive anti-surface warfare

The Hypersonic Air-Launched Offensive anti- surface warfare effort seeks to develop an air- launched hypersonic weapon for the Navy's F-18 E/F fighter aircraft.

Source: U.S. Navy. | GAO-24-106792.

Standard Missile 6 Block IB

The Standard Missile 6 (SM-6) Block IB is the latest upgrade to the Navy's existing SM-6 family currently fired from surface ships.

According to SM-6 Block IB officials, the upgrade would increase the range, speed, and capability of the missile by replacing the current second stage rocket motor with a larger, more powerful one capable of achieving hypersonic speeds.

Source: U.S. Navy. | GAO-24-106792.

Hypersonic Efforts Are Not Fully Implementing Leading Practices for Product Development

DOD's hypersonic weapon development efforts are not fully implementing leading practices for product development, which we have found enable leading commercial companies to deliver products quickly. For example, the efforts are determining which desired capabilities are most essential and achievable and intend to develop a minimum viable product that the military departments can then deliver quickly to the warfighter. Not all the efforts, however, are soliciting user feedback to determine which capabilities to include in the minimum viable product or to inform future iterations of the design. In addition, some efforts have yet to fully adopt the use of advanced digital engineering and digital twins.

Trade-Offs to Facilitate Rapid Development Do Not Fully Incorporate User Feedback

Our leading practices for product development identify the ability of programs to cancel or defer certain capabilities as critical to on-time delivery of a minimum viable product. Leading companies then improve the minimum viable product through iterative development—the successive incorporation of new capabilities over time.[576]

All six efforts we reviewed have taken steps consistent with this approach or plan to do so:

- LRHW demonstrated this approach when dealing with an issue with achieving one of the desired capabilities identified by the Army: the ability to transport the missile by rail. During system development, Army officials determined that they would have to re-design the missile launcher to transport it by rail. LRHW had already demonstrated transportability by aircraft, and according to officials, air-transportability was considered the more important capability. Army officials decided to defer meeting the rail-transportability capability to maintain schedule, as it was not necessary for a minimum viable product. Figure 6 shows LRHW on an aircraft for transport.
- CPS delayed the integration of an advanced capability into its initial design. While the Navy desires this capability, officials told us that they understood early in development that achieving it would take considerable time and effort. Therefore, they deferred this capability in favor of developing a minimal viable product that the effort can quickly field to the warfighter. Officials stated that CPS is leveraging an incremental approach to add other additional capabilities in future iterations.
- ARRW identified technologies whose inclusion would have delayed the delivery of the product beyond the 5-year rapid prototyping time frame. The effort deferred those technologies for inclusion in a future increment, should funding for that increment become available.

[576] GAO-23-106222.

Source: U.S. Army. | GAO-24-106792.

Figure 6. Demonstration of Air Transportability of Long-Range Hypersonic Weapon.

The HACM, HALO, and SM-6 Block IB efforts are also employing a minimum viable product approach, but given their current phase of development there have been fewer opportunities to make similar decisions regarding specific capabilities.

- HACM officials affirmed that they intend to refine the minimum set of capabilities to be included in the minimum viable product and are adopting software architectures that can facilitate future updates.
- HALO's acquisition strategy states that the effort will culminate with a demonstration of a minimum viable product that meets the range and speed requirements to engage a target.
- SM-6 Block IB officials described the current effort as aimed at delivering a minimum viable product. Officials acknowledged that future capabilities could be added through software. They also said that aspects of the acquisition strategy are not finalized and still subject to change.

Our leading practices also state that direct and timely collaboration with users is essential to determining the most essential capabilities for including in the minimum viable product.[577] Importantly, our leading practices state that users are individuals that directly interact with the system, such as operators and maintainers. This does not include high- level officials who may determine requirements, nor other leaders who may make decisions about whether to continue development or to buy a product. While these individuals are important, we have found that leading companies prioritize user needs in defining and iteratively improving their minimum viable product.

Our review of the six offensive hypersonic efforts showed that officials from two efforts—LRHW and CPS—are soliciting and incorporating user feedback or have plans to do so.

[577] GAO-23-106222 and GAO-22-104513.

- LRHW officials have solicited extensive feedback from operators and maintainers in the design phase. These officials told us the effort fielded the LRHW launcher and other ground equipment to soldiers over 2 years in advance of the official deployment of the system for familiarization and to obtain feedback. This feedback was then used to make changes to the launchers and associated systems. LRHW officials told us the effort intends to obtain additional feedback from soldiers after the delivery of the first battery of missiles. They will then use this feedback to inform iterative development of the system and validate future capabilities for the second and third LRHW batteries.
- According to CPS officials, they have also been soliciting and incorporating user feedback during system development. For example, officials placed a prototype weapon control system and simulated missile on the USS Zumwalt and allowed users to operate the system in a simulated environment. They then collected user feedback on the design, installation, and ability to interface with the ship's systems. Officials further stated that they will continue to solicit and incorporate user feedback during development.

We determined that officials from the remaining four efforts—SM-6 Block IB, HACM, HALO, and ARRW—do not currently solicit and incorporate direct and timely user feedback or have plans to do so in the future. While officials from these efforts are soliciting and collecting feedback from a variety of stakeholders, these stakeholders do not constitute "users" in the sense of individuals who physically interact with or directly operate the system.

- SM-6 Block IB officials told us the effort is not soliciting and obtaining feedback directly from users and has no plans to do so. Officials stated that generally their weapon development efforts rely on input from Navy leadership.
- HACM officials do not have plans to directly solicit and incorporate user feedback during the design phase. Officials stated that, while no official channel for doing so is planned, users could provide feedback to their respective requirements officers following testing.
- HALO officials told us they do not have plans for soliciting or incorporating user feedback at this time. Instead, the effort is soliciting feedback from mission planners rather than end users.
- ARRW officials also stated that they did not have plans to obtain direct, timely feedback from end users. According to officials, they have involved maintainers, but their focus was primarily on the development of training orders for maintaining the weapon rather than input on its design or features.

At present, offensive hypersonic weapon efforts are not required to solicit this type of feedback throughout development. We previously reported that DOD's software acquisition policy requires solicitation of feedback from users, but the policies for other acquisitions do not contain a similar requirement. As a result, there is no policy outlining the mechanisms to solicit and incorporate user feedback that extends to all DOD acquisition programs and efforts.

In 2022, we recommended DOD implement incorporating feedback from users of initial capabilities throughout development, but the department has yet to incorporate this into its policies.[578] Such a policy would better position programs and efforts to understand user needs, develop capabilities to meet those needs, and better plan for future iterative development. The absence of such a policy does not preclude efforts from soliciting user feedback, which could help them meet user needs in a timely fashion.

Efforts Are Not Fully Incorporating Other Practices that Could Speed Up Development

Our prior work has found that leading companies benefit from developing a variety of models using digital engineering tools, such as digital twinning, during the design modeling and simulation phase.[579] A digital twin is a virtual model that simulates the configuration, performance, and behavior of a system within a computer. This virtual model can be updated when new features are added. A high-fidelity digital twin, coupled with high-resolution simulations of the operating environment, can be used for testing the system to validate that it meets requirements. This reduces the need to build physical prototypes each time the design changes. In addition, digital twins are also useful in the sustainment phase. These digital design tools are useful in the design and validation process as they can enable more rapid iterative design cycles and facilitate stakeholder and user feedback at earlier stages.

We determined that the HALO and CPS efforts are either currently using digital engineering tools, including digital twins, or have plans to do so in the future. Specifically:

- HALO's acquisition plan states that the prime contractor will need to establish a digital engineering ecosystem—i.e., a single, integrated computer environment—as well as the use of digital engineering tools. According to HALO officials, vendors are required to provide high- fidelity digital models of the components and subsystems in the design, which DOD officials can then assemble and test as a digital prototype.
- CPS also has a digital engineering plan that includes a transition to a computing environment similar to HALO's. Program officials stated that CPS has used a 3-D model in design, which they have found to be useful in development. Officials plan to use digital engineering tools and techniques in the future to prepare for production and eventual sustainment. According to officials, the challenge with the approach is bringing the various models of the missile and missile components, such as the CHGB, together to create a full digital representation.

Officials from the remaining four efforts—LRHW, SM-6 Block IB, ARRW, and HACM—are either not using digital engineering or are using it only to a limited degree and lack any plans to further implement these tools in the future. Specifically:

- LRHW officials used some digital engineering tools to create a virtual reality model of the launcher that users could interact with to identify potential design flaws. From

[578] GAO-22-104513. DOD concurred with this recommendation and stated it would consider the integration of key product development principles, to include incorporating user feedback, when it updates its overarching acquisition policy, the individual acquisition pathways, and functional acquisition policies. DOD anticipates completing this effort by June 2024.

[579] GAO-23-106222 and GAO-22-104513.

these interactions, users were able to identify potential maintenance challenges based on the placement of certain components, such as a generator. Officials were then able to make modifications to the digital and physical design, incorporating this feedback. While this simulation is useful for some applications, it is not a high-fidelity, dynamic digital model necessary for implementing digital engineering. LRHW officials said that they do not plan to develop digital models of this kind.

- SM-6 Block IB officials reported they do not plan to use digital engineering tools or digital twins. Instead, they stated they intend to continue to leverage modeling and simulation tools used in previous Standard Missile programs. This falls short of the advanced digital engineering tools our leading practices identified. Program officials acknowledged that effort could leverage a digital twin, but they currently plan to focus their attention on developing underlying technologies for the system without it.
- ARRW officials reported they do not use digital twinning. they stated that they do not have specific plans to increase their use of digital engineering tools beyond the typical use of models and simulations.
- According to HACM officials, they attempted to leverage digital design tools, up to and including fully digital design reviews. They stated, however, that there are challenges to conducting these reviews, including the number of tools in use, licensing restrictions, limited computing power, and the logistics of conducting reviews in a way that is accessible to the large number of Air Force stakeholders. HACM officials are not planning to use a digital twin and will instead continue to use 3D models, which are less dynamic and not as easily updated in real time.

DOD issued a policy in December 2023 requiring efforts initiated after the date of issuance to incorporate digital engineering.[580] The policy also states that existing efforts, such as the offensive hypersonic efforts, may incorporate digital engineering when it is practical, beneficial, and affordable, but are not required to do so under the policy. Our past work highlighted that private companies use digital engineering tools to allow for faster iterative design cycles than what would be possible with physical prototypes alone.[581] While fully utilizing digital engineering tools can pose certain challenges, efforts that do not may be missing opportunities to take advantage of the efficiencies they can provide. This includes the ability to anticipate potential design flaws, optimize manufacturing, and reduce costs. A formal, documented assessment of whether and how to implement these tools would better position these efforts to take full advantage of the benefits they can provide.

DOD Has Identified Its Limited Experience with Hypersonic Weapons as a Risk in Accurately Estimating Costs

According to multiple DOD and program officials, DOD's limited experience designing, producing, and testing hypersonic weapons provides a limited basis to determine the cost of prototyping while also complicating efforts to project those costs into the future. We assessed the cost estimate for the CPS effort, one of the more mature estimates among those we identified

---

[580] Department of Defense, Digital Engineering, DOD Instruction 5000.97 (Dec. 21, 2023).
[581] GAO-23-106222 and GAO-22-104513.

and found it was affected by these uncertainties. Reducing or otherwise accounting for uncertainties will limit the risk of DOD allocating resources without a full understanding of the costs for its hypersonic weapon efforts. We found that while the estimate meets many of our leading practices for cost estimating, the Navy could improve its cost estimating methodologies and how it updates the estimate to account for actual costs. For example, we found the estimate relied heavily on the input of subject matter experts, whose inputs can be prone to bias unless estimators further analyze and account for that bias.

Limited Experience with Hypersonic Weapons Contributes to Uncertainty over Costs

According to DOD officials and contractor representatives, DOD's limited experience overseeing hypersonic development efforts, particularly those with the range and performance of those systems currently under development, is a key driver for near-term costs, while also complicating efforts to estimate future costs. Program officials noted that, to a large extent, rapid prototyping efforts do not lend themselves to creating affordable designs, as their priority is to demonstrate a capability quickly.

These designs require additional work in the future to make them easier to produce or to incorporate less expensive materials or processes.

Further, many critical components for hypersonic weapons do not have commercial equivalents, and therefore rely on unique components or materials. In many cases, only one or two vendors may be capable of producing these items, limiting opportunities for competition. Some of these unique items also rely on labor-intensive processes with significant lead times. As DOD and industry gains experience designing, building, and testing hypersonic weapons, officials expect the designs will become easier to manufacture at scale. Until this occurs, however, the full cost of these programs is difficult to estimate with confidence.

Current estimates of unit cost are also a consideration in these procurement decisions. For example, LRHW officials stated that missiles for the second and third batteries are estimated to cost over $40 million each. This results in a total potential cost of over $1.28 billion to fully equip those two batteries with missiles once the launchers and other ground support equipment have been procured.

DOD officials also acknowledged that high unit costs could result in DOD procuring fewer weapons, potentially at levels below those needed to meet a combatant command's operational requirements. At the same time, DOD officials said that military departments have been reluctant to plan and budget for significant quantities until they know which offensive hypersonic weapons will meet warfighters' needs. These uncertainties for production complicate the calculation of costs for any subsequent programs to field these weapons.

Despite these difficulties, most hypersonic weapon efforts have some form of cost estimate, even if it only includes costs associated with the current phase. Table 2 provides key details for these estimates.

Table 2. Details of hypersonic weapon efforts' cost estimates

Name	Year conducted	Total (then-year dollars in billions)	Scope
Long-Range Hypersonic Weapon	2023	$10.3	Army lifecycle cost estimate for ground support equipment and missiles
Conventional Prompt Strike	2021	$30.1	Navy lifecycle cost estimate, including development and production
Air-launched Rapid Response Weapon	2019	$1.2	Air Force lifecycle cost estimate for rapid prototyping
Hypersonic Attack Cruise Missile	2022	$1.9	Air Force cost estimate for rapid prototyping
Hypersonic Air-Launched Offensive anti-surface warfare	2023	$1.5	Navy cost estimate for research and development
Standard Missile-6 Block IB	TBD	TBD	No cost estimate available due to program restructuring

TBD = to be determined.
Source: GAO analysis of Department of Defense documentation. | GAO-24-106792
Note: This table should not be used for comparison purposes, due to the differences in the scope of the estimates.

CPS Cost Estimate Could Improve Its Methodologies and Better Incorporate Actual Costs

Our analysis of the CPS effort's most recent cost estimate showed that while it substantially met three of the five elements of an accurate cost estimate, it could improve its methodologies to better account for DOD's limited experience with hypersonic weapons and its approach to updating the estimate with actual costs.[582] According to our leading practices, an accurate cost estimate is based on a detailed cost model—a build-up of program costs for each cost element necessary for accomplishing all work and delivering the end product. Validating that a cost estimate is accurate requires thoroughly understanding and investigating how the cost model was constructed.

While most of the offensive hypersonic weapon efforts we assessed have some form of cost estimate, we assessed CPS because it was among the more mature of the six efforts we identified and approaching a decision to enter the rapid fielding phase of the MTA pathway. According to CPS officials, this meant that the estimate was relatively up-to-date, but was still technically in draft form. We assessed the CPS estimate against five leading practices for the accurate characteristic: (1) the methodology used; (2) the adjustments made for inflation; (3) whether it

[582] GAO-20-195G also identifies "Contains few, if any, minor mistakes" as an element of the "accurate" characteristic. The Navy declined to provide to us the full cost model for review due to the draft nature of the estimate. As a result, we did not score this element. Not having the full cost model complicates our ability to determine if all calculations are accurate and account for all costs. Westill reviewed the supplemental documentation provided and solicited information on the CPS effort's quality control process and found few, if any, mistakes. Additionally, we are not providing the total figure in this report because doing so would limit our ability to publicly release the report.

reflected changes made to the program; (4) the treatment of variances in planned and actual costs; and (5) the use of historical records from analogous programs.[583]

Methodology used (partially met). We found that the cost estimate only partially met the requirement that estimators ensure it is based on the best methodology available. Given the unique nature of the CPS effort there are few, if any, direct analogies to offensive hypersonic weapons.

To account for this in their methodology, cost estimators leveraged analogous historical data from other missile programs. For example, the cost estimators used the Trident II D-5 program as a point of comparison, making an adjustment based on the different weights of the missiles.

However, D-5 outweighs the CPS missile by a considerable factor. According to our leading practices, when choosing analogous programs, the technical parameters compared should be roughly similar to the program for which the estimate is performed. When the technical parameter is as different as the D-5 analogy, the cost adjustment may not reasonably explain the potential cost differences between the programs.

In cases where analogous programs were not available, or deemed not appropriate, we found the estimate relied on the opinion of subject matter experts. The expert opinion method should be considered subjective unless estimators further analyze and account for the assumptions within those experts' opinions. Our leading practices state that expert opinion should be used sparingly, and the estimate should account for the possibility that bias influenced the results. We found that, while the estimate relied heavily on subject matter expertise, the estimate's methodology did not capture the full extent of potential bias in these inputs.

While the estimate accounts for the uncertainty stemming from the limited availability of analogous or actual data by increasing the margins within the estimate, estimators did not employ cross-checks to further account for uncertainties when using historical analogies. Cost estimators apply cross-checks by using alternative estimating methodologies to see if they produce similar results, ensuring greater credibility in the resulting estimate. However, for the CPS cost estimate, the Navy did not document its cross-checks in the model or provide supplemental documentation for crosschecks during our review. When asked about the lack of cross- checks on areas such as the D-5 analogy, CPS cost estimators stated they were not necessary because the element they estimated was a small percentage of the total estimate. While the analogy may only represent a portion of the estimate, our concerns with the lack of cross-checks remain.

Additionally, the estimate used an atypical methodology to update engineering costs based on schedule delays. Rather than updating the entire estimate's assumptions for the time (and thus expense) necessary to complete this engineering, the estimate added a new element to account for them. Because the cost of each element of the estimate is linked to specific schedule assumptions, not performing a global update to the estimate's schedule risks invalidating estimates for other elements of the overall effort. CPS cost estimators stated that they used a new element because it was too burdensome to update the entire model. In addition, they stated

[583] GAO-20-195G. Our leading practices also include three other characteristics of a reliable cost estimate: comprehensive (meaning it reflects the current schedule and technical baselines), well-documented (meaning all parts can easily be repeated or updated and can be traced to original sources), and credible (meaning it discusses and documents any limitations, including the possibility of uncertainty or bias). We chose the accurate characteristic for this assessment because it was the most relevant to the scope of this report.

that the costs of this specific delay would soon be realized and thus "sunk." Even if these costs are sunk, however, the risk of invalidating other aspects of the model remains.

Adjustments for inflation (substantially met). We found that the Navy substantially met the criteria for properly adjusting for inflation in both its cost model and the inputs for the estimate. Cost estimators used a database that provides cost data normalized to a specific base year, and they used a software package that has built-in functions to account for inflation. For the development and production phases, cost estimators used 2023 indexes from the National Reconnaissance Office since they were more applicable to CPS than other Navy indexes.

Updates to reflect program changes and actual costs (partially met). We found that the cost estimate partially met the criteria to update the results of the cost estimate to reflect program changes and actual costs. Navy estimators received cost data from contractors and subcontractors and used these data to update their estimate. However, due to data quality issues with the prime contractor's reports and the unique nature of the first efforts to build the system, cost estimators did not input this information directly into the estimate. Instead, estimators used the actual cost data to cross-check their existing estimate. These crosschecks are not documented in the cost model, however. In addition, as noted above, the estimate accounted for the costs of schedule delays by placing them as a separate entry in the estimate rather than updating the model's schedule and examining its effects across the original estimate.

According to our leading practices, not updating the estimate makes it more difficult to analyze changes in program costs and hinders efforts to collect cost and technical data to support future estimates.

Document, explain, and review variances between planned and actual costs (substantially met). We found that the estimate substantially met the criteria for documenting, explaining, and reviewing variances between planned and actual costs. We found that cost estimators kept an estimating change log that documented the changes made within the model as well as the resulting variances between different iterations of the estimate. Estimators demonstrated that costs within the estimate are updated as actual cost data arrive, though not at the most detailed levels of the estimate.

Use of historical records from analogous programs (substantially met). We found that the cost estimate substantially met the criteria for using historical records from analogous programs. As stated previously, DOD does not have significant experience overseeing the development or fielding of hypersonic weapons. As a result, estimators had limited historical data applicable to the effort. In spite of that limitation, we found that the limited historical data sources the Navy used for the CPS cost estimate are applicable to the program. We found that CPS cost estimators tried to draw from a range of missile systems, scaling to the extent possible to the relevant technical parameters. As noted above, the estimate's methodology did not completely account for the disparity between the specifications and requirements of the CPS system and the chosen analogous programs. The system's unique design, however, means that the data used in the CPS estimate are among the best available.[584]

While the cost estimate substantially meets most of our leading practices associated with the accurate characteristic of cost estimating, there are further steps cost estimators could take.

[584] GAO-20-195G also identifies "Contains few, if any, minor mistakes" as an element of the "accurate" characteristic. Due to the draft nature of the Navy's estimate, we were unable to review the full cost model. As a result, we did not score this element. Not reviewing the full cost model complicates our ability to determine if all calculations are accurate and account for all costs. Westill reviewed the supplemental documentation provided and solicited information on the CPS program's quality control process and found few, if any, mistakes.

CPS cost estimators said that they were aware of these steps and that they did not take them as they would be burdensome to execute on all parts of the estimate. Adjusting the methodology of the estimate and better integrating updated schedule assumptions and real-world data could be worth this effort as it will provide Navy decision-makers a more accurate cost estimate and could act as a model to cost estimators for future offensive hypersonic weapons.

DOD Is Not Fully Implementing Enterprise Risk Management for Hypersonic Weapons

In addition to cost risks, DOD has identified a number of challenges and risks at the enterprise level for managing hypersonic weapons. DOD also identified a number of actions intended to address them. These actions implement most of the essential elements of an effective enterprise risk management framework as described in our leading practices.

Enterprise risk management is a forward-looking management approach that allows an agency to assess risks that could prevent the achievement of its goals. In this sense, "risks" can be both threats to achieving goals as well as opportunities identified by the organization as important to meeting its goals. We previously identified six essential elements for an agency to effectively manage risks through this framework.[585] These elements are: (1) aligning the risk management process with the agency's goals and objectives; (2) identifying risks; (3) assessing risks; (4) selecting risk responses; (5) monitoring risks; and (6) communicating and reporting on risks. While DOD is meeting the first five elements, it is inconsistently reporting progress on addressing risks and the costs to do so to Congress.

DOD Is Substantially Identifying, Assessing, Addressing, and Monitoring Risks

DOD completed the first essential element of an enterprise risk management framework, which is to align the risk management process for offensive hypersonic weapons to the agency's goals and objectives. The 2022 National Defense Strategy, which describes the military challenges facing the country and the means to address them, states that DOD will develop hypersonic systems to bolster the military capabilities of the United States.[586] The strategy also states that DOD will support hypersonic research and development as part of building an enduring foundation for future military advantage. Furthermore, the Under Secretary of Defense for Research and Engineering identified hypersonics as one of DOD's 14 critical technology areas in 2022. These areas guide investments to accelerate transitioning key capabilities to the military services and the combatant commands.[587]

The Under Secretary designated a senior DOD official, the Principal Director for Hypersonics, to serve as the focal point to coordinate hypersonic research and engineering

---

[585] GAO-17-63.

[586] Department of Defense, 2022 National Defense Strategy of the United States of America Including the 2022 Nuclear Posture Review and the 2022 Missile Defense Review (Washington, D.C.: Oct. 27, 2022).

[587] Department of Defense, Under Secretary of Defense for Research and Engineering, USD(R&E) Technology Vision for an Era of Competition (Washington, D.C.: Feb. 1, 2022); and National Defense Science and Technology Strategy 2023.

activities across the department.[588] The Principal Director is also to lay out a strategy for achieving both the first generation of hypersonic weapons, as well as subsequent generations covering more missions and with expanded capabilities. The activities coordinated by the Principal Director's office include managing enterprise risks that affect the development and fielding of offensive hypersonic weapons. The director coordinates with other offices with responsibilities for elements of hypersonic development, including the Joint Hypersonics Transition Office, Office of Manufacturing Capability Expansion and Investment Prioritization, and Test Resource Management Center.

DOD has substantially implemented the second, third, fourth, and fifth essential elements of an enterprise risk management framework by identifying risks, assessing them, selecting responses to address them, and monitoring them. We categorized these risks and responses into four enterprise-level areas: technology, industrial base, test and evaluation, and workforce. We have previously reported on these risks as well as some of DOD's responses to address them.[589]

Technology. DOD stakeholders in offensive hypersonic weapon efforts have identified technology risks for both present and future systems and assessed their importance. These risks include current technology challenges as well as enabling technologies that, if developed successfully, could increase production rates or provide additional capabilities for future systems. Specific technology risks include materials for thermal protection systems, alternative navigation systems, propulsion systems, and sensors, among others. Figure 7 illustrates components within hypersonic boost-glide and cruise weapons.

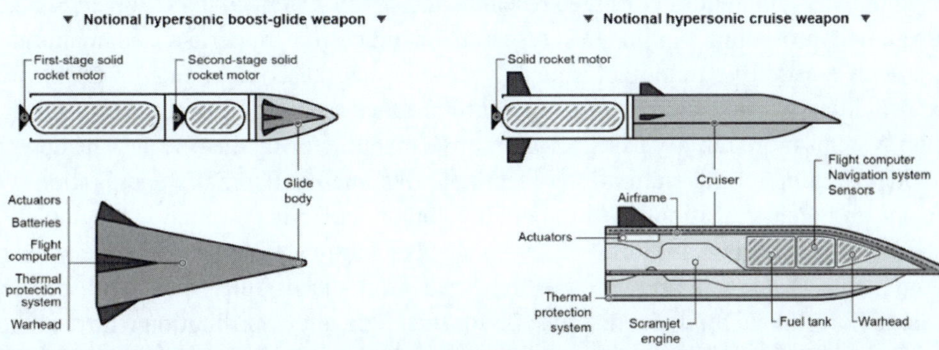

Source: GAO analysis of Department of Defense and contractor information. GAO-24-106792.

Notes: Actuators are hardware that take command signals and move control surfaces, such as fins, enabling a weapon to maneuver in the atmosphere.

Scramjet engines are propulsion systems that use shock waves to compress incoming oxygen from the atmosphere and slow it to supersonic speeds. The engine then mixes the compressed oxygen with fuel, combusts the mixture, and expels the combusted products from the cruiser to generate thrust.

Warheads are a part of a weapon's payload and contain explosives and other components designed to damage a target.

Figure 7. Notional Hypersonic Boost-Glide and Hypersonic Cruise Weapons.

- Thermal protection systems insulate offensive hypersonic weapon systems against air heated to high temperatures by shock waves and aerodynamic friction, and are a

[588] William M. (Mac) Thornberry National Defense Authorization Act for Fiscal Year 2021, Pub. L. No. 116-283, § 217.

[589] GAO-21-378.

limiting factor for a system's range and maneuverability. These systems are difficult to manufacture and are a cost and schedule driver, according to DOD officials and industry representatives.

- Alternative navigation systems maintain the course of offensive hypersonic weapons in contested areas where GPS signals may bejammed or spoofed.[590] Navigation systems need to survive and function in hypersonic flight conditions.
- Propulsion systems include solid rocket motors and scramjet engines. The development of solid rocket motors is a large technical challenge, according to DOD officials, and their production is a cost and schedule driver. While DOD has successfully demonstrated scramjet engines, the demonstration vehicles were not designed to be operational weapons or mass-produced. Furthermore, the scramjet engine is a major cost driver for hypersonic cruise missiles.
- Sensors can increase the capabilities and effectiveness of offensive hypersonic weapons. DOD officials said that additional technology development is needed for sensors to survive and function in hypersonic flight conditions, to integrate them into weapon systems, and to improve their manufacturability.

To address these risks, the Joint Hypersonics Transition Office and other DOD offices prioritize and provide funding for technology investments.

They provide funding for multiple areas of hypersonic research at government labs, academia, industry, and federally funded research and development centers. For example, DOD offices reported providing funding for research of alternative materials and manufacturing methods, such as additive manufacturing for scramjet engine components.[591] DOD officials reported providing partial funding for a project to produce carbon- carbon material for thermal protection systems more rapidly using an alternative manufacturing method.[592] The department also provides funding for the maturation of technologies, such as alternative navigation systems and sensors, and their transition into weapon development efforts.

In terms of monitoring, the Joint Hypersonics Transition Office oversees the science and technology projects it funds and whether they transition successfully. For every project, the office assigns a technical manager responsible for oversight and coordination with stakeholders to transition technologies, according to office officials. Officials also said that the office monitors projects by conducting quarterly technical and programmatic reviews to validate performance measures related to schedule and budget as well as deliverables to an assigned transition partner. The office briefs the Principal Director for Hypersonics on the progress and transition status of its projects on a quarterly basis. The office is working with the Principal Director and other stakeholders to revise its hypersonic science and technology roadmap and

---

[590] We have previously reported on DOD's efforts to develop positioning, navigation, and timing technologies to complement GPS. GAO, *Defense Navigation Capabilities: DOD Is Developing Positioning, Navigation, and Timing Technologies to Complement GPS*, GAO-21-320SP (Washington, D.C.: May 10, 2021); and *GPS Alternatives: DOD Is Developing Navigation Systems but Is Not Measuring Overall Progress*, GAO-22-106010 (Washington, D.C.: Aug. 5, 2022).

[591] Additive manufacturing, also known as 3-D printing, is a computer-controlled process that creates physical objects by depositing materials, usually in layers. Additive manufacturing can produce complex objects that would be difficult or impossible to produce using traditional manufacturing methods.

[592] Carbon-carbon is a composite material composed of carbon fibers weaved together and infused with a carbon-based resin. The composite material is cured in a process called densification that requires high temperatures and months of time, according to DOD officials and industry representatives. The alternative manufacturing method, called rapid densification, could significantly reduce the densification time.

align the community's investments based on the capabilities needed for DOD's hypersonic weapon systems.

Industrial base. DOD, through multiple studies, has identified and assessed risks with the industrial base that affect the manufacturing and cost of offensive hypersonic weapons and can impede increased and affordable production. Specific risks include those listed below.

- There is a limited number of suppliers for critical components including solid rocket motors and thermal protection materials. In some cases, only one supplier exists for the specialized components DOD needs, which reduces opportunities for competition. Additionally, with multiple concurrent hypersonic weapon development efforts, some components come from the same limited pool of suppliers, further straining the industrial base. Some components, such as solid rocket motors and carbon-carbon materials, are also in demand for other weapon systems, further complicating these problems.
- Some critical components require long lead times for production due to the processes used to manufacture them. For example, it takes months to create the specialized carbon-carbon material used for some weapons' thermal protection systems. This is because producing these materials currently requires a significant amount of manual labor as opposed to automated machinery, according to industry representatives. Other examples include solid rocket motors, specialty aluminum airframes, positioning systems, actuators, complex electrical wiring harnesses, and warheads. These long lead times can slow production rates if not considered in a timely manner and contribute to higher costs.

DOD is addressing industrial base risks for offensive hypersonic weapons through several investments made by the Office of Manufacturing Capability Expansion and Investment Prioritization. The office reported that it awarded contracts to increase the number of suppliers in the industrial base and that base's capacity and capability for manufacturing. For example, to increase the capacity and robustness of the industrial base for solid rocket motors, the office reported awarding a $64 million contract to a second supplier of solid rocket motors for LRHW and CPS. Another investment was to reduce the manufacturing time for carbon- carbon materials by using robotic additive manufacturing techniques.

The office is monitoring its industrial base investments and has established performance measures and goals, though it is too early to track progress with those measures. Office officials said that they conduct review meetings for their investments and brief the Principal Director for Hypersonics about every 6 months. The investments have performance measures and goals for specific production rates and percent reductions in costs for producing hypersonic weapon components. However, the investments have not produced components yet, since they are currently building new facilities.

Test and evaluation. DOD has identified and assessed test and evaluation capacity and capability as major risks to the development of offensive hypersonic weapons. The risks stem, in part, from the age and insufficient sustainment and improvement of ground- and flight-test facilities due to the inconsistent focus on hypersonic systems in previous decades. In addition, test facilities and assets that support the concurrent development of multiple offensive hypersonic weapon systems also support the development of missile defense and strategic systems. As a result, the demand for ground- and flight-test facilities is much greater than what can be supported by what a DOD official said is available in federal agencies, industry, and academia. This high demand leads to schedule conflicts and could ultimately delay hypersonic weapon efforts and increase costs.

Additionally, existing test facilities and ranges, such as those to test airbreathing propulsion systems, do not reproduce all of the physical conditions associated with hypersonic flight, according to DOD officials. Without sufficient capacity to test or the capability to replicate certain physical conditions, hypersonic weapon development efforts may lack accurate information to adequately lower technical risk or validate design and performance models. Hypersonic weapon development efforts then might have to choose between adopting more conservative designs, which can limit a weapon's performance, or having lower confidence in a weapon's performance.

To address test and evaluation risks, the Test Resource Management Center is making investments to act on the highest priority needs. A center official said that the center has $1.5 billion in investments planned from fiscal years 2023 through 2028 to upgrade and expand test infrastructure. The center is making investments in ground facilities, such as building a new arc-heated facility and a new airbreathing propulsion test facility that will better replicate hypersonic flight conditions compared to existing test facilities. The center is also making several investments for flight tests, including airborne sensors and data collection platforms, more frequent and affordable flight testing, and additional test corridors. For example, the SkyRange effort converts Global Hawk and Reaper uncrewed aerial systems to collect data on test assets in flight. These aircraft can be deployed more agilely and collect certain types of data with higher quality than the crewed surface ships that traditionally support flight tests, according to DOD officials. In addition, the center is investing in the Multi-Service Advanced Capability Hypersonic Test Bed, which DOD officials expect to provide affordable and more frequent flight-test opportunities for a broad set of advanced technologies and systems across different operational hypersonic conditions. Finally, the center plans to establish long-range flight-test corridors over the Atlantic Ocean, over land in the western United States, and over the Pacific Ocean into Alaska. The U.S. is also leveraging its partnership with Australia to use its test ranges.

DOD officials said that they have established the frequency of hypersonic flight testing as a relevant performance measure to monitor test and evaluation risk and have set a goal of conducting one hypersonic flight test per week. Officials have said that the Multi-Service Advanced Capability Hypersonic Test Bed is the main method to achieving this goal. The Principal Director for Hypersonics said that he typically receives quarterly briefs from the Test Resource Management Center with less formal updates and communication occurring frequently between the more formal briefings. Center officials said that they monitor how well they address test and evaluation risks by annually reassessing shortfalls in needed capabilities and updating the hypersonic test infrastructure roadmap. Figure 8 illustrates some of DOD's risk responses for hypersonic boost-glide and cruise weapons.

Workforce. DOD has identified the potential of a limited workforce for hypersonics as a risk. The sensitive nature of hypersonic technologies and systems limits the available workforce as it requires that the workforce has the appropriate technical expertise and the ability to access classified materials. Additionally, much of the workforce with technical knowledge of hypersonics is nearing retirement. Many sectors of government and industry that need similar expertise as hypersonics, including the commercial space industry, compete with hypersonics to attract and retain the limited supply of talent, according to a DOD official. It is especially challenging to recruit talent for some high-demand occupations, such as software engineers.

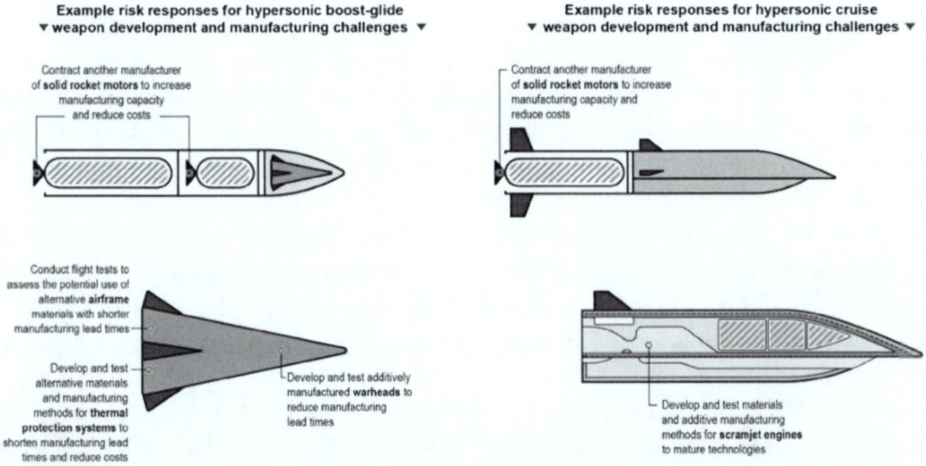

Source: GAO analysis of Department of Defense and contractor information. GAO-24-106792.

Figure 8. Example Risk Responses for Hypersonic Boost-Glide and Hypersonic Cruise Weapons.

To address workforce risks, the Joint Hypersonics Transition Office has taken several actions, including developing a roadmap to create a pipeline of candidates graduating from academic or technical training programs to work on hypersonics. This roadmap targets workforce gaps in and disciplines related to priority technology challenges and areas of high competition for talent. The office has also established a university consortium to give undergraduate and graduate students opportunities to conduct sensitive applied hypersonic research. Finally, it has funded internship opportunities at national laboratories, government agencies, and universities for students to gain experience in hypersonics.

The Joint Hypersonics Transition Office's workforce development team monitors and measures students' progression and its outreach with various groups. The office collects information about students working on projects it funded, including milestones such as internships, graduation, and job acquisition. The office's workforce development roadmap lists performance measures on outreach with university students, community college and technical school students, industry, mid-career professionals, educators and students, and underrepresented groups. The office briefs the Principal Director for Hypersonics on student development, training courses, and outreach activities on a quarterly basis.

DOD Is Not Comprehensively Reporting Progress on Managing Risks

DOD has not implemented the sixth, and final, essential element of an effective enterprise risk management framework: communicating and reporting on risks both internally and externally. DOD offices communicate and report on risks internally, but DOD does not do so externally in a comprehensive fashion. Sharing risk information and incorporating feedback from internal and external stakeholders can help organizations identify and better manage risks. In a federal setting, communicating risk is also important because of the additional transparency expected by Congress, taxpayers, and other relevant stakeholders. Communicating management of enterprise-level risks through a dedicated report or integrating risk information into existing enterprise-level performance management reports are useful ways of sharing progress on managing risks.

While DOD stakeholders involved in offensive hypersonic weapon efforts regularly coordinate and communicate among each other, DOD is not comprehensively communicating and reporting to Congress its progress on managing risks. Reporting comprehensive information would enable Congress to better understand and oversee DOD's progress in managing risks at the enterprise level for offensive hypersonic weapons.

DOD does provide some information to Congress on these risks. The Under Secretary of Defense for Research and Engineering must report annually on the activities and statuses of DOD's critical technology areas.[593] The Under Secretary must also report quarterly on DOD's forward-looking hypersonic science and technology development strategy as well as investments and progress toward achieving DOD's hypersonic science and technology goals.[594] Additionally, DOD has been required to report information on different test and evaluation subjects on a nonrecurring basis since at least 2019. An official from one DOD office said that the office briefs congressional staff members on investments in the industrial base through the annual budgeting process. This information, however, does not include progress on managing enterprise- level risks since the official said that the information contains limited data on individual projects.

More recently, Section 218 of the National Defense Authorization Act for Fiscal Year 2024 established new periodic reporting requirements for DOD, including an "Annual Report on Funding and Investments in Hypersonic Capabilities."[595] The statute requires DOD to include in this report information on funding and investments in "procurement, research, development, test, and evaluation, and operation and maintenance of offensive and defensive hypersonic weapons." The act, however, does not explicitly require reporting on enterprise-level risks and progress managing them.

While the reports to meet prior reporting requirements include information on DOD's activities in different areas, they do not provide comprehensive, enterprise-level information on DOD's progress across all four risk areas.

DOD has not been required to report its progress on managing enterprise-level risks in this way, because existing reporting requirements have not mandated it. Without comprehensive enterprise-level reporting Congress will have an incomplete or fragmented perspective on the risks DOD has identified, the actions it is taking to address them, or the costs that these actions entail.

Conclusion

Years of effort and billions of dollars spent on hypersonic weapon development have yielded considerable progress, but DOD has yet to field its first operational hypersonic weapon system.

---

[593] William M. (Mac) Thornberry National Defense Authorization Act for Fiscal Year 2021, Pub. L. No. 116-283, § 217.

[594] Department of Defense Appropriations Bill, 2020, H.R. Rep. No. 116-84, at 227 (2019).

[595] National Defense Authorization Act for Fiscal Year 2024, Pub. L. No. 118-31, § 218.

Yet even fielding these first prototypes will not ensure an effective or affordable capability. Direct and continuous feedback from potential users on product development efforts assists developers in prioritizing and delivering the most essential capabilities quickly and ensuring that the insertion of capabilities in the future meets the warfighters' needs. Expanded use of modern digital engineering tools will help these efforts to iterate more quickly on designs than is possible with physical prototyping alone. Given the risks identified by DOD in estimating costs for hypersonic weapons, improvements to cost estimating based on leading practices will help DOD plan better and allocate resources. Finally, by providing additional information to Congress, DOD will provide additional transparency on its progress implementing enterprise-level solutions for hypersonics and the costs associated with addressing identified risks.

Recommendations for Executive Action

We are making the following 10 recommendations to DOD:

The Secretary of the Navy should ensure the Standard Missile 6 (SM-6) Block IB effort solicits and incorporates relevant user feedback throughout development. (Recommendation 1)

The Secretary of the Air Force should ensure the Hypersonic Attack Cruise Missile effort solicits and incorporates relevant user feedback throughout development. (Recommendation 2)

The Secretary of the Navy should ensure the Hypersonic Air-Launched Offensive anti-surface warfare effort solicits and incorporates relevant user feedback throughout development. (Recommendation 3)

The Secretary of the Air Force should ensure the Air-launched Rapid Response Weapon effort, if further development and production is planned, solicits and incorporates relevant user feedback throughout development. (Recommendation 4)

The Secretary of the Army should ensure the Long-Range Hypersonic Weapon effort assesses the practicality, benefits, and affordability of implementing digital engineering, including digital twinning, and whether to incorporate these tools into the effort. (Recommendation 5)

The Secretary of the Navy should ensure that the Standard Missile 6 (SM-6) Block IB effort assess the practicality, benefits, and affordability of implementing digital engineering, including digital twinning, and whether to incorporate these tools into the effort. (Recommendation 6)

The Secretary of the Air Force should ensure the Air-launched Rapid Response Weapon effort, if further development and production is planned, assesses the practicality, benefits, and affordability of implementing digital engineering, including digital twinning, and whether to incorporate these tools into the effort. (Recommendation 7)

The Secretary of the Air Force should ensure the Hypersonic Attack Cruise Missile effort assesses the practicality, benefits, and affordability of implementing digital engineering, including digital twinning, and whether to incorporate these tools into the effort. (Recommendation 8)

The Secretary of the Navy should ensure the Conventional Prompt Strike program office improves its cost model in accordance with the "accurate" characteristic described in GAO's Cost Estimating and Assessment Guide, including by estimating each element using the best methodology from the data collected and updating the estimate to reflect program changes and actual costs. (Recommendation 9)

The Secretary of Defense should ensure that the Under Secretary of Defense for Research and Engineering periodically provides information on DOD's progress in managing enterprise-level risks to congressional decision-makers. (Recommendation 10)

Agency Comments and Our Evaluation

We provided a draft of this report to DOD for review and comment. DOD provided an official comment letter (reproduced in appendix II) which concurred with our recommendations. They also provided technical comments which we incorporated as appropriate.

We are sending copies of this report to the appropriate congressional committees, the Secretary of Defense, the Secretary of the Air Force, the Secretary of the Army, the Secretary of the Navy, and other interested parties. In addition, the report is available at no charge on the GAO website at https://www.gao.gov/.

If you or your staff have any questions concerning this report, please contact us at (202) 512-4841 or ludwigsonj@gao.gov and at (202) 512- 6888 or bothwellb@gao.gov. Contact points for our Offices ofCongressional Relations and Public Affairs may be found on the last page of this report. Key contributors to this report are listed in appendix III.

Jon Ludwigson, Director
Contracting and National Security Acquisitions

Brian Bothwell, Director
Science, Technology Assessment, and Analytics

List of Committees

The Honorable Jack Reed
Chairman
The Honorable Roger Wicker
Ranking Member
Committee on Armed Services
United States Senate

The Honorable Mike Rogers
Chairman
The Honorable Adam Smith
Ranking Member
Committee on Armed Services
House of Representatives

The Honorable Jon Tester
Chair
The Honorable Susan Collins
Ranking Member
Subcommittee on Defense Committee on Appropriations
United States Senate

The Honorable Ken Calvert
Chair
The Honorable Betty McCollum
Ranking Member
Subcommittee on Defense Committee on Appropriations
House of Representatives

Appendix I: Objectives, Scope, and Methodology

The Joint Explanatory Statement accompanying the James M. Inhofe National Defense Authorization Act for Fiscal Year 2023 included a provision that GAO conduct a review of the offensive hypersonic weapons programs of the Department of Defense. We defined hypersonic vehicles in a previous report as those that (1) travel at speeds of Mach 5 or greater, (2) spend a majority of their flight path inside the atmosphere, and (3) are capable of maneuvering in flight.[596] For this report, we added a fourth element: that the system is designed to provide an operational, offensive strike capability.

Using this definition, we identified six offensive hypersonic weapon development efforts for inclusion in this review: (1) the Army's Long- Range Hypersonic Weapon (LRHW), (2) the Navy's Conventional Prompt Strike (CPS), (3) the Air Force's Air-launched Rapid Response Weapon (ARRW), (4) the Air Force's Hypersonic Attack Cruise Missile (HACM), (5) the Navy's Hypersonic Air-Launched Offensive anti-surface warfare weapon (HALO), and (6) the Navy's Standard Missile 6 (SM-6) Block IB. To validate this selection, we consulted with DOD's Principal Director for Hypersonics, who confirmed that the six efforts met our definition and no other efforts within his portfolio met these criteria.

This report examines (1) the acquisition approaches DOD is using for offensive hypersonic weapons, (2) the extent to which DOD is employing leading practices for product development to address risks associated with the schedule and performance of these weapons, (3) the extent to which DOD has identified and analyzed cost risks for these weapons, (4) the extent to which DOD is managing enterprise-level risks for the development of hypersonic systems.

To identify the acquisition approaches used by DOD for its offensive hypersonic development efforts, we interviewed officials from each of the identified efforts about the status and history of these efforts. We reviewed key documents from these efforts, including acquisition strategies, acquisition decision memorandums, and briefings.

To determine the extent to which DOD employed leading practices for product development to address risks associated with the schedule and performance of these weapons, we reviewed documents provided by the Army, Air Force, and Navy and interviewed officials from the six efforts'offices. These documents included schedules, requirements, and other documentation. We also reviewed DOD instructions and guidance for the middle tier of acquisition and major capability acquisition pathways. We assessed the efforts by comparing this information to our leading practices for product development.[597] We focused on the principles most relevant to the six identified efforts based on their phase of product development: prioritizing schedule by off-ramping capabilities when necessary, soliciting user feedback to inform a minimum viable product, and the use of digital engineering tools such as digital twinning.

To assess DOD's efforts to identify and analyze cost risks, we interviewed DOD officials and industry representatives regarding cost issues. We reviewed high-level cost estimates for each effort. We then performed a more detailed assessment of the CPS effort's rapid fielding cost estimate based on its adherence to our leading practices for the "accurate" characteristic.

---

[596] GAO, Hypersonic Weapons: DOD Should Clarify Roles and Responsibilities to Ensure Coordination across Development Efforts, GAO-21-378 (Washington, D.C.: Mar. 22, 2021).

[597] GAO, Leading Practices: Agency Acquisition Policies Could Better Implement Key Product Development Principles, GAO-22-104513 (Washington, D.C.: Mar. 10, 2022); and Leading Practices: Iterative Cycles Enable Rapid Delivery of Complex, Innovative Products, GAO-23-106222 (Washington, D.C.: July 27, 2023).

We selected the CPS cost estimate for assessment because the effort was among the most mature of the six identified and the Navy prepared the cost estimate to inform a major acquisition decision. We chose the "accurate" characteristic because, among the four characteristics, it was most relevant to the scope of this report.

To perform this assessment, we obtained and analyzed portions of the draft CPS rapid fielding cost estimate and documentation supporting the Navy's cost estimating practices. For the portions we could not access, we interviewed CPS cost estimators as they guided us through a copy of the estimate. We assessed the CPS cost estimate by comparing the estimates and supporting documentation to the leading practices discussed in our Cost Estimating and Assessment Guide.[598] These practices have been found to be the basis for reliable cost estimates. In making these assessments, we sorted our findings into categories of fully met, substantially met, partially met, minimally met, or not met.[599] Accurate cost estimates are developed by the following: estimating each cost element using the best methodology from the data collected; using appropriate adjustments for inflation; ensuring there are few, if any, minor mathematical mistakes; being based on a historical record of cost estimating and actual experiences from comparable programs; being updated regularly to reflect significant changes in the program; and ensuring any variances between estimated and actual costs are documented, explained, and reviewed.

To determine the extent to which DOD is identifying and managing enterprise-level risks related to hypersonic systems, we interviewed DOD officials at both the enterprise level and within offices responsible for executing risk responses. We reviewed documentation from the Offices of the Under Secretary of Defense for Research and Engineering and the Under Secretary of Defense for Acquisition and Sustainment on identifying and responding to risks. These documents included technology development roadmaps, enterprise strategies, risk analyses, and other reports describing accomplishments, setbacks, and risks. In addition, we conducted a site visit to Dynetics' Common Hypersonic Glide Body manufacturing facility, the Lockheed Martin CPS/LRHW missile assembly facility, and the LRHW Systems Integration Lab. We then assessed this information against the six elements listed in our Enterprise Risk Management framework: (1) align process to goals and objectives, (2) identify risks, (3) assess risks, (4) select risk response, (5) monitor risks, and (6) communicate and report on risks.[600]

We conducted this performance audit from April 2023 to July 2024 in accordance with generally accepted government auditing standards. Those standards require that we plan and perform the audit to obtain sufficient, appropriate evidence to provide a reasonable basis for our findings and conclusions based on our audit objectives. We believe that the evidence

[598] GAO, *Cost Estimating and Assessment Guide: Best Practices for Developing and Managing Program Costs,* GAO-20-195G (Washington, D.C.: Mar. 12, 2020). The *Cost Estimating and Assessment Guide* also articulates three other characteristics of a reliable cost estimate: comprehensive (meaning it reflects the current schedule and technical baselines), well-documented (meaning all parts can easily be repeated or updated and can be traced to original sources), and credible (meaning it discusses and documents any limitations, including the possibility of uncertainty or bias).

[599] "Fully met" means the Navy provided complete evidence that satisfied the entire criterion. "Substantially met" means the Navy provided evidence that satisfied a large portion of the criterion. "Partially met" means the Navy provided evidence that satisfied about half of the criterion. "Minimally met" means the Navy provided evidence that satisfied a small portion of the criterion. "Not met" means the Navy provided no evidence that satisfies any of the criterion.

[600] GAO, *Enterprise Risk Management: Selected Agencies' Experiences Illustrate Good Practices in Managing Risk,* GAO-17-63 (Washington, D.C., Dec. 1, 2016).

obtained provides a reasonable basis for our findings and conclusions based on our audit objectives.

Appendix II: Comments from the Department of Defense

OFFICE OF THE ASSISTANT SECRETARY OF DEFENSE
3600 DEFENSE PENTAGON
WASHINGTON, DC 20301-3600

ACQUISITION

JUN 2 7 2024

Mr. Jon Ludwigson
Director, Contracting and National Security Acquisitions
U.S. Government Accountability Office
441 G Street, N.W.
Washington, DC 20548

Dear Mr. Ludwigson:

The Department has completed a security and accuracy review of Government Accountability Office (GAO) Draft Hypersonic Weapons Report 24-106792, 'DOD Could Reduce Cost and Schedule Risks by Following Leading Practices.'

The Department finds that the draft report is UNCLASSIFIED and cleared for public release, pending the GAO addressing the security and sensitivity concerns provided. Enclosed is a copy of the Department's official security review. Additionally, comments and recommended changes to the content of the draft report are enclosed.

Department concurs with all ten of the GAO recommendations provided in the draft report.

My point of contact is Mr. Brandon Sweat who can be reached at brandon.c.sweat.civ@mail.mil and phone 703-697-8183.

Sincerely,

Gary A. Ashworth
Performing the Duties of the Assistant Secretary of Defense for Acquisition

Enclosures:
As stated

GAO DRAFT REPORT DATED APRIL 29, 2024
GAO-24-106792 (GAO CODE 106792)

"HYPERSONIC WEAPONS: DOD COULD REDUCE COST AND SCHEDULE RISKS BY FOLLOWING LEADING PRACTICES"

DEPARTMENT OF DEFENSE COMMENTS TO THE GAO RECOMMENDATION

RECOMMENDATION 1: The Secretary of the Navy should ensure the Standard Missile 6 (SM-6) Block IB efforts solicits and incorporates relevant user feedback throughout development.

DoD RESPONSE: Concur: The DON recognizes the value of user feedback across all acquisition programs, particularly MTA efforts; this feedback can help enable these programs to move at speed while also providing optimal capability to users. The Joint Urgent Operational Needs statement and Nimble Fire war games were instances of user feedback which informed early conceptual design and systems trades that in turn facilitated the inception of SM-6 Block IB. The DON has instituted a framework in which the program office collaborates with the Naval Surface and Mine Warfighting Development Command, leveraging feedback from the Navy's Weapons Tactics Instructors, and is engaged with the Naval Capabilities Improvement Program (NCIP), where capability gaps and desires are discussed with the technical community, fleet representatives, and resource sponsors.

RECOMMENDATION 2: The Secretary of the Air Force should ensure the Hypersonic Attack Cruise Missile (HACM) effort solicits and incorporates relevant user feedback throughout development.

DoD RESPONSE: Concur, the HACM Program has and will continue to solicit inputs from all stakeholders throughout development. Stakeholder inputs are assessed for impact to cost, schedule, and performance prior to deciding to incorporate into the program.

RECOMMENDATION 3: The Secretary of the Navy should ensure the Hypersonic Air-Launched Offensive anti-surface warfare (HALO) effort solicits and incorporates relevant user feedback throughout development.

DoD RESPONSE: Concur. The DON recognizes the value of user feedback across all acquisition programs, particularly MTA efforts; this feedback can help enable these programs to move at speed while also providing optimal capability to users. The HALO program will solicit inputs from a diverse set of users, including immediate operators, mission planners and operational commands, maintainers, pilots, and other actors. This will be done throughout development, with a time-phased plan to incorporate feedback from each user type at the optimal time.

2

RECOMMENDATION 4: The Secretary of the Air Force should ensure the Air-launched Rapid Response Weapon (ARRW) effort, if further development and production is planned, solicits, and incorporates relevant user feedback throughout development.

DoD RESPONSE: Concur, if further development and production of the ARRW Program occurs, the program will solicit inputs from relevant users and will assess their impact to cost, schedule and performance prior to deciding to incorporate into the program.

RECOMMENDATION 5: The Secretary of the Army should ensure that the Long-Range Hypersonic Weapon (LRHW) effort asses the practicality, benefits, and affordability of implementing digital engineering, including digital twinning and whether to incorporate these tools into the effort.

DoD RESPONSE: Concur without specific comment.

RECOMMENDATION 6: The Secretary of the Navy should ensure that the Standard Missile 6 (SM-6) Block IB effort assess the practicality, benefits, and affordability of implementing digital engineering, including digital twinning, and whether to incorporate these tools into the effort.

DoD RESPONSE: Concur. The program will continue to seek opportunities to inject digital engineering processes into SM-6 Block IB development. The program incorporated digital engineering requirements into its most recent solid rocket motor efforts, and the prime contractor employes Model Based Systems Engineering with many of their current tasks. The program also leverages both Inert Operational Missiles and Missile Simulation Units in order to increase the pace of integration and reduce costs where possible.

RECOMMENDATION 7: The Secretary of the Air Force should ensure the Air-launched Rapid Response Weapon effort, if further development and production is planned, assess the practicality benefits, and affordability of implementing digital engineering, including digital twinning, and whether to incorporate these tools into the effort.

DoD RESPONSE: Concur, if further development and production of the ARRW Program occurs, the program will, as part of acquisition decision, assess the practicality and affordability of implementing digital engineering, including digital twinning, and whether to incorporate these tools into the effort.

RECOMMENDATION 8: The Secretary of the Air Force should ensure the Hypersonic Attack Cruise Missile effort asses the practicality, benefits, and affordability of implementing digital engineering, digital twinning, and whether to incorporate these tools into the effort.

DoD RESPONSE: Concur, the HACM Program will assess expanding its implementation of digital engineering activities to include digital twinning, as part of its transition to a major capability acquisition.

3

RECOMMENDATION 9: The Secretary of the Navy should ensure the Conventional Prompt Strike program office improves its cost model in accordance with the "accurate" characteristic described in GAO's Cost Estimating and Assessment Guide, including by estimating each element using the best methodology from the data collected and updating the estimate to reflect program changes and actual costs.

DoD RESPONSE: Concur. The Conventional Prompt Strike program continuously updates its cost model in alignment with the "accurate" characteristic described in GAO's Cost Estimating and Assessment Guide. This includes updates to reflect program changes and actual costs, and incorporates the best methodology possible based on improved data availability.

RECOMMENDATION 10: The Secretary of Defense should ensure that Under Secretary of Defense for Research and Engineering periodically provides information on DoD's progress in managing enterprise-level risks to Congressional decisionmakers.

DoD RESPONSE: Concur, agree to periodically providing information on DoD's Hypersonic progress in managing enterprise-level risks to Congressional decisionmakers.

Appendix III: GAO Contact and Staff Acknowledgments

GAO Contacts

Jon Ludwigson, (202) 512-4841 or ludwigsonj@gao.gov
Brian Bothwell, (202) 512-6888 or bothwellb@gao.gov

Staff Acknowledgments

In addition to the contacts named above, the following staff members made key contributions to this report: Kristopher Keener (Assistant Director), Matthew J. Ambrose (Analyst-in-Charge), Rose Brister, Nicole Daniels, Susan C. Ditto, Jennifer Leotta, Mark Luth, Chi L. Mai, Mark Oppel, Andrew N. Powell, Benjamin Wilder, and Adam Wolfe.

Chapter 10

Defense Primer: Hypersonic Boost-Glide Weapons*

Kelley M. Sayler

The U.S. Department of Defense (DOD) is pursuing two types of hypersonic weapons technologies: boost-glide systems that place a maneuverable glide vehicle atop a ballistic missile or rocket booster, and cruise missiles that would use high-speed, air-breathing engines known as scramjets to travel to hypersonic speeds. This chapter addresses only the first of these technologies.

The Pentagon's FY2025 budget request for hypersonic weapons was $6.9 billion (of a total $9.8 billion request for long-range fires), up from $4.7 billion in FY2023. The Pentagon declined to provide a breakout of funding for hypersonic weapons in FY2024, but it requested a total of $11 billion for long-range fires. Pentagon officials have stated that hypersonic weapons could attack priority targets promptly and with improved accuracy without facing defeat by an adversary's air or missile defense systems. Pentagon officials have also expressed concerns about advances in hypersonic weapons technologies in Russia and China, as well as about the potential threats to U.S. forces, allies, and territory.

Characteristics of Hypersonic Glide Vehicles

Hypersonic glide vehicles (HGVs), like all weapons delivered by medium- and longer-range rocket boosters, can travel at speeds of at least Mach 5, or about 1 mile per second. The key difference between missiles armed with HGVs and missiles armed with ballistic reentry vehicles (i.e., those that travel on a ballistic trajectory throughout their flight) is not their speed, but their ability to maneuver and change course after they are released from their rocket boosters. In addition, although it is not necessary, many concepts for the delivery of HGVs presume that the boosters will launch along a flatter, or depressed, trajectory than standard ballistic missiles, and will release their gliders at a lower altitude of flight (see *Figure 1*).

Taken together, the HGV's novel trajectory and maneuverability in flight would complicate a U.S. effort todetect, track, and defend against an attack. The United States would likely detect the booster's launch, as it would for the launch of any ballistic missile, but it would not be able to predict the HGV's flight path. In addition, although an HGV launched by a rocket booster would reach its target far more quickly than a warhead delivered by an aircraft or

---

* This is an edited, reformatted and augmented version of Congressional Research Service Publication No. IF11459, Updated November 1, 2024.

In: Emerging Military Technologies
Editor: Jerry Shields
ISBN: 979-8-89530-613-0

subsonic cruise missile (in minutes instead of hours), it would not travel faster than a ballistic reentry vehicle.

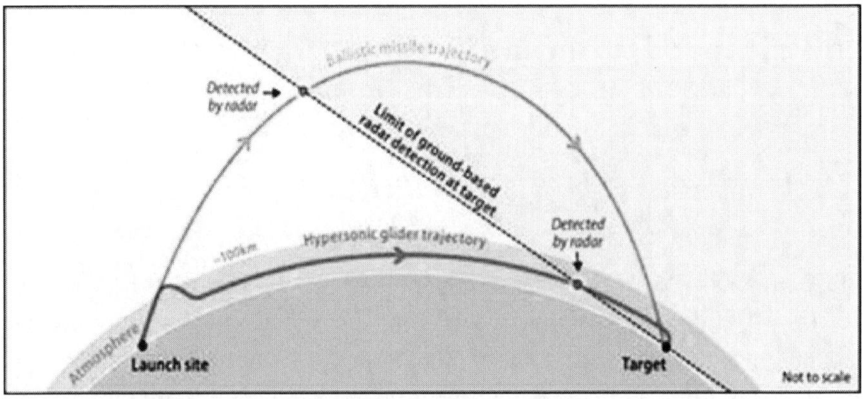

Source: CRS image based on an image in "Gliding missiles that fly faster than Mach 5 are coming," The Economist, April 6, 2019.

Figure 1. Flight Trajectory of Ballistic Missiles vs. Hypersonic Boost-Glide Weapons.

However, it would be more difficult to predict the intended target and to direct missile defense interceptors toward the attacking HGV.

Hypersonic Boost-Glide Programs

United States

When the United States began to assess the military utility of hypersonic boost-glide weapons in the early 2000s, it sought to develop longer-range systems that could reach deep into an adversary's territory to attack defended, hardened, and time-urgent targets. More recently, the Pentagon has increased funding and accelerated development programs for shorter and intermediate-range boost-glide systems that could conduct prompt attacks against heavily defended targets in regional conflicts. The Navy and Army are known to maintain hypersonic boost-glide programs.

Unlike Russia and China, the United States is not developing HGVs for use with nuclear warheads. As a result, U.S. gliders may require greater accuracy and will be more technically challenging to develop than nuclear-armed Chinese and Russian systems.

Russia

Russia's HGV, known as Avangard, is equipped with a nuclear warhead and deployed on SS-19 long-range land- based ballistic missiles. While SS-19 missiles can carry six nuclear warheads on a ballistic trajectory, reports indicate that they will deploy with only one Avangard HGV. Avangard reportedly features onboard countermeasures and will be able to maneuver in flight so that it can evade U.S. ballistic missile defenses. Russia conducted successful tests of

Avangard in 2016 and 2018. On December 27, 2019, the Russian military announced that it had activated two SS-19 missiles equipped with Avangard.

Russia views the Avangard system as a part of its nuclear retaliatory capability, ensuring that Russian missiles could penetrate U.S. ballistic missile defenses. According to the Pentagon's 2022 Missile Defense Review, the United Statesrelies on nuclear deterrence, not ballistic missile defenses, to counter Russia's long-range missile forces. Nevertheless, in March 2018, President Putin stated that Russia had pursued HGV technologies in response to the U.S. 2002 withdrawal from the 1972 Anti-ballistic Missile Treaty.

Some U.S. analysts have noted, however, that Russia could use the Avangard as part of a first strike, even in the absence of U.S. ballistic missile defenses, to attack critical targets quickly. Others have assessed that Russia is likely to deploy Avangard in small numbers, so it will add little to Russia's existing nuclear force structure.

China

China has developed an HGV known as the DF-ZF (previously referred to as the WU-14) and has tested it at least nine times since 2014. U.S. defense officials have stated that the HGV may be capable of performing "extreme maneuvers" during flight, which would allow it to evade U.S. ballistic missile defenses. Unclassified reports indicate this glider would likely be equipped with conventional warheads, and when mated with the DF-17 booster, could travel to ranges of 1,800-2,500 kilometers.

China is also developing the DF-41 long-range intercontinental ballistic missile, which former commander of U.S. Northern Command General Terrence O'Shaughnessy seemed to confirm could carry a nuclear- capable HGV. Reports additionally indicate that China has tested an HGV on a fractional orbital bombardment system (FOBS)—a ballistic missile booster that carried the HGV into orbit before the HGV de-orbited and approached its target. Experts assert that this type of system might provide China with the ability to launch HGVs over the South Pole, thus evading U.S. early warning assets that track threats over the North Pole and further reducing the amount of warning time prior to a strike.

Some contend that China has prioritized HGV development to counter "specific security threats from increasingly sophisticated U.S. military technology." This includes both U.S. hypersonic weapons that could threaten strikes against China's nuclear arsenal and supporting infrastructure, and U.S. missile defense deployments that could then limit China's ability to conduct a retaliatory strike against the United States. In this framework, nuclear-armed HGVs on long-range missiles would ensure that China had the ability to retaliate after a U.S. attack, even if the United States were to expand its ballistic missile defense capabilities. HGVs on medium-range missiles would aid China's efforts to threaten U.S. assets in the Indo-Pacific region, particularly when faced with growing U.S. regional missile defense capabilities.

Arms Race Dynamics?

Many analysts have characterized the ongoing U.S., Russian, and Chinese development of HGVs and boost- glide weapons as an arms race because each nation seems to be competing to be the first to deploy these systems. Some argue that the United States is falling behind in this contest because Russia and China have both displayed operational systems. Others argue that the United States is leading, and may be escalating the race because it has accelerated its programs and expanded them to include short-, medium-, and long-range systems.

When asked about this dynamic in February 2020, Admiral Charles Richard, then-commander of U.S. Strategic Command, noted that there "is a competition, just like any other military competition." But he did not characterize it as an arms race. He noted that the United States was developing technologies needed to meet U.S. national objectives and that he expected the United States to be successful in meeting its security goals. Others have noted that the same is true for Russia and China; each is developing HGVs to meet its own security interests, not to counter or match the U.S. development of HGVs.

Specifically, both seem to be responding to concerns about U.S. ballistic missile defense programs.

Game-Changing Technology?

Some analysts have asserted that the speed, accuracy, and maneuverability of hypersonic boost-glide weapons will fundamentally change the character of warfare. Former acting Secretary of the Navy Thomas Modly made this case in January 2020 when he noted that these technologies "have already changed the nature of the battlespace" and that they "can destabilize the global security environment and pose an existential threat to our nation." Others question this assessment. They note that boost-glide systems can reach their targets more quickly than other maneuverable systems, like aircraft and subsonic cruise missiles. But adversaries armed with ballistic missiles have long been able to attack U.S. forces, allies, and territory, even without maneuvering warheads. Consequently, they argue that there is nothing new about the threat from nuclear-armed HGVs, when compared with other nuclear- armed missiles, and nothing existential about a threat from conventionally armed HGVs.

Crisis Instability?

Boost-glide systems could accelerate the pace of warfare and create incentives to strike first in a crisis. If the United States and potential adversaries develop these systems to strike promptly against high-value targets protected by missile defenses, each side may believe it has to strike first, and strike fast, to achieve its objectives. This dynamic— often referred to as crisis instability—could provoke the start of a conflict even if neither party to the crisis initially planned to strike first.

DOD is seeking to address the potential threats posed by hypersonic boost-glide weapons by developing defensive systems to track and engage them. Experts disagree on the cost and technological feasibility of this approach.

Policymakers may also consider mechanisms to ban or limit the deployment of these weapons to avoid the crisis instabilities created by their short time of flight. On the other hand, nations might be unwilling to agree to limit these weapons without corresponding limits on missile and air defenses.

Amy F. Woolf, former Specialist in Nuclear Weapons Policy, originally coauthored this report.

Chapter 11

Hypersonic Missile Defense: Issues for Congress*

Jennifer DiMascio
and Kelley M. Sayler

The Missile Defense Agency (MDA) and Space Development Agency (SDA) are currently developing elements of a hypersonic missile defense system to defend against hypersonic weapons and other emerging missile threats. These elements include the tracking and transport layers of the Proliferated Warfighter Space Architecture (PWSA) and various interceptor programs. As MDA and SDA continue to develop these systems, Congress may consider implications for oversight and defense authorizations and appropriations.

Background

Hypersonic weapons, like ballistic missiles, fly at speeds of at least Mach 5, or roughly 1 mile per second. Unlike ballistic missiles, hypersonic weapons do not follow a ballistic trajectory and can maneuver en route to their target. Russia reportedly fielded its first hypersonic weapons in December 2019, while some experts believe that China fielded hypersonic weapons as early as 2020.

The United States does not have any fielded hypersonic weapons.

The maneuverability and low flight altitude of hypersonic weapons could challenge existing detection and defense systems. For example, most terrestrial-based radars cannot detect hypersonic weapons until late in the weapon's flight due to line-of-sight limitations of radar detection. This leaves minimal time for a defender to launch interceptors that could neutralize an inbound weapon. Figure 1 depicts the differences in terrestrial-based radar detection timelines for ballistic missiles versus hypersonic weapons.

U.S. defense officials have stated that both existing terrestrial- and space-based sensor architectures are insufficient to detect and track hypersonic weapons; former Under Secretary of Defense for Research and Engineering Mike Griffin has noted that "hypersonic targets are 10 to 20 times dimmer than what the U.S. normally tracks by satellites in geostationary orbit."

---

* This is an edited, reformatted and augmented version of Congressional Research Service Publication No. IF11623, updated June 24, 2024.

In: Emerging Military Technologies
Editor: Jerry Shields
ISBN: 979-8-89530-613-0

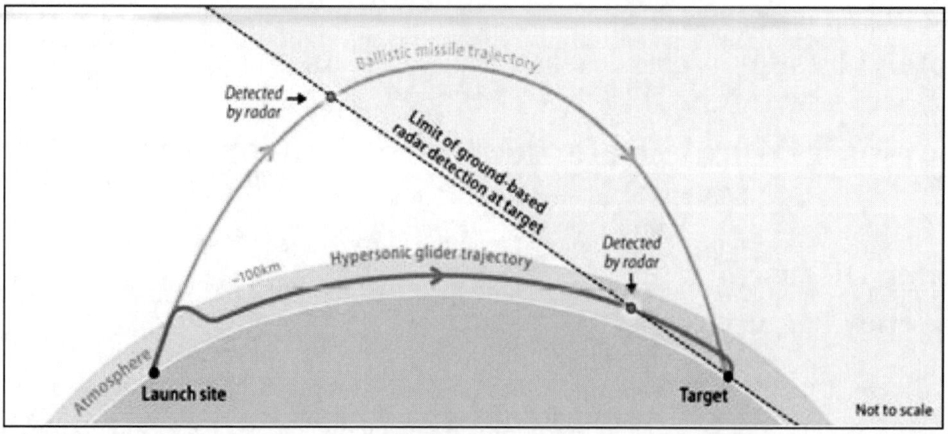

Source: CRS image based on an image in "Gliding missiles that fly faster than Mach 5 are coming," The Economist, April 6, 2019, https://www.economist.com/science-and-technology/2019/04/06/gliding-missiles-that-fly-faster-than-mach-5-are-coming.

Figure 1. Terrestrial-Based Detection of Ballistic Missiles vs. Hypersonic Weapons.

Proliferated Warfighter Space Architecture

SDA developed the PWSA, formerly known as the National Defense Space Architecture, to "unify and integrate next generation capabilities across [the Department of Defense (DOD)] and industry." According to SDA, the PWSA aims to be a "single, coherent proliferated space architecture with seven layers," which include the data tracking and transport layers depicted in Figure 2 and discussed below. Other layers include the custody layer to support the targeting of mobile ground assets; the battle management layer to provide space-based command and control; the navigation layer to provide "alternate positioning, navigation, and timing for potential GPS-denied environments"; the deterrence layer to detect potentially hostile actions in deep space; and the support layer to facilitate satellite operations for the other PWSA layers. Once fully fielded, the PWSA is to include 550 satellites and provide full global coverage.

Tracking Layer

SDA states that the tracking layer is to "provide global indications, warning, tracking, and targeting of advanced missile threats, including hypersonic missile systems." As part of this layer, SDA is developing an architecture of Wide Field of View (WFOV) satellites, which are to eventually provide global coverage. SDA requested $108.7 million for Tranche 0 tracking activities in FY2025 and $1.5 billion for Tranche 1 tracking activities (also known as Resilient Missile Warning Missile Tracking - Low Earth Orbit).

Working in tandem with the SDA's tracking satellites will be the Hypersonic and Ballistic Tracking Space Sensor (HBTSS), previously known as the Space Sensor Layer, which is being developed by MDA in collaboration with SDA and the U.S. Space Force. HBTSS is to provide more sensitive, but more limited (or Medium Field of View [MFOV]) coverage, compared to WFOV. For this reason, WFOV is intended to provide cueing data to HBTSS, which could then provide more specific, target quality data to a ground-based interceptor. MDA requested $76 million for HBTSS in FY2025.

The Space Force's Space Systems Command (SSC) is developing a third set of tracking satellites called Resilient Missile Warning Missile Tracking - Medium Earth Orbit (MEO). According to SDA, the MEO satellites will add "low-latitude coverage and track custody" and enhance resilience in the nation's missile defense architecture. The Space Force requested $846.3 million for Resilient Missile Warning Missile Tracking - MEO in FY2025.

Section 1682 of the FY2020 NDAA (P.L. 116-92) tasks the director of the Missile Defense Agency to "develop a hypersonic and ballistic missile tracking space sensor payload." Section 1645 of the FY2021 NDAA (P.L. 116- 283) requires that integration of the sensor payload into the SDA's broader space-based sensor architecture begin "as soon as technically feasible." In 2022, the Space Force established a Combined Program Office to coordinate missile warning and tracking efforts, including SDA's PWSA, MDA's HBTSS, and SSC's MEO satellites.

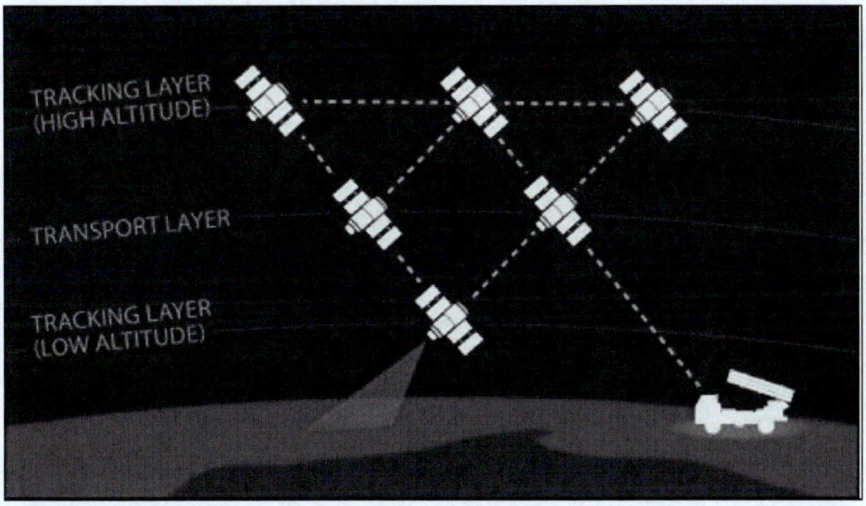

Source: CRS image; not to scale.

Figure 2. Selected Elements of the PWSA.

Transport Layer

SDA has stated that the PWSA's transport layer, which is intended to connect the tracking layer to interceptors and other weapons systems on the ground, will "enhance several mission areas including missile defense." According to DOD, SDA has awarded prototype agreements for Tranches 1 and 2 of the transport layer. The transport layer is to eventually consist of a constellation of approximately 300-500 satellites. SDA requested $1.4 billion for "the data transport layer, sensor capabilities, and alternate position, navigation, and timing capabilities" in FY2025.

Interceptors

MDA has explored options for neutralizing hostile hypersonic weapons, including interceptor missiles, hypervelocity projectiles, directed energy weapons, and electronic attack systems. In January 2020, MDA issued a draft request for prototype proposals for a Hypersonic Defense Regional Glide Phase Weapons System interceptor. This program was intended to "reduce interceptor key technology and integration risks." In April 2021, MDA then shifted to the Glide Phase Interceptor (GPI), which is to be integrated with the Aegis Weapon System. Although GPI was to notionally provide a hypersonic missile defense capability in FY2034, Section 1666 of the FY2024 NDAA (P.L. 118-31) directs DOD to achieve initial operational capability for the program by December 31, 2029, and full operational capability by December 31, 2032. Lockheed Martin, Northrop Grumman, and Raytheon Missiles and Defense have been awarded contracts for the "accelerated concept design" phase of the GPI. In May 2024, DOD signed a formal agreement with Japan to cooperatively develop GPI.

In addition, Section 1664 of the FY2022 NDAA (P.L. 117- 81) grants the director of MDA "the authority to budget for, direct, and manage directed energy programs applicable" for hypersonic missile defense, while Section 1662 of the FY2023 NDAA (P.L. 117-263) directs MDA to produce a strategy to use asymmetric capabilities (e.g., directed energy, microwave systems) to defeat hypersonic missile threats. The Defense Advanced Research Projects Agency (DARPA) is also working on a program called Glide Breaker, which is to "develop critical component technology to support a lightweight vehicle designed for precise engagement of hypersonic threats at very long range." DARPA requested $38 million for Glide Breaker in FY2025. Overall, MDA requested $182 million for hypersonic defense in FY2025.

Issues for Congress

Some analysts have suggested that space-based sensor layers—integrated with tracking and targeting systems to direct high-performance interceptors or directed energy weapons—could theoretically present viable options for defending against hypersonic weapons. The 2019 Missile Defense Review notes that "such sensors take advantage of the large area viewable from space for improved tracking and potentially targeting of advanced threats, including hypersonic [weapons]." Other analysts have questioned the affordability, technical feasibility, and/or utility of hypersonic weapons defense. In addition, some analysts have argued that the United States' current command and control architecture would be incapable of "processing data quickly enough to respond to and neutralize an incoming hypersonic threat."

Some analysts have also questioned the current division of labor and level of coordination between the SDA and MDA on hypersonic missile defense. SDA director Tournear has previously responded to criticisms of potential redundancies between the two agencies, stating that both report to the Under Secretary of Defense for Research and Engineering. Since October 2022, SDA director Tournear instead reports to the Assistant Secretary of the Air Force for Acquisition and Integration. Also in 2022, the Space Force created the combined program office overseeing missile warning and tracking efforts to address concerns about integration.

Congress may monitor the implications of this reporting structure and new combined program office for efficiency and efficacy.

Potential Questions for Congress

- Is an acceleration of research on hypersonic missile defense options both necessary and technologically feasible? Does the technological maturity of hypersonic missile defense options warrant current funding levels?
- How are SDA, MDA, and SSC collaborating on various elements of hypersonic missile defense? Are their current roles increasing or decreasing costs and the speed and efficiency of technology development?
- Does DOD have the enabling capabilities, such as adequate command and control architectures, needed to execute hypersonic missile defense?

Chapter 12

The U.S. Army's Long-Range Hypersonic Weapon (LRHW): Dark Eagle*

Andrew Feickert

What Is the Army's Long-Range Hypersonic Weapon?

The Army's Long-Range Hypersonic Weapon (LRHW), also known as Dark Eagle (*Figure 1*), with a reported range of 1,725 miles, consists of a ground-launched missile equipped with a hypersonic glide body and associated transport, support, and fire control equipment. According to the Army,

> This land-based, truck-launched system is armed with hypersonic missiles that can travel well over 3,800 miles per hour. They can reach the top of the Earth's atmosphere and remain just beyond the range of air and missile defense systems until they are ready to strike, and by then it's too late to react.

Source: https://www.popularmechanics.com/military/weapons/ a36421213/army-hypersonic-weapon-1700-mile-range/, accessed November 18, 2021.

Figure 1. Artist Rendition of a Notional LRHW Unit.

---

* This is an edited, reformatted and augmented version of Congressional Research Service Publication No. IF11991, updated July 2, 2024.

In: Emerging Military Technologies
Editor: Jerry Shields
ISBN: 979-8-89530-613-0

The Army further notes,

> The LRHW system provides the Army a strategic attack weapon system to defeat Anti-Access/Area Denial (A2/AD) capabilities, suppress adversary long-range fires, and engage other high payoff/time critical targets. The Army is working closely with the Navy in the development of the LRHW. LRHW is comprised of the Common Hypersonic Glide Body (C-HGB), and the Navy 34.5-inch booster.

LRHW Components

Missile

The missile component of the LRHW is reportedly being developed by Lockheed Martin and Northrop Grumman. When the hypersonic glide body is attached, it is referred to as the Navy-Army All Up Round plus Canister (AUR+C). The missile component serves as the common two-stage booster for the Army's LRHW and the Navy'sConventional Prompt Strike (CPS) system, which can be fired from both surface vessels and submarines.

Common Hypersonic Glide Body (C-HGB)

The C-HGB is reportedly based on the Alternate Re-Entry System developed by the Army and Sandia National Laboratories. Dynetics, a subsidiary of Leidos, is currently under contract to produce C-HGB prototypes for the Army and Navy. The C-HGB uses a booster rocket motor to accelerate to well above hypersonic speeds and then jettisons the expended rocket booster. The C-HGB, which can travel at Mach 5 or higher on its own, is planned to be maneuverable, potentially making it more difficult to detect and intercept.

LRHW Organization and Units

The LRHW is organized into batteries. According to the Army "a LRHW battery consists of four Transporter Erector Launchers on modified M870A4 trailers, each equipped with two AUR+Cs (eight in total), one Battery Operations Center (BOC) for command and control, and a BOC support vehicle."

The 5th Battalion, 3rd Field Artillery Regiment at Joint Base Lewis-McChord, Washington, was designated to operate the first battery of eight LRHW missiles. The battalion, also referred to as a Strategic Long-Range Fires battalion, is part of the Army's 1st Multi Domain Task Force (MDTF), a unit in the Indo Pacific-oriented I Corps stationed at Joint Base Lewis-McChord, WA. Other LRHW batteries are planned for Strategic Long-Range Fires battalions in the remaining MDTFs scheduled for activation.

LRHW Testing and Program Activities

According to a 2023 Congressional Budget Office (CBO) Study, "U.S. Hypersonic Weapons and Alternatives," "Extensive flight testing is necessary to shield hypersonic missiles' sensitive electronics, to understand how various materials perform, and predict aerodynamics at sustained temperatures as high as 3,000° Fahrenheit." The Army originally planned for three flight tests of the LRHW before the first battery fielding in FY2023. On October 21, 2021, the booster rocket carrying the C-HGB vehicle reportedly failed a test flight, resulting in what defense officials characterized as a "no test" as the C-HGB had no chance to deploy. Reportedly, a June 2022 test of the entire LRHW missile also resulted in failure.

Flight Test Delays

In October 2022, it was reported the Department of Defense (DOD) delayed a scheduled LRHW test in order to "assess the root cause of the June [2022] failure." Reportedly, the delayed test would be rescheduled to the first quarter of FY2023.

March 2023 LRHW Test Scrubbed

On March 10, 2023, it was reported,

> On March 5, DOD was preparing to execute Joint Flight Campaign-2 featuring the Army version of the prototype weapon launched at Cape Canaveral Space Force Station, FL, when the countdown was halted as a result of pre-flight checks during that event, the test did not occur.

Cancelled September 2023 LRHW Test and Program Delay

On September 6, 2023, it was reported,

> The DOD planned to conduct a flight test at the Cape Canaveral Space Force Station, Florida, to inform hypersonic technology development. As a result of pre-flight checks, the test did not occur.

On September 14, 2023, in an Army statement to Bloomberg News, the Army reportedly acknowledged it would not be able to meet its goal of deploying the LRHW by the end of FY2023.

Change in LRHW Testing Pathway

In late November 2023, Navy and Army acquisition executives reportedly decided to "revamp efforts to prepare for [LRHW] flight test following three flight test attempts this year that were scrubbed because of problems with the Lockheed Martin-produced launcher." The Army's new testing approach will feature subcomponent testing.

LRHW Fielding Delayed Until FY2025 According to a June 2024 Government Accountability Office (GAO) report to Congress,

> The Army missed its goal of fielding its first LRHW battery—including missiles— by fiscal year 2023 due to integration challenges. Based on current test and missile production plans, the Army will not field its first complete LRHW battery until fiscal year 2025. Before the Army can field an operational system, it must conduct a successful end-to-end missile flight test using the Army's launch system.

GAO further notes,

> The LRHW integration issues discovered during testing also affect missile production. The Army cannot complete the missiles for the first battery until a successful test demonstrates that the current design works. LRHW officials stated that once a successful flight test is achieved, the first production missile will be delivered within approximately six weeks and the first battery of eight missiles will be delivered within approximately 11 months. If the Army discovers issues with missile performance in flight testing, missile deliveries and the fielding of the first operational LRHW system could be further delayed.

Successful LRHW Flight Test

On June 28, 2024 DOD announced,

> The U.S. Navy and U.S. Army recently completed an end-to-end flight test of a hypersonic missile from the Pacific Missile Range Facility, Kauai, HI. The test provided data on the end-to-end performance of the Conventional Prompt Strike and Long- Range Hypersonic Weapon All Up Round.

Reportedly, the two-stage missile was launched from a ground stand from Hawaii across the Pacific Ocean more than 2,000 miles to a test range in the Marshall Islands, with the missile flying its intended course and releasing the glide body, which flew to the target.

Considerations for Congress

Possible oversight considerations for Congress could include the following:

Additional LRHW Testing

Now that the Army has achieved a successful end-to-end test flight of the LRHW, the Army has stated that "once a successful flight test is achieved, the first production missile will be delivered within approximately six weeks and the first battery of eight missiles will be delivered within approximately 11 months." If the Army proceeds with its stated course of action, it is possible that the first production missiles would be delivered to the Army in mid- August 2024. Given this short timeline between missile production and fielding to the first LRHW battery, Congress might examine what additional testing is planned to be conducted with the newly produced missiles before they are delivered to the first LRHW battery.

FY2025 LRHW Budgetary Information

Table 1. FY2025 LRHW Budget Request

Funding Category	Total Request ($M)
RDT&E	$538
Procurement (Missiles)	$744

Source: Assistant Secretary of the Army (Financial Management and Comptroller), U.S. Army FY2025 Budget Highlights, March 2024, p. 33.

Notes: RDT&E = Research, Development, Test & Evaluation; $M = U.S. dollars in millions.

LRHW Missile Costs

According to a January 2023 CBO study, "U.S. Hypersonic Weapons and Alternatives," purchasing 300 Intermediate- Range Hypersonic Boost-Glide Missiles (similar to the LRHW) was estimated to cost $41 million per missile (in 2023 dollars). A January 2023 Center for Strategic and International Studies report, "The First Battle of the Next War: Wargaming a Chinese Invasion of Taiwan," contends, regarding hypersonic weapons, "their high costs limits inventories, so they lack the volume needed to counter the immense numbers of Chinese air and naval platforms." Given concerns about how LRHW missile costs could influence LRHW inventories, policymakers might decide to further examine LRHW missile costs as well as quantities of LRHW missiles needed to support potential sustained combat operations in various theaters of operations.

Chapter 13

Defense Primer: U.S. Policy on Lethal Autonomous Weapon Systems*

Kelley M. Sayler

Lethal autonomous weapon systems (LAWS) are a special class of weapon systems that use sensor suites and computer algorithms to independently identify a target and employ an onboard weapon system to engage and destroy the target without manual human control of the system.

Although these systems are not yet in widespread development, it is believed they would enable military operations in communications-degraded or -denied environments in which traditional systems may not be able to operate.

Contrary to a number of news reports, U.S. policy does not prohibit the development or employment of LAWS. Although the United States is not known to currently have LAWS in its inventory, some senior military and defense leaders have stated that the United States may be compelled to develop LAWS in the future if U.S. competitors choose to do so. At the same time, a growing number of states and nongovernmental organizations are appealing to the international community for regulation of or a ban on LAWS due to ethical concerns.

Developments in both autonomous weapons technology and international discussions of LAWS could hold implications for congressional oversight, defense investments, military concepts of operations, treaty-making, and the future of war.

U.S. Policy

Then-Deputy Secretary of Defense Ashton Carter issued DOD's policy on autonomy in weapons systems, Department of Defense Directive (DODD) 3000.09 (the directive), in November 2012. DOD has since updated the directive—most recently in January 2023.

Definitions. There is no agreed definition of lethal autonomous weapon systems that is used in international fora. However, DODD 3000.09 provides definitions for different categories of autonomous weapon systems for the purposes of the U.S. military. These definitions are principally grounded in the role of the human operator with regard to target selection and engagement decisions, rather than in the technological sophistication of the weapon system.

DODD 3000.09 defines LAWS as "weapon system[s] that, once activated, can select and engage targets without further intervention by a human operator." This concept of autonomy

* This is an edited, reformatted and augmented version of Congressional Research Service Publication No. IF11150, updated February 1, 2024.

In: Emerging Military Technologies
Editor: Jerry Shields
ISBN: 979-8-89530-613-0

is also known as "human out of the loop" or "full autonomy." The directive contrasts LAWS with human- supervised, or "human on the loop," autonomous weapon systems, in which operators have the ability to monitor and halt a weapon's target engagement. Another category is semi-autonomous, or "human in the loop," weapon systems that "only engage individual targets or specific target groups that have been selected by a human operator." Semi-autonomous weapons include so-called "fire and forget" weapons, such as certain types of guided missiles, that deliver effects to human-identified targets using autonomous functions.

The directive does not apply to autonomous or semi- autonomous cyberspace capabilities; unarmed platforms; unguided munitions; munitions manually guided by the operator (e.g., laser- or wire-guided munitions); mines; unexploded explosive ordnance; or autonomous or semi- autonomous systems that are not weapon systems, nor subject them to its guidelines.

Role of human operator. DODD 3000.09 requires that all systems, including LAWS, be designed to "allow commanders and operators to exercise appropriate levels of human judgment over the use of force." As noted in an August 2018 U.S. government white paper, "'appropriate' is a flexible term that reflects the fact that there is not a fixed, one-size-fits-all level of human judgment that should be applied to every context. What is 'appropriate' can differ across weapon systems, domains of warfare, types of warfare, operational contexts, and even across different functions in a weapon system."

Furthermore, "human judgment over the use of force" does not require manual human "control" of the weapon system, as is often reported, but rather broader human involvement in decisions about how, when, where, and why the weapon will be employed. This includes a human determination that the weapon will be used "with appropriate care and in accordance with the law of war, applicable treaties, weapon system safety rules, and applicable rules of engagement."

To aid this determination, DODD 3000.09 requires that "[a]dequate training, [tactics, techniques, and procedures], and doctrine are available, periodically reviewed, and used by system operators and commanders to understand the functioning, capabilities, and limitations of the system's autonomy in realistic operational conditions." The directive also requires that the weapon's human-machine interface be "readily understandable to trained operators" so they can make informed decisions regarding the weapon's use.

Weapons review process. DODD 3000.09 requires that the software and hardware of covered semi-autonomous and autonomous weapon systems, be tested and evaluated to ensure they

> Function as anticipated in realistic operational environments against adaptive adversaries taking realistic and practicable countermeasures, [and] complete engagements within a timeframe and geographic area, as well as other relevant environmental and operational constraints, consistent with commander and operator intentions. If unable to do so, the systems will terminate the engagement or obtain additional operator input before continuing the engagement.

Systems must also be "sufficiently robust to minimize the probability and consequences of failures." Any changes to the system's operating state—for example, due to machine learning—would require the system to go through testing and evaluation again to ensure that it has retained its safety features and ability to operate as intended. The directive also notes that "the use of AI capabilities in autonomous or semi-autonomous systems will be consistent with the DOD AI Ethical Principles."

Senior-level review. In addition to the standard weapons review process, a secondary senior-level review is required for covered autonomous and semi-autonomous systems.

This review requires the Under Secretary of Defense for Policy (USD[P]), the Vice Chairman of the Joint Chiefs of Staff (VCJCS), and the Under Secretary of Defense for Research and Engineering (USD[R&E]) to approve the system before formal development. USD(P), VCJCS, and the Under Secretary of Defense for Acquisition and Sustainment (USD[A&S]) must then approve the system before fielding. In the event of "urgent military need," this senior-level review may be waived by the Deputy Secretary of Defense. DODD 3000.09 additionally establishes the Autonomous Weapon System Working Group—composed of representatives of USD(P); USD(R&E); USD(A&S); DOD General Counsel; the Chief Digital and AI Officer; the Director, Operational Test and Evaluation; and the Chairman of the Joint Chiefs of Staff—to support and advise the senior-level review process.

Congressional notification. Per Section 251 of the FY2024 NDAA (P.L. 118-31), the Secretary of Defense is to notify the defense committees of any changes to DODD 3000.09 within 30 days. The Secretary is directed to provide a description of the modification and an explanation of the reasons for the modification.

International Discussions of LAWS

Since 2014, the United States has participated in international discussions of LAWS, sometimes colloquially referred to as "killer robots," under the auspices of the United Nations Convention on Certain Conventional Weapons (U.N. CCW). In 2017, these discussions transitioned from an informal "meeting of experts" to a formal "Group of Governmental Experts" (GGE) tasked with examining the technological, military, ethical, and legal dimensions of LAWS. In 2018 and 2019, the GGE has considered proposals by states parties to issue political declarations about LAWS, as well as proposals to regulate them.

In addition, approximately 30 countries and 165 nongovernmental organizations have called for a preemptive ban on LAWS due to ethical concerns, including concerns about operational risk, accountability for use, and compliance with the proportionality and distinction requirements of the law of war. The U.S. government does not currently support a ban on LAWS and has addressed ethical concerns about the systems in a March 2018 white paper, "Humanitarian Benefits of Emerging Technologies in the Area of Lethal Autonomous Weapons." The paper notes that "automated target identification, tracking, selection, and engagement functions can allow weapons to strike military objectives more accurately and with less risk of collateral damage" or civilian casualties.

Although the U.N. CCW is a consensus-based forum, the outcome of its discussions could hold implications for U.S. policy on lethal autonomous weapons.

Potential Questions for Congress

- What is the status of U.S. competitors' development of LAWS? Is the United States adequately investing in counter-autonomy capabilities?
- To what extent, if at all, should the United States initiate or accelerate its own development of LAWS?
- How should the United States balance LAWS research and development with ethical considerations? What, if any, restrictions should there be on DOD's development or employment of LAWS?
- If the United States chooses to develop LAWS, are current weapons review processes and legal standards for their employment in conflict sufficient?
- What role should the United States play in U.N. CCW discussions of LAWS? Should the United States support the status quo, propose a political declaration, or advocate regulation of or a ban on LAWS?

Other Resources

Department of Defense Directive 3000.09, "Autonomy in Weapon Systems," Updated January 25, 2023, https://www.esd.whs.mil/portals/54/documents/ dd/issuances/dodd/300009p.pdf.

U.S. Government, "Humanitarian Benefits of Emerging Technologies in the Area of Lethal Autonomous Weapons," March 28, 2018.

U.S. Government, "Human-Machine Interaction in the Development, Deployment and Use of Emerging Technologies in the Area of Lethal Autonomous Weapons Systems," August 28, 2018.

United Nations Office at Geneva, "Background on Lethal Autonomous Weapons Systems in the CCW."

Chapter 14

International Discussions Concerning Lethal Autonomous Weapon Systems*

Kelley M. Sayler

Lethal autonomous weapon systems (LAWS), or weapons designed to independently select and engage targets without the need for manual human control, could enable military operations in communications-degraded or -denied environments where traditional systems may not be able to operate. LAWS are not yet in widespread development.

However, as technology advances—particularly artificial intelligence (AI)—a larger number of countries may consider developing and operating LAWS. This could hold potential implications for congressional oversight, defense investments, military concepts of operations, treaty-making, and the future of warfare.

Furthermore, incorporation of new technology into weapons systems could create a number of potential legal, ethical, strategic, and operational challenges. For this reason, some members of the international community seek through international discussions to constrain—if not ban— LAWS.

What Are LAWS?

Definitions. No single, universally accepted definition of LAWS is used in international discussions. However, Department of Defense Directive 3000.09, which establishes U.S. policy on autonomy in weapons systems, defines LAWS as "weapon system[s] that, once activated, can select and engage targets without further intervention by an operator." This definition's principal characteristic is the role of the operator with regard to target selection and engagement decisions.

Other countries, however, have grounded their definition of LAWS on different characteristics, in particular the technological sophistication of the weapon system, such that LAWS are considered to be weapon systems capable of human-level cognition. Still others do not believe that a definition of LAWS is required—or desirable—for international discussions. Despite these differences, most parties to the LAWS discussions generally agree that the defining features of LAWS include full autonomy (no manual human control of the system) and the potential to produce lethal effects.

---

* This is an edited, reformatted and augmented version of Congressional Research Service Publication No. IF11294, updated February 14, 2023.

In: Emerging Military Technologies
Editor: Jerry Shields
ISBN: 979-8-89530-613-0

Status. Although the pursuit of LAWS is not yet widespread, some analysts have argued that Israel's Harpy loitering munition—which the weapon's manufacturer, IAI, describes as being fully autonomous—qualifies. Israel has exported the Harpy to Chile, China, India, South Korea, and Turkey. Similarly, former Secretary of Defense Mark Esper has noted that Chinese manufacturer Ziyan has advertised a fully autonomous system, the Blowfish A3 helicopter drone, which it has reportedly exported to the Middle East. In addition, according to a report by the Defense Innovation Board, the United States developed LAWS during the 1980s but no longer has LAWS in its inventory.

International Forum for LAWS Discussions

The international community examines the implications of LAWS in discussions held primarily under the auspices of the United Nations Convention on Certain Conventional Weapons (CCW), a multilateral arms control agreement to which the United States became a party in 1982. The CCW's purpose is to "ban or restrict the use of specific types of weapons that are considered to cause unnecessary or unjustifiable suffering to combatants or to affect civilians indiscriminately."

Since 2014, the CCW has convened annual meetings of States Parties, observers, and members of civil society to discuss the legal, ethical, technological, and military facets of LAWS. These meetings were elevated in 2017 from informal Meetings of Experts to a formal Group of Government Experts (GGE). After each session of the GGE, the session's chair produces a draft report that details session proceedings and offers conclusions and recommendations for future work. States Parties then adopt the final report by consensus.

In 2019, States Parties additionally agreed to a set of "guiding principles" for LAWS. States Parties agreed that international humanitarian law (IHL) would apply to LAWS, that humans must remain responsible for decisions about the use of force, and that states must consider the risks of LAWS acquisition by, or proliferation to, terrorists.

Table 1. State stances on preemptive LAWS ban

Support		Oppose[b]	Other[c]
Algeria Argentina	Holy See Iraq Jordan	Australia France	China
Austria Bolivia	Mexico Morocco	Germany India Israel	
Brazil Chile	Namibia	Russia	
Colombia Costa	New Zealand Nicaragua	South Korea Spain	
Rica Cuba Djibouti	Pakistan Panama	Turkey United Kingdom	
Ecuador Egypt[a]	Peru Uganda Venezuela	United States	
El Salvador	Zimbabwea		
Ghana[a] Guatemala			

Source: CRS consolidation of data from multiple sources.

Notes: CCW discussions on LAWS exclude existing weapons systems. Therefore, States Parties consider any potential LAWS ban to be preemptive.

[a] State is not party to the CCW.

[b] States that oppose a preemptive LAWS ban do not necessarily share the same alternative approach to managing LAWS.

[c]See section below on China.

Debate About Approaches to Managing LAWS

The CCW has considered proposals to ban LAWS, as well as proposals to regulate or issue political declarations about them. CCW States Parties have not reached a consensus on which of these approaches, if any, they will adopt.

Arguments Supporting Preemptive LAWS Ban

In addition to the states listed in Table 1, approximately 165 nongovernmental organizations have called for a preemptive ban on LAWS due to ethical concerns. These concerns include a perceived lack of accountability for use and a perceived inability to comply with the proportionality and distinction requirements of IHL. Some analysts have also raised concerns about LAWS's potential operational risks. For example, Center for a New American Security analyst Paul Scharre has noted that risks could arise from "hacking, enemy behavioral manipulation, unexpected interactions with the environment, or simple malfunctions or software errors" and could potentially result in civilian or noncombatant casualties. Although such risks could be present in automated systems, they could be heightened in autonomous systems, in which the human operator would be unable to physically intervene to terminate engagements—potentially resulting in wider-scale or more numerous instances of fratricide, civilian casualties, or other unintended consequences.

Those supporting a preemptive ban on LAWS have additionally appealed to the Martens Clause, which appears in the 1899 Hague Convention preamble and states that weapons use should conform to the "principles of humanity and the dictates of the public conscience." These analysts believe that LAWS contravene that requirement; however, others have noted that the Martens Clause has not been used previously to ban a weapons system and, furthermore, that the legal status of the Martens Clause is questionable and instead constitutes "merely a recognition of 'customary international law.'"

Arguments Opposing Preemptive LAWS Ban

A number of countries have voiced their opposition to a preemptive ban on LAWS. These countries have noted the potential military utility of LAWS, which could operate in communications-degraded or -denied environments where traditional systems may not be able to operate. LAWS could also potentially enable new concepts of operations, such as *swarming*, in which large formations of autonomous vehicles could be deployed to overwhelm defensive systems. (Some proponents of a ban have argued that swarms of autonomous vehicles could alternatively provide states or terrorist organizations with comparatively inexpensive weapons of mass destruction.)

Countries opposing a preemptive ban have additionally noted the potential humanitarian benefits of LAWS, which may be able to "strike military objectives more accurately and with less risk of collateral damage or civilian casualties" than traditional systems. These countries contend that human operators will remain accountable for the deployment of the systems and for ensuring that the systems' use complies with IHL. Finally, some countries are concerned that a preemptive ban on LAWS could inhibit research into technologies that may provide civilian benefits (e.g., elder care robots).

Alternative Approaches to Managing LAWS

Other states have proposed that the CCW instead focus on enhancing transparency in weapons development and sharing best practices for weapons review processes. France and Germany, for example, have proposed issuing a non- legally binding political declaration that would affirm that international humanitarian law applies to LAWS and that "[States Parties] share the conviction that humans should continue to be able to make ultimate decisions with regard to the use of lethal force and should continue to exert sufficient control over lethal weapons systems they use." Similarly, the United States has proposed a nonbinding Code of Conduct to "help States promote responsible behavior and compliance with international law." France and Germany have additionally proposed the establishment of a consultative committee of technical experts to advise the CCW on relevant technological developments.

Positions of the United States, Russia, and China

Although the CCW operates by consensus, the United States, Russia, and China—as leading military powers—are likely to be particularly influential in determining the trajectory of international discussions of LAWS.

United States. The U.S. delegation to the CCW has consistently opposed any preemptive ban on LAWS, arguing that LAWS could potentially provide a humanitarian benefit and that existing IHL is sufficient to govern the development and use of LAWS.

Russia. The Russian delegation to the CCW has also opposed a preemptive ban on LAWS, noting that LAWS could "ensure the increased accuracy of weapon guidance on military targets, while contributing to lower rate of unintentional strikes against civilians and civilian targets." It has also argued there is no proper legal precedent for a preemptive international ban on an entire class of weapons.

China. The Chinese delegation has stated that China supports a ban on the use—but not development—of LAWS, which it defines to be indiscriminate, lethal systems that do not have any human oversight and cannot be terminated. China is the only country that defines LAWS in this manner, and analysts note that such a weapon would be unable to comply with IHL and therefore would be inherently illegal. Some analysts have argued that China is maintaining "strategic ambiguity" about its position on LAWS.

Potential Questions for Congress

- Is the executive branch keeping Congress adequately informed about developments, both international and domestic, concerning LAWS and their regulation?
- What role should the United States play in UN CCW discussions of LAWS? Should the United States support the status quo, propose a political declaration, or advocate regulation of or a ban on LAWS?
- To what extent are potential U.S. adversaries developing LAWS? How, if at all, should this affect U.S. LAWS and counter-LAWS research and development or the United States' UN CCW position on LAWS?

Chapter 15

Navy DDG(X) Next-Generation Destroyer Program: Background and Issues for Congress*

Ronald O'Rourke

Introduction

The Navy's DDG(X) program envisages procuring a class of next-generation guided-missile destroyers (DDGs) to replace the Navy's Ticonderoga (CG-47) class Aegis cruisers and older Arleigh Burke (DDG-51) class Aegis destroyers. The Navy wants to procure the first DDG(X) in FY2032. The Navy's proposed FY2025 budget requests $102.8 million in research and development funding for the program.

Navy Large Surface Combatants (LSCs)

Force-Level Goal

The Navy refers to its cruisers and destroyers collectively as large surface combatants (LSCs). The Navy's current 355-ship force-level goal, released in December 2016, calls for achieving and maintaining a force of 104 LSCs. The Navy's FY2025 30-year (FY2025-FY2054) shipbuilding plan summarizes Navy and OSD studies outlining potential successor Navy force-level goals that include 72 to 96 LSCs.

Existing LSCs

The Navy's CG-47s and DDG-51s are commonly called Aegis cruisers and destroyers because they are equipped with the Aegis combat system, an integrated collection of sensors and weapons named for the mythical shield that defended Zeus. The Navy procured 27 CG-47s between FY1978 and FY1988. The ships entered service between 1983 and 1994. The first five, which were built to an earlier technical standard, were judged by the Navy to be too

* This is an edited, reformatted and augmented version of Congressional Research Service Publication No. IF11679, updated August 5, 2024.

In: Emerging Military Technologies
Editor: Jerry Shields
ISBN: 979-8-89530-613-0

expensive to modernize and were removed from service in 2004-2005. The Navy began retiring the remaining 22 ships in FY2022 and wants to retire all 22 by the end of FY2027.

The first DDG-51 was procured in FY1985 and entered service in 1991. The version of the DDG-51 that the Navy is currently procuring is called the Flight III version. The Navy also has three Zumwalt (DDG-1000) class destroyers that were procured in FY2007-FY2009 and are equipped with a combat system that is different than the Aegis system.

LSC Industrial Base

All LSCs procured for the Navy since FY1985 have been built at General Dynamics/Bath Iron Works (GD/BIW) of Bath, ME, and Huntington Ingalls Industries/Ingalls Shipbuilding (HII/Ingalls) of Pascagoula, MS. Lockheed Martin and Raytheon are major contractors for Navy surface ship combat system equipment. The surface combatant industrial base also includes hundreds of additional component and material supplier firms.

DDG(X) Program

Program Designation and Lead Ship Procurement

In the program designation DDG(X), the X means the precise design for the ship has not yet been determined. As mentioned earlier, the Navy wants to procure the first DDG(X) in FY2032, though the date for procuring the first ship has changed before and could change again.

Procurement of DDG-51s—the type of LSC currently being procured by the Navy—would end sometime after procurement of DDG(X)s begins. Navy officials have stated that they would like to see a three-year overlap between the start of DDG(X) procurement and the end of DDG-51 procurement.

Navy's General Concept for the Ship

Figure 1 shows a Navy rendering of a notional DDG(X) design. The Navy approved the DDG(X)'s top-level requirements (i.e., its major required features) in December 2020. An October 2023 Congressional Budget Office (CBO) report on the Navy's FY2024 30-year shipbuilding plan states that "the Navy has indicated that the initial [DDG(X)] design prescribes a displacement of 13,500 tons," which would be about 39% greater than the 9,700- ton Flight III DDG-51 design.

The Navy envisages the DDG(X) as having (1) Flight III DDG-51 Aegis combat system elements; (2) more growth margin than the Flight III DDG-51 design, meaning more space, weight-carrying capacity, electrical power, and cooling capacity (aka SWAP-C) for accepting additional or higher-power equipment and weapons (including directed- energy weapons) over the ship's service life; (3) an integrated power system (IPS); (4) reduced vulnerability due to reduced infrared, acoustic, and underwater electromagnetic signatures; (5) increased cruising range and time on station; and (6) increased weapon capacity.

Source: Navy rendering of notional DDG(X) design accompanying Sam LaGrone, "Navy Wants 3-Year Overlap Between Arleigh Burkes and DDG(X), Considering Propulsion System," USNI News, January 10, 2024.

Figure 1. Navy Rendering of Notional DDG(X) Design.

The Navy states that the baseline DDG(X) design, like the Fight III DDG-51 design, is to include 96 standard Vertical Launch System (VLS) cells, with an ability to incorporate 12 large missile launch cells in place of 32 of the 96 standard VLS cells. It is also to include two 21-cell Rolling Airframe Missile (RAM) launchers, and possibly also an ability to be built with an additional mid-body hull section, called the Destroyer Payload Module, that would provide additional payload capacity. The Navy states that

> The Future Naval Force Study (FNFS) and the Future Surface Combatant Force Analysis of Alternatives (FSCF AoA) identified the requirement for future large surface combatants (LSCs) to be capable of hosting directed energy (DE) weapons, larger missiles for increased range and speed, increased magazine depth, growth in organic sensors, and an efficient integrated power system to manage the dynamic loads. [The] DDG 51 Flight (FLT) III [design] is highly capable, but after over 40 years in production and 30 years of upgrades the [DDG-51] hull form does not provide sufficient space and center of gravity margin to host these future capabilities. To reset these design allowances for the future of naval warfare, requirements tradeoff and design studies were performed from FY 2018 to FY 2020.... These studies concluded that DDG(X) is required to deliver the necessary margins and flexibility to succeed the DDG 51 Class as the next enduring LSC combining the DDG 51 FLT III combat system elements with new hull form, an efficient Integrated Power System (IPS) and greater endurance reducing the Fleet logistics burden.

(Source: Department of Defense Fiscal Year (FY) 2025 Budget Estimates, Navy, Justification Book, Volume 2 of 5, Research, Development, Test & Evaluation, Navy, March 2024, p. 498.)

Procurement Quantities and Procurement Cost. The Navy's FY2025 30-year shipbuilding plan projects LSCs being procured in FY2032 and subsequent years in annual quantities of generally one to two ships per year.

The October 2023 CBO report estimates the DDG(X)'s average procurement cost in constant FY2023 dollars at $3.2 billion to $3.5 billion—about 33% to 40% more than the Navy's estimate (shown in the CBO report) of $2.4 billion to $2.5 billion. The CBO and Navy estimates are about 45% to 59%, and 9% to 14%, respectively, more than the DDG-51's procurement cost of about $2.2 billion. The CBO report states that "the Navy's estimates imply that the DDG(X) would cost about 14 percent more than the DDG-51 Flight III but would have a full-load displacement that is 40 percent greater. Such an outcome, however, seems unlikely given the history of the Zumwalt class DDG-1000 guided missile destroyer."

Issues for Congress

Issues for Congress regarding the DDG(X) program include the following: (1) Would a new LSC larger than the Flight III DDG-51 design be consistent with the Navy's desire to shift to a more distributed fleet architecture that includes a larger number of smaller ships? (2) The Navy in the past has studied options for a lengthened version of the DDG-51 that would displace between 11,000 and 12,000 tons. Would the DDG(X) be more cost-effective than a lengthened DDG-51? (3) Has the Navy accurately identified the DDG(X)'s required operational capabilities? (4) Why is there a 35% to 43% difference between the CBO and Navy estimates of the DDG(X)'s average procurement cost? (5) Would future Navy budgets permit the procurement of DDG(X)s in desired numbers while adequately funding other Navy priorities? (6) Has the Navy taken adequate steps to mature DDG(X) technologies and mitigate technical, schedule, and cost risk in the program? (7) Has the Navy planned adequately for the transition from DDG-51 procurement to DDG(X) procurement, and for resulting impacts on the shipbuilding industrial base?

Funding Request

The Navy's proposed FY2025 budget requests $28.3 million for Project 0411 (DDG[X] Concept Development) within Program Element (PE) 0603564N (Ship Preliminary Design & Feasibility Studies), which is line 46 in the Navy's FY2025 research and development account, and $74.5 million for "DDG(X) Power & Propulsion Risk Mitigation & Demonstration," which forms part of Project 2471 (Integrated Power Systems [IPS]) within PE 0603573N (Advanced Surface Machinery Systems), which is line 48.

The House Armed Services Committee, in its report (H.Rept. 118-529 of May 31, 2024, page 463) on the FY2025 National Defense Authorization Act (NDAA) (H.R. 8070); the Senate Armed Services Committee, in its report (S.Rept. 118-188 of July 8, 2024, page 484) on the FY2025 NDAA (S. 4638); the House Appropriations Committee, in its report (H.Rept. 118-557 of June 17, 2024, page 186) on the FY2025 DOD Appropriations Act (H.R. 8774); and the Senate Appropriations Committee, in its report (S.Rept. 118-204 of August 1, 2024, page 214) on the FY2025 DOD Appropriations Act (S. 4921), recommended approving the Navy's funding requests.

H.Rept. 118-529 directs the Government Accountability Office (GAO) to review and provide a briefing on the Navy's large surface combatant programs, with one or more GAO reports to follow (pages 24-25), and directs the Navy to provide a briefing on propulsion motor technology for the DDG(X) program (page 65). Section 214 of S. 4638 would require full-scale testing of a minimum of two electric propulsion motor technologies for the DDG(X) program.

Chapter 16

The U.S. Army's Typhon Strategic Mid-Range Fires (SMRF) System*

Andrew Feickert

Reported improvements to Russian and Chinese artillery systems present a challenge to the U.S. Army. Improved longer-ranged artillery systems, new employment techniques for unmanned aerial vehicles (UAVs), and the proliferation of special munitions (such as precision, thermobaric, loitering, and top-attack munitions) have renewed concerns about the potential impact of Russian and Chinese fires on U.S. combat operations and ground combat systems. In response, the U.S. Army is seeking to improve its ability to deliver what it refers to as long-range precision fires (LRPF) by upgrading current artillery and missile systems, developing new longer-ranged systems and hypersonic weapons, and modifying existing air- and sea-launched missiles for ground launch.

Source: The Drive: https://www.thedrive.com/the-war-zone/army- fires-tomahawk-missile-from-its-new-typhon-battery-in-major- milestone, accessed July 6, 2023.

Figure 1. Typhon Launchers and Battery Operations Center.

Originally known as the Mid-Range Capabilities (MRC) System, SMRF is part of the Army's LRPF modernization portfolio. It is intended to hit targets at ranges between the

---

* This is an edited, reformatted and augmented version of Congressional Research Service Publication No. IF12135, updated April 16, 2024.

In: Emerging Military Technologies
Editor: Jerry Shields
ISBN: 979-8-89530-613-0

Army's Precision Strike Missile (PrSM) and the developmental Long-Range Hypersonic Weapon (LRHW) system. The SMRF Weapon System leverages existing Raytheon-produced SM-6 missiles and Raytheon-produced Tomahawk cruise missiles and modifies them for ground launch. The SMRF system is also known as the "Typhon" missile system (Figure 1).

SMRF Weapon System Components

According to the Army, the prototype SMRF battery is planned to consist of four launchers and a battery operations center (BOC) (Figure 1). Reportedly, a decision has not been made on how many missiles each battery will have. SMRF batteries are to be equipped with a number of prime movers, trailers, generators, and support vehicles. Numbers of soldiers assigned to each battery is presently unknown. The Army originally planned for the first prototype SMRF battery to be fielded no later than the fourth quarter of FY2023 and three additional batteries were to be fielded on an annual basis thereafter. It is not clear at this time if the Army will field more than four SMRF batteries or if any of the batteries will be Army National Guard units.

SMRF Unit Organization

The Army plans to field one SMRF battery in the Strategic Fires Battalion of the Army's regionally aligned Multi- Domain Task Force (MDTF) (Figure 2).

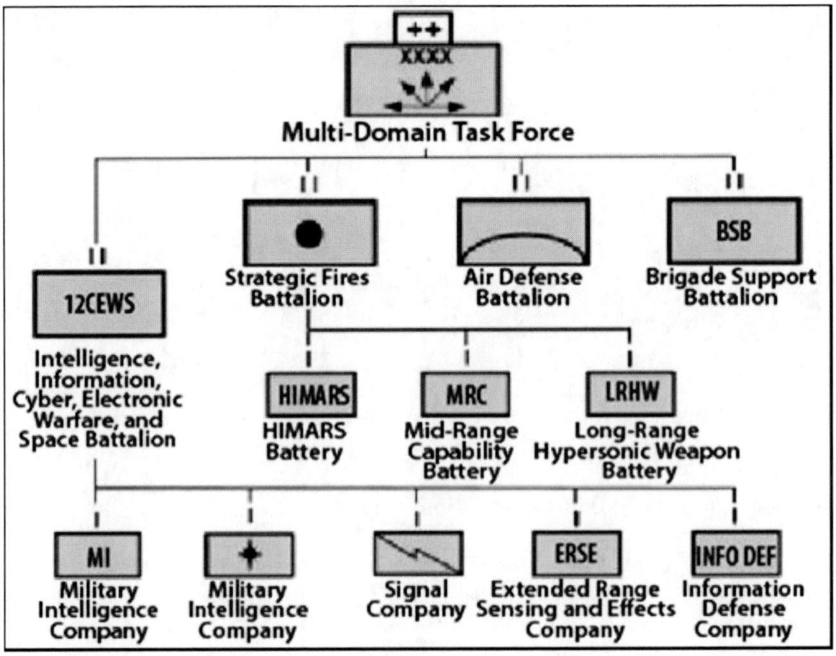

Source: Chief of Staff Paper #1 Army Multi-Domain Transformation Ready to Win in Competition and Conflict, March 16, 2021, p. 12.

Note: MRC Battery depicted above is now known as the SMRF Battery.

Figure 2. Army Multi-Domain Task Force (MDTF) Organization.

The Army describes MDTFs as "theater-level maneuver elements designed to synchronize precision effects and precision fires in all domains against adversary anti- access/area denial (A2/AD) networks in all domains, enabling joint forces to execute their operational plan (OPLAN)-directed roles."

Program Status

Reportedly, Lockheed Martin delivered the first of four prototype Typhon systems to the Army on December 2, 2022. The Army originally planned to field its first prototype SMRF battery no later than the fourth quarter of FY2023.

> What Is Anti-Access/Area Denial (A2/AD)?
>
> Anti-Access (A2) is an action, activity, or capability, usually long-range, designed to prevent an advancing enemy force from entering an operational area.
>
> Area Denial (AD) is an action, activity, or capability, usually short-range, designed to limit an enemy force's freedom of action within an operational area.
>
> Source: Department of Defense Dictionary of Military and Associated Terms, November 2021.

SMRF Test Launches and Full Operational Capability

On June 27, 2023, the Army reported

> The Army's Rapid Capabilities and Critical Technologies Office's Mid-Range Capability Project Office successfully demonstrated the launch of a Tomahawk missile from the Army's prototype Mid-Range Capability system. Soldiers assigned to the 1st MDTF conducted this live-fire event. This test follows the successful launch of an SM-6 missile from the Mid-Range Capability system earlier this year, confirming the full operational capability of the system.

Plans to Deploy SMRF in 2024

Reportedly, in November 2023, U.S. Army Pacific Commander General Charles Flynn noted

> We have tested [Typhon] and we have a battery or two of them today. In 2024, we intend to deploy that system in the region. I'm not going to say where and when, but I will just say that we will deploy them in the region.

SMRF Battery Deployed to the Philippines for Exercise Salaknib 2024

According to U.S. Army Pacific, the Joint Base Lewis- McChord's 1st MDTF deployed a SMRF battery to Northern Luzon, Philippines, on April 11, 2024, as part of Exercise Salaknib 24. Reportedly, the deployment of the SMRF battery is temporary for exercise purposes, and it is to be used during Exercise Salaknib 24 and the upcoming Exercise Balikatan 24.

SMRF Battery Activation

Reportedly, the Army activated its second SMRF battery— D Battery, 5th Battalion, 3rd Field Artillery Regiment—as part of Joint Base Lewis-McChord's 1st MDTF in January 2024. The article notes that "while no statement can be found on the creation of the first battery," it is assumed that "the second battery activated was part of the Army's 1st Long-Range Fires Battalion, implying that that the 1st MDTF now has at least two Typhon batteries."

FY2025 SMRF Budgetary Information

Table 1. FY2025 SMRF budget request

Funding Category	Total Request ($ Million)	Total Request (Qty.)
RDT&E	$183	—
Procurement	$233	—

Source: Assistant Secretary of the Army (Financial Management and Comptroller), FY2025 President's Budget Highlights, March 2024. p. 33.

Notes: RDT&E = Research, Development, Test & Evaluation;
Qty. = FY2025 procurement quantities.

According to the *Army's FY2025 President's Budget Highlights*,

> Base funding also allows for purchasing and receiving hardware and materials to implement prototype fabrication, and to support component- level and system-level qualification, adding additional capabilities to the batteries. Procures thirty-two Tactical Tomahawks (TACTOMs) and MK14 cannisters It also supports the procurement of a MRC Battery, Ground Support Equipment to include one Battery Operations Center (BOC), four launcher Payload Deployment Systems (PDS), one Reload Capability, and one BOC Support Vehicle, [and] associated Government Furnished Equipment, and program management costs.

Potential Issues for Congress

Planned SMRF Units

The Army's original plans called for fielding four SMRF batteries. The Army, however, plans for five MDTFs, with each MDTF having one organic SMRF battery. Reports suggest the 1st MDTF now has two SMRF batteries instead of one. The Army has reportedly stated that while "each [MDTF] is supposed to have one battery," that "MDTFs can be adjusted to their combatant commander's requirements with more or fewer units." If this is the case, it suggests the Army may eventually field more than the four originally planned SMRF batteries. Congress might seek clarification on the Army's current plans for total number of SMRF batteries, including if any batteries will be fielded in the Army National Guard.

Overseas Stationing of SMRF Units

As previously noted, the Army reportedly planned to deploy SMRF in 2024 in support of U.S. Army Pacific at undisclosed locations in the Pacific region. In April 2024, a SMRF battery was temporarily deployed to the Philippines for exercises. It is not known if this temporary deployment could eventually be made permanent. It is possible that SMRF units may be stationed elsewhere in the Indo-Pacific region or at other overseas locations. Given the importance and issues that often affect securing overseas basing, Congress might examine ongoing efforts to secure Army long-range precision fires unit basing in the Indo-Pacific region and other overseas regions.

Chapter 17

U.S. Army's Maneuver Short-Range Air Defense (M-SHORAD) System*

Andrew Feickert

Background

The Army is developing a new maneuver short-range air defense system, or M-SHORAD, to perform short-range air defense (SHORAD). The Army defines SHORAD as capabilities that provide air defense against low-altitude air threats. In the early 2000s, the Army divested SHORAD units from Army force structure to free up personnel to create other types of units deemed more mission-critical at the time. After 2005, SHORAD force structure was reduced to two battalions of Active Component Avenger systems and Counter-Rocket, Artillery and Mortar (C-RAM) batteries and seven Army National Guard Avenger battalions.

Since 2005, there has been a dramatic increase in air and missile platforms that could threaten U.S. ground forces. The use of unmanned aerial systems (UASs) has increased exponentially and UASs have been used successfully in a variety of conflicts, including the current Ukrainian conflict. Given the increase in threat and limited air defense assets available to Army divisions, the Army decided to improve the air defense posture of its maneuver forces.

M-SHORAD Requirement

In response to the growing aerial threat, the Army originally planned to field 144 M-SHORAD Increment 1 systems to four battalions and could field additional battalions in the future. Each M-SHORAD battalion would consist of 40 M- SHORAD systems, support vehicles and equipment, and about 550 soldiers. In April 2021, the 5th Battalion, 4th Air Defense Artillery Regiment received the first four of its M- SHORAD systems, becoming fully equipped in late 2022. In addition to the 144 systems designated for operational units, the Army planned to procure 18 additional systems for training, operational spares, and testing, for a total of 162 systems. Reportedly, as of June 2024, the Army envisions building about 312 M-SHORAD systems but that number could increase to as many as 361 "based on future Army decisions," but, at present, the Army only has approval for 162 systems.

* This is an edited, reformatted and augmented version of Congressional Research Service Publication No. IF12397, updated June 25, 2024.

In: Emerging Military Technologies
Editor: Jerry Shields
ISBN: 979-8-89530-613-0

M-SHORAD Variants/Increments

Initially, there were three M-SHORAD variants, or "Increments," planned by the Army.

Source: https://www.leonardodrs.com/what-we-do/products-and- services/m-shorad/.

Figure 1. M-SHORAD Increment I.

M-SHORAD Increment 1

M-SHORAD Increment 1 (*Figure 1*) was developed under the Other Transaction Authority contracting process. M- SHORAD uses the M-1126 Stryker combat vehicle as its chassis. The weapons and radar packages are configured by Leonardo DRS and then installed on the Stryker by General Dynamics Land Systems (GDLS), the vehicle's original manufacturer. Leonardo DRS reports the multipurpose unmanned turret includes

- two AGM-114L Longbow Hellfire missiles capable of hitting ground targets;
- four FIM-92 Stinger missiles for aerial targets in a launcher (configured by Raytheon);
- an XM914 30 mm automatic cannon;
- an M-240 7.62 mm machine gun; and
- a multi-mission radar capable of tracking both ground and air targets.

Prohibition on Hellfire Missile Use on M-SHORAD Increment 1

Reportedly, the Army is planning to replace the Longbow Hellfire missile launcher on M-SHORAD Increment 1 vehicles because the Hellfire mounted on the side of the vehicle "created wear and tear on the missile, leading to potential safety concerns." The Army reportedly plans to retrofit M-SHORAD vehicles "by swapping out the Hellfire launcher and for a second Stinger pod," which would give the modified vehicles eight total FIM-92 Stinger rounds.

M-SHORAD Increment 2

M-SHORAD Increment 2 is also referred to as DE (Directed Energy) M-SHORAD and incorporates a 50 kilowatt (kW) laser as its primary armament to defend against a variety of air and artillery threats. Efforts to develop the 50 kW laser began in 2019 and, in 2021, Raytheon was awarded a $123 million developmental contract after a competitive shoot-off against Northrop Grumman. The Army reports additional testing of the 50 kW laser has enjoyed success against a variety of drones, but according to Army program officials, "challenges remain" in terms of defending against rockets, artillery, and mortars. Army plans called for M-SHORAD Increment 2 to start a user assessment beginning in the fourth quarter of FY2023, running through the first quarter of FY2024. The Army also plans for an additional final contract competition prior to FY2025.

Soldier Observations on DE M-SHORAD Deployed to the Middle East for Testing

Reportedly, according to Army officials, in February 2024 the Army deployed four Stryker-mounted 50 kW laser M- SHORAD prototypes to the Middle East for soldiers to test the system's performance against aerial threats. Initial soldier feedback was characterized as "not overwhelmingly positive," with officials suggesting "results from the lab environment and test ranges are very different from the tactical environment."

M-SHORAD Increment 3

The Army reportedly plans for M-SHORAD Increment 3 to incorporate the FIM-92 Stinger replacement missile—the Next Generation Short Range Interceptor—into the Increment 1 system. In addition, these plans call for the Increment 1 30 mm automatic cannons to receive the XM 1223 Multi-Mode Proximity Airburst munition (MMPA), which features a multipurpose munition that can be employed against air, ground, and personnel targets. In March 2023, the Army reportedly selected Lockheed Martin and Raytheon Technologies to develop competing prototypes of a Next-Generation Short-Range Interceptor. The Army has indicated it is planning for a technology demonstration in FY2024, an operational demonstration in FY2026, and a production decision by FY2027.

Army Issues a Request for Information for M-SHORAD Increment 4

On May 8, 2024, the Army issued a request for information (RFI) for M-SHORAD Increment 4, stating that this capability is to focus on delivering "air defense capability to support dismounted maneuver." M-SHORAD Increment 4 is to include capabilities that can be transported on a C- 130 aircraft and that are air-droppable and sling load capable. It is also to be capable of integration onto platforms such as the Joint Light Tactical Vehicle (JLTV) and/or a robotic vehicle. The Army has requested RFI responses from industry by July 16, 2024, and that industry provide solutions that would deliver capabilities in the near term (FY2027 to FY2028), mid term (FY2030 through FY2032), and the far term (beyond FY2035).

Army Names M-SHORAD Vehicle and Program Update

On June 14, 2024, the Army reportedly renamed the M-SHORAD the "SGT STOUT" in honor of Sergeant Mitchell William Stout, an Air Defense Artillery soldier who won the Congressional Medal of Honor during the Vietnam War. In addition to the Army's first two M-SHORAD battalions—the 5th Battalion, 4th Defense Artillery in Germany and the 4th Battalion, 60th Air Defense rtillery at Fort Sill, OK—Army officials reportedly noted that the Army is currently in the process of fielding to the 6th Battalion, 56th Air Defense Artillery at Fort Cavazos, TX, and plans to activate a fourth M-SHORAD battalion at Fort Liberty, NC, in the third quarter of FY2025.

FY2025 Budgetary Information

Table 1. FY2025 M-SHORAD budget request

	Procurement ($M)	RDT&E ($M)	Quantity
M-SHORAD Increment 1	$69.091	—	—
M-SHORAD Increment2	—	$88.480	—
M-SHORAD Increment 3	—	$204.880	—

Sources: Procurement: Department of Defense Fiscal Year (FY) 2025 Budget Estimates, March 2024, Army Justification Book, Volume 1 of 1 Missile Procurement, Army. RDT&E: Department of Defense Fiscal Year (FY) 2025 Budget Estimates, March 2024, Army Justification Book Volume 2b of 2 Research, Development, Test & Evaluation, Army RDT&E – Volume II, Budget Activity 4B.

Notes: $ M = million dollars; RDT&E = Research Development Test & Evaluation; Quantity = procurement quantity.

Considerations for Congress

Oversight questions Congress could consider include the following.

Lessons Learned from the Russo-Ukraine Conflict

The ongoing Ukrainian conflict has featured the use of a variety of military and commercial UASs employed in both kinetic and non-kinetic roles. In addition to both fixed and rotary wing air threats, loitering munitions have also been employed, reportedly with considerable effect. Lessons learned related to the employment of the aforementioned systems may inform current and future development of all four M-SHORAD Increments. Congress may consider what efforts have been undertaken by the Army to incorporate lessons learned into M-SHORAD design.

2024 Army Force Structure Transformation Initiative and M-SHORAD

On February 27, 2024, the Army released a white paper, *Army Force Structure Transformation*, outlining plans to transform the force. Regarding M-SHORAD, the Army stated it would add four additional M-SHORAD battalions.

Reportedly, Army officials noted on June 14, 2024, that the four additional battalions would be activated in the Army National Guard "assuming they are funded across the Program Objective Memorandum (POM) between FY2026 and FY2030." Given the Army's new transformation plans, Congress might seek to clarify the Army's plans for the four new Army National Guard M-SHORAD battalions including the timeline for unit activations and where the units will be stationed.

Index